T5-BZM-475

2nd World Congress on Superconductivity

Progress in High Temperature Superconductivity – Vol. 28

2nd World Congress on Superconductivity

Houston, Texas 10 – 13 September 1990

Editor

Calvin G. Burnham

Published by

World Scientific Publishing Co. Pte. Ltd.
P O Box 128, Farrer Road, Singapore 9128
USA office: Suite 1B, 1060 Main Street, River Edge, NJ 07661
UK office: 73 Lynton Mead, Totteridge, London N20 8DH

2ND WORLD CONGRESS ON SUPERCONDUCTIVITY

ISBN 981-02-0618-6
ISBN 981-02-1999-7 (pbk)

Printed in Singapore by JBW Printers and Binders Pte. Ltd.

In memory of Dr. Chiu T. Lam.

We are deeply sadden by the passing of Dr. Chiu T. Lam. Dr. Lam, a research scientist for Akzo Chemicals, volunteered his time and energy in serving the World Congress on Superconductivity as Program Chairman for the 2nd WCS. He was an energetic, outgoing person, with a genuine interest in his fellow man. He was a good friend and will be greatly missed.

We have established a memorial scholarship fund in his memory. Donations may be sent to:

Lam Memorial Fund
World Congress on Superconductivity
P.O. Box 27805
Houston, Texas 77227-7805

World Congress on Superconductivity

Introduction

In today's highly sophisticated, complex, competitive world, with super communications, technical and scientific advancements can no longer remain isolated by distance and geography. Discoveries and developments in superconductivity occurring in rapid succession and sometimes simultaneously in different laboratories and different countries bear testament to this fact. Superconductivity has the potential to trigger dramatic changes over the entire globe. Its development offers new opportunities and challenges to not only the technological and business communities, but the world at large. Probably the single most important challenge to the widespread application of superconductivity is the need to create the means and the environment to encourage coordination among the individuals and groups pursuing a variety of efforts in the scientific, commercial and social aspects of this technology. The World Congress on Superconductivity offers the international community a forum in which to openly share progress, ideals and concerns thereby hastening the technology's translation to useful products which benefit all mankind.

Goals and Objectives

The World Congress on Superconductivity has been established to foster the development and commercial application of superconductivity technology on a global scale. This will be achieved by providing an organized, constructive, non-adversarial, non-advocacy forum where scientists, engineers, businessmen, and governmental personnel can freely exchange information and ideas on recent developments and directions for the future.

The objectives of the World Congress on Superconductivity are:

To draw together pre-eminent superconductivity researchers and technologists from around the world to discuss developments and directions,

To involve industry leaders directly with technologists to accelerate the translation of discoveries into products which benefit all mankind,

To interface international government agencies with research and business communities to promote globally harmonious development and application,

To encourage and facilitate the exchange of ideas, goals and needs among research, industry and government communities,

To function as an information and networking hub, providing data and referrals to those seeking knowledge and contacts, and

To focus attention on the development of a worldwide professional community dedicated to superconductivity research and business development.

Implementation

The World Congress on Superconductivity held its first international meeting in February, 1988 in Houston, Texas. It was attended by over 700 people from 13 countries. Presentations were made on superconductivity materials, processes and fabrication, on applications in energy, space, electronics, transportation, medicine and physics, and on business aspects in the fields of technology, funding and marketing. Keynote speakers and industry leaders participating included Richard Blaugher, Paul Chu, Sadeg Faris, Laurie Gavin, Yoshihiro Kyotani, Yuri Osipiyan, Roger Poeppel, Zhang Qirui, Bernard Raveau, Congressman Donald Ritter, Thomas Rona, Carl Rosner, Rustom Roy, Arthur Sleight, Shoji Tanaka, Edward Teller and Sergio Trindade.

The second congress was held in September 1990 and continued the building process of the organization and the advancement of its objectives. Key leaders in the field from government, academia, business, and science representing sixteen nations participated in a technology update of broad proportions and mapped a strategy for cooperative ventures to hasten development and application.

The objectives of the World Congress will be implemented in the future by:

Holding major international congresses biennially to which everyone involved or interested in superconductivity is invited. Leaders from all aspects of the field will present the latest developments and trends in physics and materials research and technology, and in applications of the technology. Projections will be made of the timing and nature of the commercial availability of superconductivity applications. Government and business leaders will discuss government support and interaction, technology transfer, patent status, and business opportunities and funding. World-renowned figures will present keynote addresses. Apart from the formal exchange of ideas through session speakers, there will be ample opportunities for informal exchanges throughout these conferences.

Sponsoring and co-sponsoring topical workshops, seminars, short courses and other educational activities related to the physics, chemistry, engineering and application of superconductivity technology development.

Providing scholarships in superconductivity and related research in university facilities dedicated to the field.

Building a technical library of superconductivity documents for the use of all participants in and friends of the technology.

Interfacing and cooperating with individuals and organizations having superconductivity interests throughout the world.

Steering Committee

The World Congress on Superconductivity is a non-profit (501.c.3) organization managed by a team of volunteers.

1,2 Calvin G. Burnham, P.E.
Houston Lighting & Power Company
1,2 Kumar Krishen, Ph.D.
NASA Johnson Space Center
1,2 Jack Randorff, Ph.D.
Randorff and Associates
1,2 John L. Margrave, Ph.D.
Rice University; H.A.R.C.
1,2 Robert W. Temple
TMR, Inc.
2 Jeffrey King
Tensar
2 Jimmy Salinas
Southwestern Bell
2 Shiou-Jyh Hwu, Ph.D.
Rice University
2 Jim Beckman
Business Development
Pat Harper
Brown & Root
Glenn Carraux
CEI Inc.
3 Cynthia King
Putman & Associates

1=officer, 2=director, 3=staff

Honorary Advisory Board

The World Congress on Superconductivity is assisted by an Honorary Advisory Board of noted national and international representatives from the scientific, engineering, academic, business and government communities. Board Members are called on for their expertise, to act as WCS ambassadors, and to provide a network of information and resources. The Advisory Board consists of:

Samuel Kramer
National Bureau of Standards
Dr. John Margrave
Rice University; HARC
Dr. Valeri Ozhogin
Kurchatov Atomic Institute - USSR
Dr. Frank Patton
Defense Advanced Research Projects Agency
Jack E. Randorff, Ph.D.
Randorff and Associates
The Honorable Donald L. Ritter (R-PA)
U.S. House of Representatives
Dr. Alan Schriesheim
Argonne National Laboratory
Dr. Z. Z. Sheng
University of Arkansas
Dr. Martin Sokoloski
NASA Headquarters
Dr. Shoji Tanaka
International Superconductivity Technology Center
Al Thumman
Association of Energy Engineers
Dr. Roy Weinstein
University of Houston

Sponsorship

It is the fervent hope of the World Congress on Superconductivity that it will ultimately be self-supporting through fees from its conferences and other activities. Meanwhile, the ability of the World Congress on Superconductivity to continue operation and maintain its independence will depend on the monetary funding and donated services which individuals and institutions benefitting from the activities of the World Congress or supporting its objectives will provide. Sponsors to date have included:

American Chemical Society
American Institute of Chemical Engineers
American Physical Society
Arnold, White & Durkee
Association of Energy Engineers
Coopers & Lybrand
Cortest Laboratories, Inc.
Defense Advanced Research Projects Agency
Engineers Council of Houston
Electric Power Research Institute
E.I. Du Pont de Nemeurs & Company
Hawker Siddeley Power Engineering, Inc.
Houston Advanced Research Center
Houston Chamber of Commerce
Houston Economic Development Council
Houston Lighting & Power Company
Institute of Electrical & Electronic Engineers
Japan External Trade Organization
Krishen Foundation
Leybold Inficon, Inc.
Materials Research Society
MIT Enterprise Forum of Texas
Motorola Corporation
NASA
National Association of Manufacturers
Rice University
Southwestern Bell Corporation
Technova, Inc.
Texas A&M University
Texas Society of Professional Engineers
Thomas Alva Edison Foundation
Trahan & Associates
Turner, Collie & Braden, Inc.
University of Houston
University of Texas
Westinghouse Electric Corporation

The 3rd International Conference and Exhibition of the World Congress on Superconductivity is scheduled for September 14-18, 1992 at the prestigious Hotel Bayerischer Hof in Munich, Germany. The event promises to be both exciting and productive in all respects.

For more information, please contact:

World Congress on Superconductivity
P.O. Box 27805
Houston, Texas 77227-7805 USA

telephone (713) 895-2500
telefax (713) 469-5788

Table of Contents

4. CHARACTERIZATIONS

5. FABRICATIONS

6. MONOLITHICS

7. MICROWAVES AND SQUIDS

8. BUSINESS, MARKETS AND LEGAL

9. POSTER SESSIONS

1. LUNCHEON ADDRESS

2nd World Congress on Superconductivity
Luncheon Speaker
September 10, 1990

Dr. Edward Teller
Lawrence Livermore National Laboratory

Thank you very, very much. I would like to add to the description of my history. I was there on two very interesting occasions. One is I started to study physics in Leipzig with Heisenberg, whose name is associated with the Uncertainty Principle.

Quantum mechanics was brand new. The behavior of atoms was understood and that explained, in principle, all the properties of matter. We could calculate the behavior of matter, and we did not need any more chemists - if only we could solve differential equations in thousands of dimensions. But that looks like an insuperable task. High-temperature superconductivity which brings us here together is a property of matter that went unpredicted.

The second memorable occasion took place in Los Alamos where we planned the first nuclear explosion. This would have been impossible without the initiative and advice of the very wonderful Hungarian, Johnny von Neumann. He introduced us to early computing machines. They were of a very clumsy variety. I would call them monoflops, meaning that they could take action once a second. They used mechanical motion programmed by electric impulses. Now Johnny proposed to go over to electronics and very soon we went from monoflops to megaflops, a million actions per second. And this in a technical way is an important part of what I want to tell you. There is another part less technical and no less important.

A few days ago I came back from a trip to an international conference and also to the Middle East, and right now as I stand here I am not yet quite sure which way is up. It is not only jet lag. I am seeing the world change in the last weeks. Of course everybody knows about the danger, everybody knows about the excitement.[1] I feel

---

[1]This was September, 1990, a few weeks after Saddam Hussein invaded Kuwait.

that one point is less noticed. It is precisely that point that an international gathering of scientists should remember.

There is now international corporation, and there has been for the last few weeks. It is on a scale with such determination that I have a hard time to find a parallel in history. When Hitler came to power, I was 25 years old. If at that time there had been as much cooperation between those who wanted peace, the Second World War never would have occurred and 50 million lives would not have been lost. What is being done now is something quite remarkable.

During my trip in Israel, people asked again and again, will it last? Will the coalition continue to work together? The way I see it, there is a good chance that it will. There is a common purpose between incredibly divergent interests who have never before cooperated. And what is the common interest? It is the determination that aggression must not happen; that aggression must be stopped. There are all kinds of other interests, but this is the powerful one and the common one.

And I want to share with you a peculiar idea I have about one man, the President of the United States. A very careful man, an experienced man, a patient man. He reminds me of what I learned in school about a famous Roman counsel. I want to be a little extravagant and call our President a reincarnation of Quintus Fabius Maximus Cunctator. He is the man who saved Rome from Hannibal. And there was a peculiar way in which he did it, which earned him the name Cunctator, the Delayer. He delayed pitched battles and maintained pressure. Therefore Hannibal could never beat him. And in the end Hannibal had to leave. The methods of our President doing things gently but building for permanence give me hope for success.[2]

Our international meeting here gives all of us a new opportunity for close collaboration in science. I believe that unity in resisting aggression is very important. Building common interest through collaboration and through more rapid progress in the interest of everyone. As time goes on, everybody will understand - that what is good for my neighbor is good for me; that what is good for Japan is good for the United States and what is good for the United States is good for the Soviet Union.

I would now like to apply this to the question before us. The development of high-temperature superconductivity which makes experimentation with superconductivity much more easy. A committee on the applications of superconductivity on which I served has just completed its task.

We recommended among other things that we make more of an effort to provide high schools with apparatus for playing with superconductivity because these games can and do lead to surprises, and because science is built on such surprises. We also reviewed

---

[2]The time was the beginning of Desert Shield. What followed in Desert Storm is more reminiscent of the victory of Scipio Africanus over Hannibal.

the hopeful field of superconductivity and in this wide field there are many special projects of interest, like for instance, the levitated train or the superconducting supercollider and many others.

But among all of these, in my mind, two plans stand out which are more general and are the most important developments in superconductivity. One is to make wires out of this new superconductors which can carry a strong current. Let me mention very briefly why this is difficult. The reason for the difficulty is that the new superconductors are layer cakes with a high maximum current along the layers and a much lower maximum current perpendicular to the layers. To get these materials in bulk we produce them inexpensively in a powder form. If the crystallites in the powder are unoriented, then high currents will not be produced. One needs to press the wires to cycle them at appropriate temperatures with the hope and with some experience, that orientation of the crystallites occurs. Strong currents, even in high magnetic fields, can be reached, and the high magnetic field is to a great extent what counts.

High magnetic fields mean that you can make more powerful motors in a small volume. High magnetic fields are needed in NMR (nuclear magnetic resonance) where the spins of atomic nuclei precess in the magnetic field. The frequency of precession is a little different according to the chemical surrounding. Using high magnetic fields, one gets a much more differentiated view of the big molecules and their chemical surroundings in our organisms. So we get a magnetic picture similar to an x-ray picture. Except that an x-ray picture really shows the bones which in a way are the least interesting part of us. NMR tells us more of the actively living substance.

Another possibility is to produce these high temperature superconductors in very thin layers. Then they are nicely orientated. You get very big current densities but the layers are so thin you don't get very high currents. Nevertheless, they are most useful in electronic equipment, particularly in computers.

One may think of a computing machine as an abacus in which beads can occupy one of two positions. Only the bead can be electronic. In 1950, Johnny von Neumann could count at a megaflops, a million moves a second. Today, we have reached gigaflops, a billion counts per second. With superconduction, we are aiming at what we call a petaflops, ten to the fifteenth (10^{15} or a million billion) decisions per second. Not all "beads" are at the same place, but they are contained in a relatively small volume, tightly packed. And the tight packing is easier with superconduction because superconductors exclude magnetic fields and thereby exclude the influence of one element upon a nearby element.

If we ever we get to the petaflops, you'll have the job to write the programs for machines that act ten to the fifteenth times a second. I am very sure that the coding will become too extensive and complicated. The only way we can handle it is that we leave the program making in part to the machine itself. And with that the thinking machine will have come one step closer to the behavior of human beings. This will be a very interesting problem in software engineering where ingenuity of a mathematical kind, of an imaginative kind, will be needed. Of course, machines will go far beyond mathematics. They will translate, recognize patterns and take over a number of human like activities.

And now in the end, let me come back to the topic I really want to talk to about, international cooperation. With the possibility of small effective motors, with deeper insight into living organisms, advanced and more accurate electronics and incredible improvements in computing machines, I believe that we will need all the capable people in all countries to bring about these developments.

Let me compare this revolution with another revolution in which I was active - the revolution of atomic energy. Nuclear energy was developed under conditions of secrecy. This secrecy still continues to erect walls that impede international cooperation.

High-temperature superconductivity was discovered by a German and a Swiss physicist working in an American company in Zurich. That excluded secrecy. I don't think anybody wants it. I think all of us want openness. All of us want cooperation; yet, I hesitate to recommend plans to go ahead with big international projects. Whatever is big can be spoiled by one or two wrong decisions. Duplication and competition are important. But competition must be open so that every one of the competitors can learn from the others' mistakes. The man who is often considered the greatest physicist, Niels Bohr, had a definition for an expert - a person, who through his own painful experience has found out all the mistakes that one can commit in a very narrow field. International cooperation means that we all can become experts faster. On successes and, more importantly, our mistakes should be advertised. I believe that small international meetings should be held and bigger ones, and also organizational meetings. Western Europe, the United States, the Soviet Union and Japan should all know mutually what's going on. Exchange of personnel, different projects in different companies should be adapted to the greatest interest in that country. For instance, the United States, I would recommend, may put the greatest emphasis on petaflops. I hope we won't be the only ones, and I hope we will get help and correction from others. And in the geographically smaller countries like Japan, the levitated train may be emphasized, the United States should assist the Japanese, but in turn should be told about the progress so that the fruits can be enjoyed in every case by everybody.

I believe that a climate where the United Nations has become active, where people who, otherwise never could agree, work together will be conducive to full scientific cooperation. High-temperature superconductivity uses the discoveries of quantum mechanics. It will lead, in turn, to computers that will lead to unimagined new progress. This is the field which we are here discussing and which I hope will help bring the scientists closer together, bring the countries closer together, demonstrate the great advantages of cooperation, and make in the long run, peace much more secure.

2. PLENARY SESSIONS

The Harwell-Industry Superconducting Ceramics Club

Dr. Alan Hooper

AEA Technology
Materials and Manufacturing Technology Division
B552, The Harwell Laboratory
Oxfordshire, OX11 0RA
United Kingdom

ABSTRACT
A brief review and status report is given of an industrial consortium programme, based at the Harwell Laboratory, which forms part of the United Kingdom National Programme on High T_c Superconductivity. Related programmes at Harwell are also summarised.

The Harwell-Industry Superconducting Ceramics Club (HISCC) Programme was formally launched in October 1988. The three-year programme has a budget of £2M and is jointly sponsored by the United Kingdom Department for Trade and Industry and a consortium of industrial companies, viz : Ford, BICC, BOC, Air Products, Oxford Instruments and Johnson Matthey. A Collaboration Agreement was also signed with the U.K. Ministry of Defence, through their Admiralty Research Laboratory at Holton Heath, in November 1989.

The overall objective of the Programme is to develop fabrication technologies for the high T_c materials appropriate to the application requirements of the industrial members. To date, the majority of work has been undertaken at Harwell in the areas of ceramic fabrication (1,2), swaging and wire drawing, sol-gel processing (3), plasma-spraying (3) and theory (4). In addition, a major subcontract with Oxford University (5)was set up in May 1989 for studies of high field behaviour, flux creep and microstructure. Work has also been undertaken at the University of Bangor (6)and at the Royal Military College of Science, Shrivenham (7). A Working Party, comprising both member representatives and Harwell has been established to consider materials' property targets and component performance specifications.

Studies have been carried out of both the YBCO and BISCCO-based materials, with emphasis on the latter. Tapes, wires and bulk ceramics with superconducting properties above liquid nitrogen temperatures have been fabricated using a variety of processing routes, for both types of material, (see illustrations). Prototype multi-filamentary wires have been produced and coils have been fabricated from drawn wire and via plasma-spraying. Densities in excess of 98% theoretical have been achieved for YBCO materials via powder and plastic-ceramic processing and up to 85% for high temperature transition (2223) BISCCO material. This latter material has exhibited critical current densities in excess of a thousand Acm^{-2} at liquid nitrogen temperature in zero field and at 4.2K in fields of up to 15T. Associated work, funded by the Ministry of Defence, includes the measurement and optimisation of the mechanical properties of bulk samples and the development of practical low resistance contacts.

In addition to the HISCC Programme, there are currently a number of other high T_c superconductivity projects underway at the Harwell Laboratory. These include a smaller industrial consortium aimed at the identification of novel superconducting compositions and an internally-funded programme concerned with radiation effects. In the latter case, interests include the possible beneficial effects of irradiation in the processing of superconducting ceramics and also the consequences of using such materials in radiation environments e.g. space. Other specialist areas of study, related to the detailed characterisation of high T_c materials, include Raman spectroscopy [8] and neutron scattering. A general theoretical solid-state physics capability is also available to underpin experimental studies.

REFERENCES

1. Briggs, A., Denton, I.E., Piller R.C., Wood, T.E. and Ferguson, G., British Ceramic Proceedings, No. 40 : " Superconducting Ceramics ", 37, (1988).

2. Perks, J.M., Denton, I.E. and Briggs, A., Proceedings of the E-MRS Meeting, Strasbourg, France, May 1990.

3. Edwards, R.E., Prentice, T.C., Rush, D.F., Scott, K.T. and Segal, D.L., The British Ceramics Society - Transactions and Journal, 89(1), 32, (1990).

4. Choy,T.C. and Stoneham, A.M., J. Phys.: Condens. Matter, 2, 939, (1990) and 2, 2867, (1990).

5. Denton, I.E., Briggs, A., Lee, J.A., Moore, J., Cowey, L., Jones, H., Grovenor, C.R.M. and Hooper, A., This publication - poster paper.

6. Ferguson, G., O'Grady, K., Briggs, A. and Denton, I.E., Cryogenics, 28, 688, (1988).

7. Edwards, M.R., Rogers, K.D., Spencer, J.W.C., Denton, I.E. and Briggs, A., to be published in J. Mat. Sci. Letts. (1990).

8. Graves, P.R. and Johnston, C., Proc. R. Soc. Lond., A420, 267, (1988).

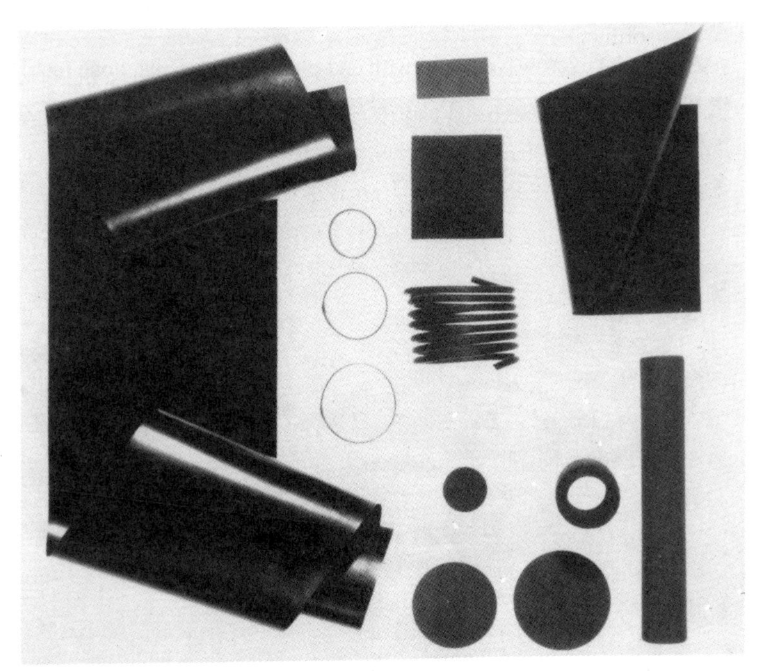

A selection of bulk YBCO components and plastic-ceramic precursors.

An YBCO microwave cavity with end-plates (approx. dia. : one inch).

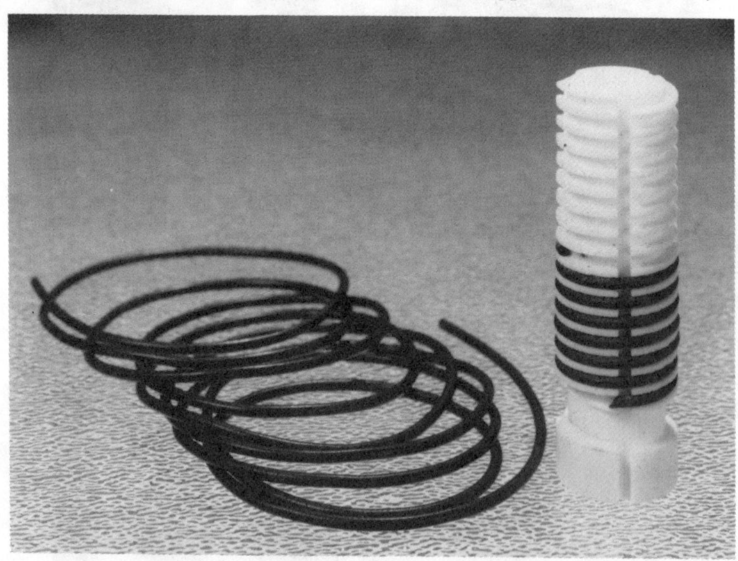

Extruded plastic-ceramic BISCCO "bootlace" precursor and sintered coil on a glass-ceramic former.

Flat and coiled BISCCO coatings deposited by spray pyrolysis onto silver substrates.

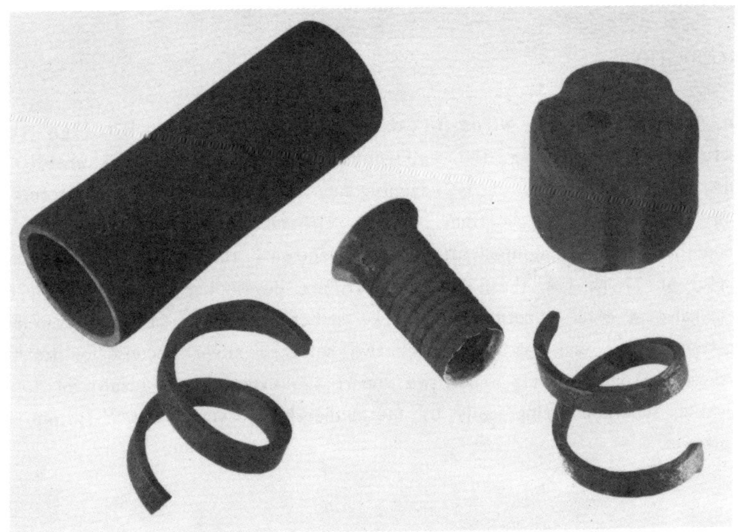

Free-standing YBCO and BISCCO components produced by plasma-spraying.

THE RECENT PROGRESS IN THE DEVELOPMENT ON HIGH-TEMPERATURE SUPERCONDUCTORS WITHIN MITSUBISHI CABLE INDUSTRIES, LTD.

K. YAMAMOTO, J. KAI, A. OKUHARA, Y. TAKADA, H. KATO and M. HIRAOKA

Central research laboratory
MITSUBISHI CABLE INDUSTRIES, LTD.

ABSTRACT : As a result of our researches on the fabrication processes of high-temperature superconducting tapes and films with high critical current density (Jc), we have obtained the tapes and the films with attractive properties by the metal sheathed process and the sol-gel process, respectively. With the metal sheathed process, 30 mm diameter coils of Ag-asheathed Bi-Pb-Sr-Ca-Cu-O superconducting tapes were prepared by the wind-and-react method. The coils have the maximum Jc of 6,300 A/cm^2 at 77 K over the full length of 60 cm. With the sol-gel process, Y-Ba-Cu-O superconducting films on Ag substrates were prepared using propionic acid solution of Y, Ba and Cu acetates. The films were about 7,000 A/cm^2 in Jc at 77 K, 0 T.

1. INTRODUCTION

In fabricating long wires of oxide superconductors with high transition temperatures (Tc), winding the superconductors into coils and stabilizing the superconducting environment, it is advantageous to make the superconductors in the forms of metal-clad tapes, films and multifilaments. In addition, to put the superconductors to practical applications, it is necessary to improve the critical current density (Jc) at 77 K. For these purposes, we are developing the fabrication processes of the metal-clad oxide superconductors, we selected Bi-Pb-Sr-Ca-Cu-O system[1)2)] and Y-Ba-Cu-O system[3)] and, as the metal, we selected silver because of its inactivity toward these systems. In this paper, we report the experimental results of fabrications of Bi-system superconducting coils by metal-sheathed process and Y-system films by sol-gel process.

2. EXPERIMENTAL

2.1 Fabrication of Bi-Pb-Sr-Ca-Cu-O Coils by Metal-sheathed Process

Bi_2O_3, PbO, $SrCO_3$, $CaCO_3$, and CuO powders were mixed in an atomic proportion of Bi:Pb:Sr:Ca:Cu=0.9:0.2:1.0:1.1:1.5 and calcined at 780 ^oC for 30 hours in air. The calcined powders were packed into the silicon-rubber molds and pressed by a CIP (Cold Isostatic Pressing) equipment. By the pressing, green rods 4.9 mm in diameter and 100 mm in length were obtained. The rods were then inserted into a silver tube 7 mm in outer diameter, 5 mm in inner diameter and 330 mm in length. Both ends of the tube were filled by Pb-Sn solder. The composite tube was cold-drawn and cold-rolled, and a tape about 10 m in length, 0.2 mm in thickness and 2.6 mm in width was obtained. The tape was cut in about 60 cm long and wound in a coil on a ceramic reel, and sintered at 830 ^oC for 160 hours in air.
In order to increase Jc, we added uniaxial pressing and re-sintering after the first sintering of this conventional wind-and-react method, because we had previously confirmed the effect of these steps in increasing Jc on 4 cm tapes. By uniaxial pressing, density of bulky superconductors increases and the c-axis of $Bi_2Sr_2Ca_2Cu_3O_y$ superconducting grains orients the direction perpendicular to the tape surface. Accordingly the grains obtain good coupling and Jc of the tapes considerably increase[5)6)], but to our knowledge, application of the steps to wind-and-react method has not been reported yet. For application of the intermediate pressing, the tapes about 60 cm in length were re-wound from the superconducting coil fabricated by the wind-and-react method and they were uniaxially pressed at 1.1×10^4 kgf/cm^2 every 6 cm length. Then they were wound into coils and sintered at 830 ^oC for 40 hours.

2.2 Fabrication of Y-Ba-Cu-O Films

As the first step, the best combination of solute and solvent was investigated to prepare homogeneous precursor in place of alkoxide, because alkoxide is unstable and expensive. As the solute, acetate, propionate, hydroxide and chloride, and as the solvent, formic acid, acetic acid and propionic acid were studied. In view of avoiding precipitation of solute during preparation of solution and coating process, the optimum combination was found to be acetate as the solute and propionic acid as the solvent[7)].

Y, Ba and Cu acetates of reagent grade were dissolved into propionic acid in a 1:2:3 molar ratio. This coating solution was spin-coated onto Ag substrates 10×10 mm^2

heated at 480 ^{o}C for 30 minutes in air in order to thermally decompose the precusors.

Various film thicknesses were obtained by repeating the process from 2 to 20 times from coating to the heat treatment. Finally the films were sintered at a temperature from 900 ^{o}C to 920 ^{o}C for 2 hours in flowing oxygen.

Tc of the films and Jc at 77K of the tapes and films were measured by standard dc four-probe method. The transport critical current (Ic) was determined with 1 μV/cm criterion, and Jc was calculated on the cross-sectional area of the specimens excluding silver. The orientation of the grains of the films was investigated by the X-ray diffractometer (XRD).

3. RESULTS AND DISCUSSION

3.1 Critical Current Density of the Coil

Figure 1 is a photograph of the coil fabricated by the wind-and-react method with intermediate pressing. Figure 2 shows the voltage-current (V-I) characteristic of the coil in liquid nitrogen, where the distance between voltage terminals was 60 cm.

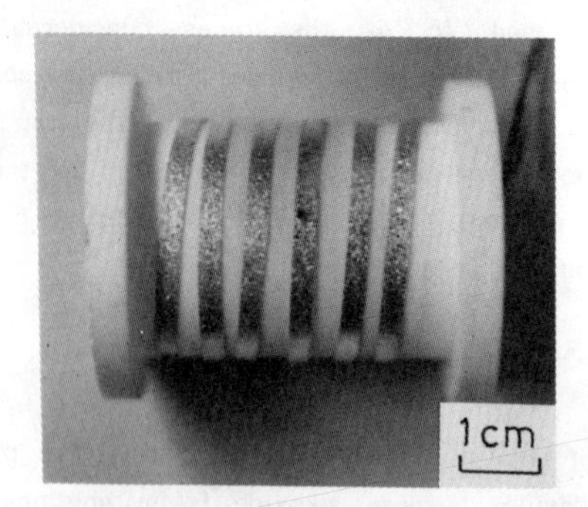

Fig.1 The coil fabricated by the wind-and-react method with intermediate pressing at 1.1×10^4 kgf/cm^2.

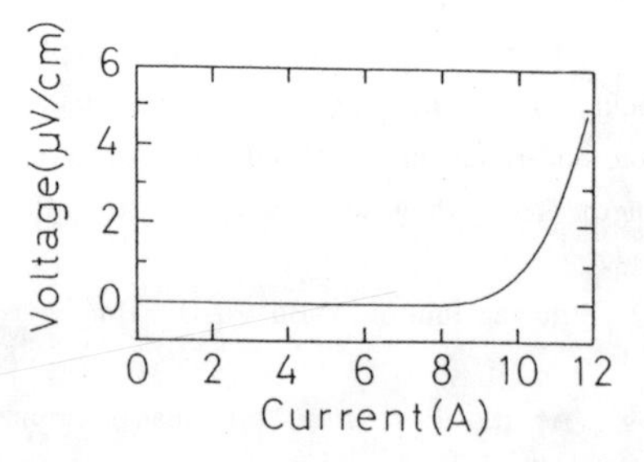

Fig.2 V-I characteristic of the coil in Fig.1. The distance between voltage terminals was 60 cm.

Figure 3 is a photograph of the polished cross-section of a tape cut from the coil. From Fig. 2, Ic of the coil was determined to be 10.1 A, and from Fig. 3, cross-sectional area of the tape excluding the Ag-sheath was 0.16 mm^2. Therefore Jc of the coil was 6,300 A/cm^2 at 77 K. The maximum magnetic field was calculated to be approximately 40 gauss.

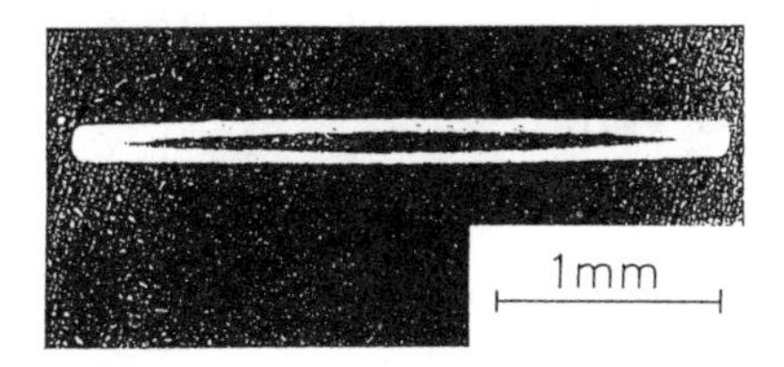

Fig.3 The polished cross-section of a tape cut from the coil in Fig.1.

The Jc, however, was lower than that obtained in our previous experiments on 4 cm -long tapes. Figure 4 shows a result of the previous experiments : low magnetic field dependence of Jc at 77 K for 4 cm-long tapes prepared under three different conditions. In this figure, S(t) denotes sintering time (hour) at 830 ^{o}C and P denotes uniaxial pressing at 1.7×10^4 kgf/cm^2. The preparing condition of S(160) + P + S(40) for a 4 cm-long tape was almost the same as that for the coil, but Jc of the tape was approximately 14,000 A/cm^2, almost twice as high as that of the coil. Figure 5 shows the measuremental results of Jc on every turn of the coil (9.8 cm in distance) to investigate the homogeneity in longitudinal direction of the coil, where Jc varied from the maximum of 8,100 A/cm^2 to the minimum of 5,700 A/cm^2. We found that the scattering of Jc over the coil decreases the total Jc. We guess that another reason for lower Jc of the coil was lower uniaxial pressure for the coil, because Y. Yamada et al.[8)] already reported that Jc increased as increasing the uniaxial pressure and saturated at 2×10^4 kgf/cm^2 and higher.

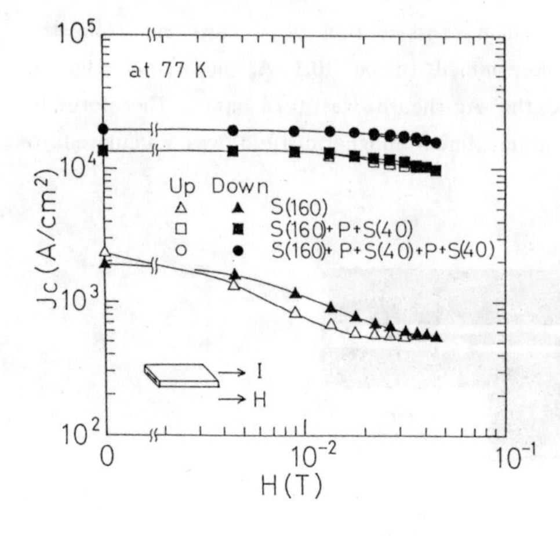

Fig.4 Low magnetic field dependence of Jc at 77 K for Ag-sheathed Bi-Pb-Sr-Ca-Cu-O superconducting tapes. This result was obtained in our previous experiments on 4 cm-long tapes. S(t) denotes sintering time(hour) at 830 ^{o}C and P denotes uniaxial pressing at 1.7×10^4 kgf/cm^2.

Fig.5 Jc at 77 K on every turn of the coil in Fig.1. The distance between voltage terminals was 9.8 cm.

3.2 Properties of the Films by Sol-gel Process

Figure 6 shows the film thickness dependence of Tc (R = 0) and Jc, where Tc increased in proportion to the film thickness and saturated around 88 K at the thicknesses from 4.0 to 6.0 μm. The maximum Jc, however, was obtained at 3.2 μm thickness. The result is not coincident with the Tc characteristics. Figure 7 shows the XRD patterns of the film surface. This figure indicates that c-axis orientation of $YBa_2Cu_3O_{7-\delta}$ grains in the films of thicknesses between 3.2 and 4.0 μm is clearer than in other films. Since it has been reported that c-axis orientation of $YBa_2Cu_3O_{7-\delta}$ grains perpendicular to the film surface increases Jc, the results mentioned above show that Jc of the films were influenced not only by Tc but also by c-axis orientation of the grains.

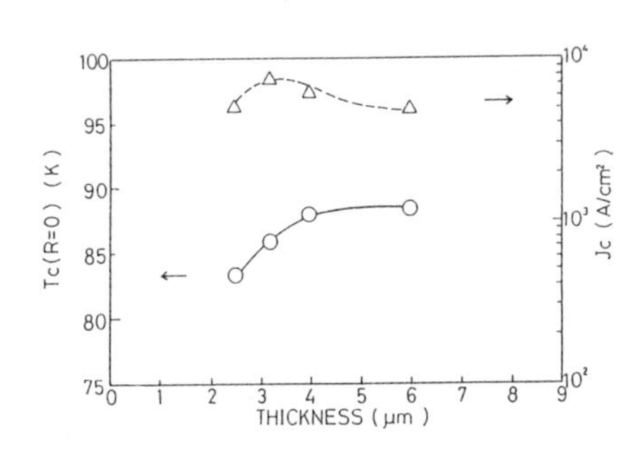

Fig.6 The film thickness dependence of Tc(R=0) and Jc(77 K,0T). The films were sintered at 900 ^{o}C for 2 hours.

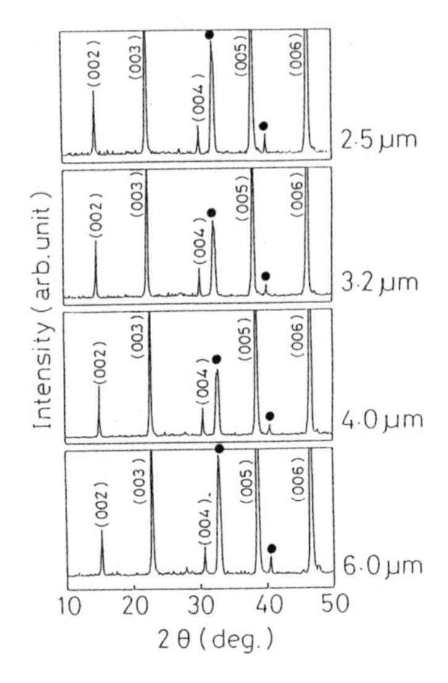

Fig.7 X-ray diffraction patterns for the $YBa_2Cu_3O_{7-\delta}$ films with various thicknesses. The closed circle denotes the (h k l) peaks other than (0 0 l) peak.

4. CONCLUSION

In order to develop high-temperature superconducting wires, we fabricated 30 mm diameter coils of Ag-sheathed Bi-Pb-Sr-Ca-Cu-O superconducting tapes by the metal-sheathed process and Y-Ba-Cu-O superconducting films on Ag substrates by the sol-gel process. The coils have the maximum Jc of 6,300 A/cm^2 over the full length of 60 cm and the films have the maximum Jc of 7,000 A/cm^2 at 77 K in zero magnetic field. The results suggest that the metal-sheathed process and sol-gel process are suitable for fabricating metal-clad tapes and films of oxide superconductors. Though these Jc (77 K) were much lower than Jc (4.2 K) of Nb-Ti and Nb_3Sn superconducting wires with low Tc, it is possible to increase Jc remarkably by improvement in fabrication processes. Our future R&D subjects include the increase in Jc in high magnetic fields as well as the fabrication of longer tapes, larger coils and multifilaments.

REFERENCES

1) H. Maeda, Y. Tanaka, M. Fukutomi and T. Asano, Jpn. J. Appl. Phys., 27, L209 (1988)

2) M. Takano, J. Takada, K. Oda, H. Kitaguchi, Y. Miura, Y. Ikeda, Y. Tomii and H. Mazaki, Jpn. J. Appl. Phys., 27, L1041 (1988)

3) M. K. Wu, J. R. Ashburn, C. J. Torng, P. H. Hor, R. L. Meng, L. Gao, Z. J. Huang, Y. Q. Wang and C. W. Chu, Phys. Rev. Lett., 58, p.908 (1987)

4) K. Yamamoto, Y. Nishikawa and M. Hiraoka, Cryogenic Engineering, 25, No. 2, p.103 (1990), (in Japanese)

5) T. Hikata, K. Sato and H. Hitotsuyanagi, Jpn. J. Appl. Phys., 28, L82 (1989)

6) K. Osamura, S. S. Oh and S. Ochiai, Supercond. Sci. Technol., 3, p.143 (1990)

7) M. Hiraoka, Y. Takada, A. Okuhara and H. Kato, Preprint of The Beijing International Conference on High Tc Superconductivity, September 4-8, (1989) Beijing, China

RESEARCH ACTIVITIES AT MATERIALS RESEARCH LABORATORIES, ITRI

Dr. Chien-Min Wang

Deputy General Director, MRL
Industrial Technology Research Institute
Chutung, Hsinchu, Taiwan 31015, R.O.C.

Ladies & Gentlemen:

It is my great pleasure to have this honor to introduce to you the Materials Research Laboratories (MRL) of Industrial Technology Research Institute (ITRI) in Taiwan, R.O.C. Before talking about MRL, I'd like to give you a little background on ITRI.

About two decades ago, Taiwan was basically an agricultural country. Her industrial production growth was just about the same as that of her agricultural production, as shown in Fig.1.

In order to feed more than 10 million people at that time and bring them better living standard & environment on this tiny island, the government of ROC devoted her every effort to industrialize the country. Industrialization, as everybody knows, is not an easy process. It needs to put the industrious and well trained labors, enormous resources and technologies together to make it work.

Relatively speaking, prominent technology is more difficult to acquire than labors and material resources. Besides transfering technology from abroad to up-grade technical level we must have our own R&D program.

In the early '70s some high ranking officials lead by former Premier of the Executive Yuan, Sun Yuen-Shian initiated to charter a national research institute in Hsinchu area, in the northern part of Taiwan. In 1973, Industrial Technology Research Institute was established with electronics and chemicals as its major research fields. Through 17 years passage, 8 more Labs or Centers were established, such as materials, machinery, energy & natural resources, electro-optics, computer & communication, measurement standards, pollution control and industrial safety, to meet the needs of local industries (Fig.2). The ITRI is the only organization for applied reserach in the ROC with the goal of raising the technology level of the industry.

MRL was established in 1983. So, we're pretty young compared with other Labs in the world. Therefore, we must work hard to win our stripes. MRL is a non-profit organiza-

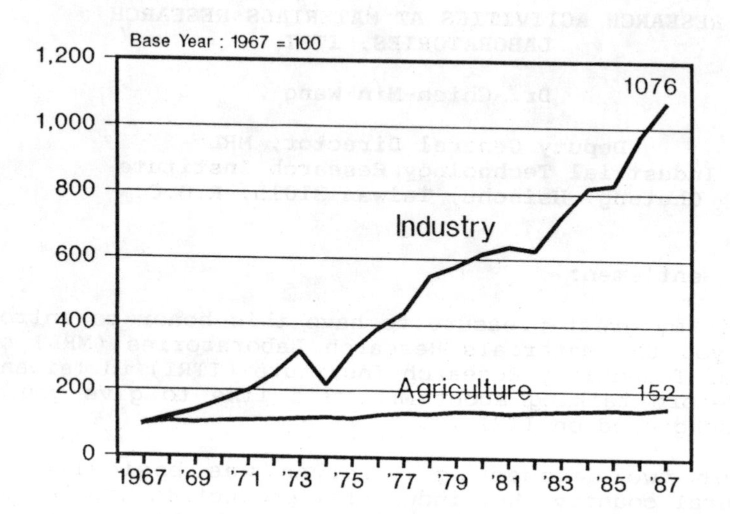

Fig. 1 The growth of industrial production v.s. agricultural production in R.O.C.

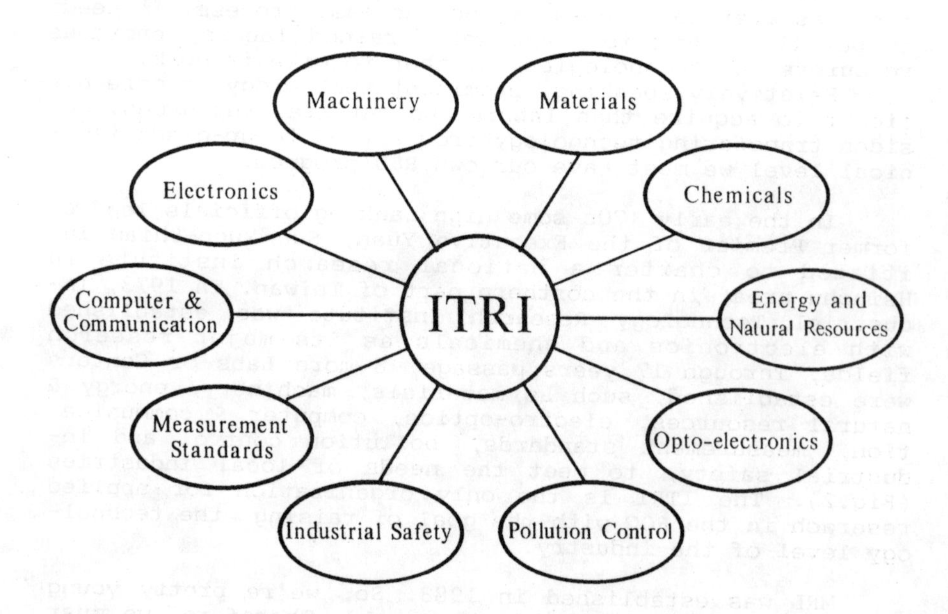

Fig. 2 Major research field.

tion in which the major sources of funding comes from both the government and the industrial sector. MRL accepts government contracts to undertake applied research relevant to national needs and to provide assistance to small and medium-size industries. MRL also accepts contracts from the industrial sector for cooperative research and development of specific products. Our goal is to keep the government/industrial sector ratio of 2 to 1. The total budget in FY1991 (started from July 1, 1990) is about 1 billion NT dollars.

At the present time, MRL has 606 employees including 75% research staff. Dr. P.T. Wu is now in charge of the MRL. Dealing with the human resources development, we choose dual ladder management system that the executive manager and R&D researcher are equally important. The salary policy is maintained above the average local job market. Even we have plenty budget and sophisticated equipment, we still need qualified people to do good research work. Therefore, MRL encourage her employees to pursue advanced degrees or to take on-job-training program. There are now 40 people in such program domestically and abroad.

In order to raise the technology level of the materials industry in the ROC, MRL set up an industrial materials R&D program to carry out the following missions:

* Indentify/solve bottlenecks in materials for existing industries.
* Increase added values of industrial products.
* Develop key materials and related technologies for new industries.
* Develop advanced materials technology and their applications.

At the stage I of Phase I Development, the establishment of fundamental capability was our primary task. At the stage II, we worked on the development of new materials technologies for the ROC industry.

MRL transfers technology to the local private companies at appropriate times through open and fair procedures. Fig.3 shows a summary of technologies developed in MRL and have been transfered to local industries.

Fig.4 shows that MRL pays much attention on the intellectural property protection. There are 52 patents applied and 27 patents approved in FY1990. Regarding publications & seminars, over 260 technical papers and 4 periodicals are published in FY1990 and 50 workshops are held annually. On a more specific level, MRL provides government, industries, and academia with a large array of technical services, information and expertise.

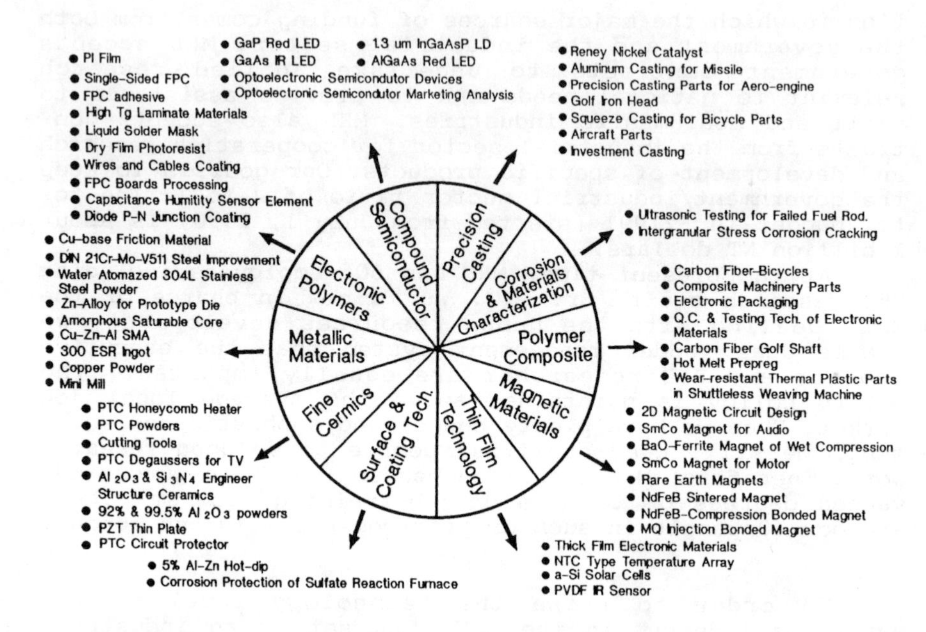

Fig.3 The contribution of MRL technologies to R.O.C. industry.

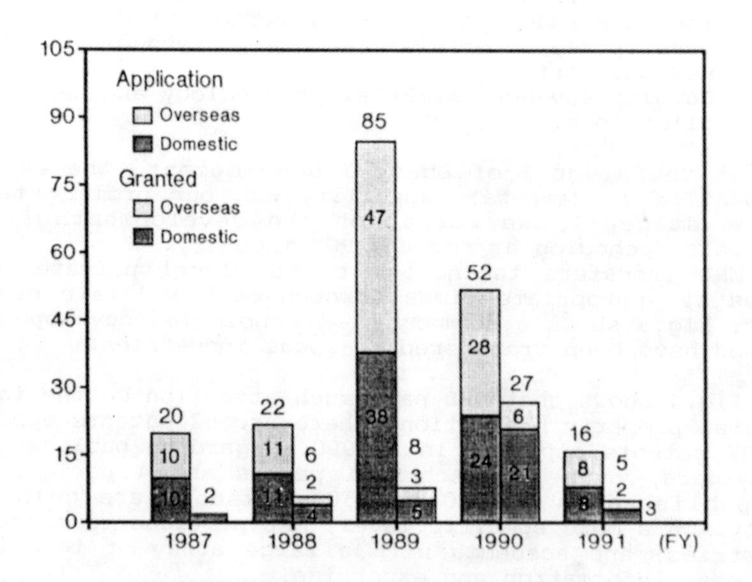

Fig.4 MRL patent application.

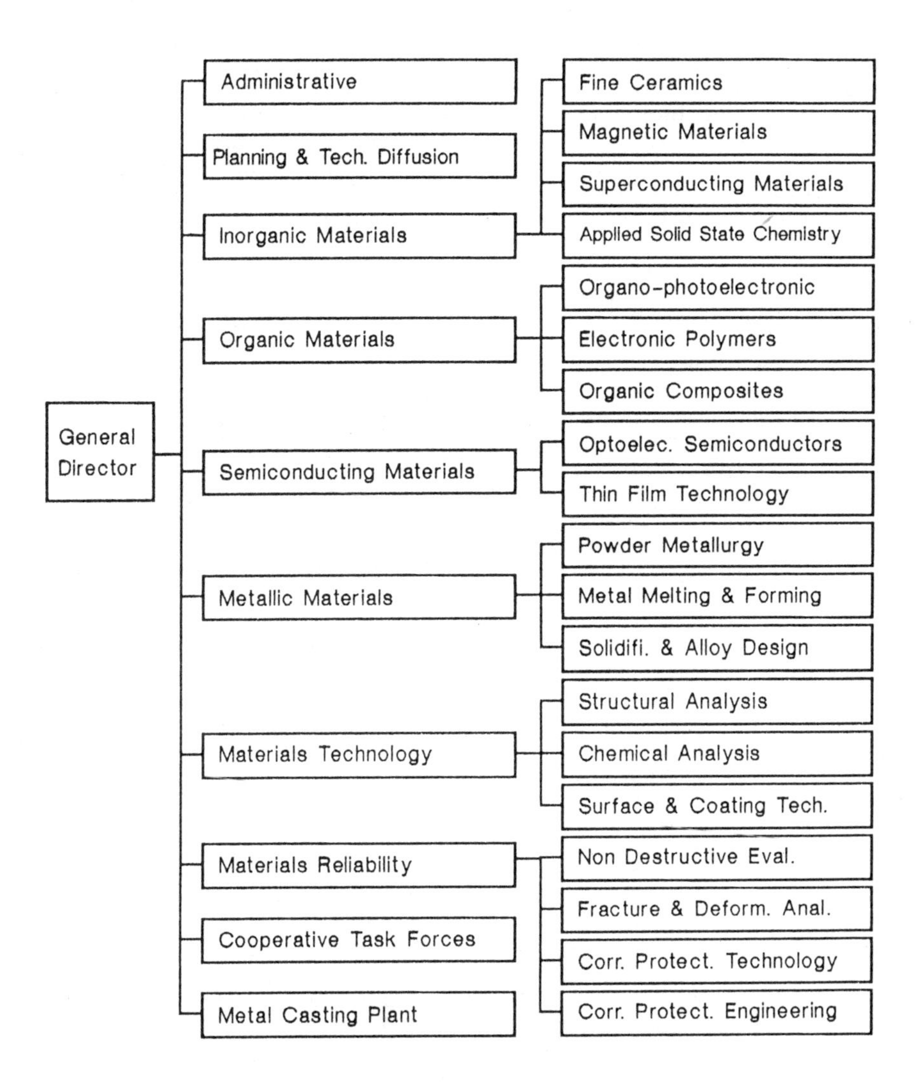

Fig.5 Organization chart of Materials Research Laboratories.

In the past 8 years, MRL grew very fast. Fig.5 shows the latest organization chart. There are 6 reserach divisions and 19 laboratories, covering inorganic materials, organic materials, semiconductors, metals, materials technology as well as materials reliability. I'll briefly introduce the research work and some highlights of each division. In the field of inorganic materials, significant progress has been made in electronic ceramics, permanent magnets, and high Tc superconductor. For example, many advanced ceramic technologies have been transfered to local industries, such as $BaTiO_3$ PTC, PZT, Al_2O_3 substrate, and Si_3N_4 guide roller; many magnetic materials & technologies have been developed to increase product value and replace import materials. As for the high Tc superconductor research, we have completed the following works:

1. The discovery of high Tc superconductivity in the new system $(Tl_{0.5}Pb_{0.5})Sr_2(Ca_{0.8}Re_{0.2})Cu_2O_y$ with Tc up to 109K.
2. Accelerated formation of the high purity (Bi-Pb)-Sr-u-O (Tc=110K) and Tl-(Pb,Bi)-Sr-Ca-Cu-O (Tc=118K) superconductors.
3. Preparation and magnetic studies on the bulk $YBa_2Cu_3O_{7-x}$ materials with transport Jc>37300A/cm^2 at 77K.
4. Magnetic recording behavior study on the reliable and stable ultra-thin film of YBCO epitaxially grown under thermodynamic equilibrium conditions.

In the field of organic materials, our research activities include the following three important areas. For organo-photoelectronic materials, we have been working on organic optical materials (e.g. optical disk, optical polymer, and ferroelectric liquid crystal) and electrophotographic materials (e.g. OPC, toner). For electronic polymers, we have made great contributions on flexible printed circuits & high Tg laminates for PCB, and now we are working on high performance plastic packaging materials (e.g. TAB, thin film packaging) and photoimageable polymer (ICPR, EDPR, liquid solder mask). For organic composites, we have been working on matrix developing (BMI, HPTP), structural design (3D, dynamic, simulation), processing (autoclave, SMC, prepreg), and product development (machinery parts, aircraft parts). Organic photoconductor (OPC) is a newly developed materials applicable to laser printer and copiers. MRL has successfully synthesized a new polymer suitable for making OPC drum and obtained 4 patents.

Optoelectronic semiconductor division including 62 scientists and technicians engaged in material growth, device design, fabrication, testing, and packaging for III-V compound semiconductor and thin film sensor technology. The following research topics on III-V compound semiconductor materials and devices have been sucessfully carried out.

* Bulk single crystals growth:
  GaP, GaAs, InP, InAs.
* Epitaxy technologies:
  LPE, MOCVD.
* Devices:
  RED LED, IR LED, LD, HP LD.

In the area of thin film technology, the progress is rapid. The materials & devices which have been developed are:

* InSb MR sensor
* SiC thermistor
* Pt resistance temperature detectors
* Amorphous Si solar cell

Metallic materials division is divided into three focused areas such as powder metallurgy, solidification & alloy design, and metal melting & forming. For powder metallurgy, we focus on friction materials, special brazing materials (RSP), and metal injection technology. For solidification and alloy design, we do researches on Al-based ceramic whisker reinforced composites, near-net shape casting technology, and computer aided casting design system. For metal melting and forming, we work on AIF melting, NiTi alloys wire-drawing, and computer aided metallic materials selection.

MRL is well equipped with various kinds of sophisticated characterization facilities and instruments in the materials technology division. It functions as a central resource for all R&D groups in MRL. For structural analysis, we have SEM, XRD, STEM, HRTEM, EDS, EELS, EPMA, SAM/ESCA, and FTIR-PAS. For chemical analysis, we have ICP-AES, AAS, C/S analyzer, spark-AES, XRF, UV/Vis, ion chromatography, electrodeposition. For surface and coating technology, we conduct researches on cermet composite plating, electroetching of aluminum foil, fully additive process (electroless copper), and surface modification of metals.

Corrosion and protection of metals is a serious problem in Taiwan under the subtropical climate. The development of methods to predict and extend the life spans of existing construction is also an important issue. In the division of materials reliability, there are 4 laboratories, such as non-destructive evaluation, fracture and deformation analysis, corrosion protection technology, and corrosion protection engineering. The materials evaluations & reliability group has made a tremendous contribution to a number of institutes including Taiwan Power Co., Chinese Petroleum Corp., Taiwan Fertilizer Co., etc.

Besides the research work done in our field, MRL has extensive and frequent contact with academic and industrial circles domestically and abroad. In this global technical interaction chart, we only list out the related foreign institutes as reference.

In Fig.6, MRL strategies for the materials technology development are divided into three milestones. The first is the Technology Oriented Phase (1985-1990), with only one large MOEA Project from our government for establishing the fundamental material capability. Substantial contributions to local industries have been accomplished as mentioned before. Since FY1991, we began to conduct 5 or more Industrial Oriented Projects simultaneously. The management system is a big concern. Basically, guidelines for Phase II Developments are

1. To upgrade the added-values of ROC's electronic and information products through the development of key materials and technologies within these industries.
2. To develop high added-value opto-electronic materials and for the furtherance of ROC's optoelectronic industry.
3. To enhance the global competitiveness of ROC' smachineries through the development of new metallic materials.
4. To restructure the material protection technologies so as to increase the reliability of the industrial infractures, to promote industrial safety and enhance the ty and reliability of the products.
5. To develop high Tc superconductor application technologies for the establishment of superconducting industry.
6. To conduct researches on materials design and processing or the support of ROC's aviation industry and for the nhancement of product qualities within the biomaterials industries.

At the time of reorganization, we perform a modified matrix management system. This means that the research manager of each laboratory plays a pivot role in the execution and control of the subproject and researchers of the related fields are regrouped in the same research team.

By the year 2000, MRL will have 800 staff members with 680 professionals. The annual budget is about 2.2 billion NT dollars with 60% coming from the government, 38% from the industry and 2% from royalties. Through the dedication of its staffs and with accumulation of experiences, MRL will become one of the best international laboratories and the centers of excellence for advanced composites, advanced powder technology, opto-electronic materials and materials design & selection. Through technology transfer, technical services and consultations, MRL will be a vital partner in ROC's industrial development.

Ladies & Gentlemen, I have just given you an overall introduction of MRL. It seems that this young research Laboratories do have done some contributions to local materials industries in Taiwan. However, one thing that I didn't mention is MRL considers its achievements based on certain values. Without those values in our heart, I don't think that we can run MRL very successfully. Those MRL values are

*** benefit to society**
*** excellence**
*** diligence**
*** industrial impact**
*** synergism**
*** respect to the individual**

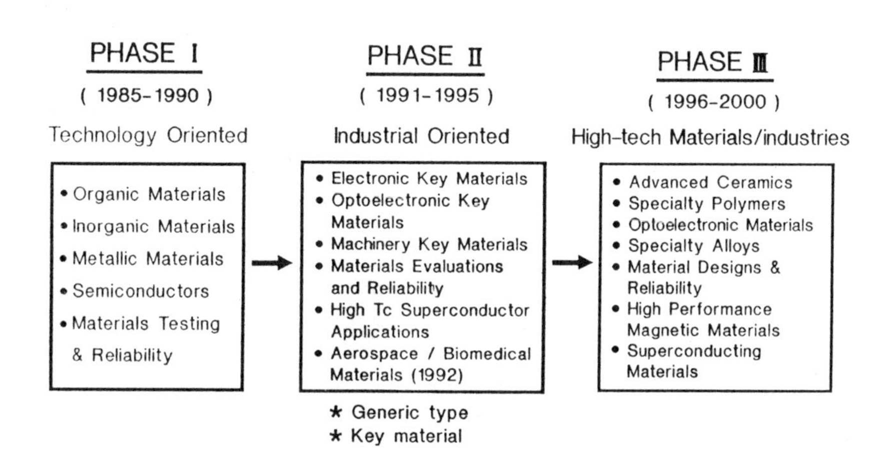

Fig.6 Strategies for the material technology development at MRL, ITRI, Taiwan, R.O.C..

Progress of Superconductivity Research and Development in Korea

Dong-Yeon Won, Hee-Gyoun Lee, Chan-Joong Kim, and Jong-Chul Park*

Korea Atomic Energy Research Institute

P.O.Box 7 Daedok Science Town, Daejun

305-353 Korea

* Korea Standards Research Institute

Abstract

Research on high temperature superconductivity in Korea started in early 1987. High Temperature Superconductivity Research Association(HITSRA) was organized in June, 1987, to promote and coordinate the research. Ministry of Science and Technology(MOST), recognizing the importance of high Tc superconductor technology choose the project of development of high temperature superconductor technology as one of the national research and development projects. In this talk, current status and future perspectives of high temperature superconductivity research in Korea will be briefly discussed. Additionally, research activity on superconductivity in Korea Atomic Energy Research Institute(KAERI) is presented.

I. Introduction

The discovery of high temperature superconductivity in copper-oxide based compounds by Bednorz and Muller[1], and followed discovery of YBaCuO type superconductors with critical temperature (Tc) higher than the boiling point of liquid nitrogen(77K) by Chu et al.[2], led to intensive research and development effort around the world. Race in quest of material with still higher Tc has taken place with unprecedented intensity in the history of physics and of modern engineering as a whole; it could be termed as ' scientific gold rush' which has continued to persist until this day.

The discovery has affected not only the scientists and engineers, but also the general public who have a little interest in the advancement of science and technology. People got informations from newspapers and television on high Tc superconductivity about possible applications such as the magnetically levitated trains, superconducting magnetic storage system, superconductive power transmission cables, superconducting motors and generators, superconducting computers and many more. Last 3 years have been truly exciting period for us who were engaged in this field of science.

In this talk, I will review the current status of research on high Tc superconductivity research in Korea, with emphasis on the national research and development project on high Tc superconductivity financed by the Ministry of Science and Technology(MOST). And the research activity on superconductivity in Korea Atomic Energy Research Institute (KAERI) will be introduced at the same time.

II. High Temperature Superconductivity Research Association (HITSRA)

In Korea, the first high temperature superconductor was synthesized by Z.G. Khim, SNU, in early March 1987[3]. He had succeeded in reaching 55K with YBaCuO, and by the end of March 123 phase was obtained. In early May, the MOST recognizing the importance of high Tc technology solicited for research proporsals on the subject. Many proposals from universities and from government supported research institutes were submitted, and a voluntary organizaiton called High Temperature Superconductivity Research Association (HITSRA) was formed to coordinate the research efforts more efficiently.

The major fuctions of the HITSRA are as follows.

- Promotion and coordination of research activities on high Tc superconductivity
- Providing expert advice to gorvenment in funding and policy making
- Organizing workshops and conferences.
- Information center.

The HITSRA had organized 14 workshops since the first one which was held on July 6, 1987 at POSTECH. At workinshop typically 200 people attended and about 30 papers were presented for a day event. There are 18 members in the HITSRA, consisting of representatives from universities, research institutes, and industries.(see footnote #1)

III. National Research and Development Project on High Temperature Superconductivity.

National R&D projects are funded by the Ministry of Science and Technology(MOST) to cope with specific national needs. The objectives of the project are broad and general, namely, development of practical high Tc supercondutor technology, buildup of research capability, and training of manpower in preparation for the coming superconductor age.

Near-term goals include

- Exploring for new material with still higher Tc
- Fabrication of bulk material with $J_c > 10^5$ A/cm^2(77K)

Footnote #1

Duk N. Yoon(KAIST), Sunggi Baik(POSTECH), Keun-Ho Chang(Gold Star), Sang-sam Choi(KIST), Min-su Jang(PNU), Zheong G. Khim(SNU), Mokhee Kim(KOSEF), Kang S. Ryou(KERI), Yoon-soo Kim(KRICT), Dong-nyung Lee(SNU), Jung-il Nam(KEPCO), Jong-chul Park(KSRI), Sin-jong Park(ETRI), Hyun-joon Shin(RIST), W.T. Son(ADD),Jae-kyun Yang(Ssangyong), and Dong-yeon Won(KAERI)

- Growth of single crystal and thin film of good quality
- Establishment of manufacturing technology of wire and ribbon
- Fundamental research on development of magnet and SQUID
- Experimental and theoretical studies on basic mechanism of high Tc superconductors.

Participating organizations are KSRI, KAIST, KRICT, KIMM, KERI, ETRI, KAERI, SNU, PNU, Yeonsei U., Sung Kwun Kwan U., and RIST(See footnote #2). They are devided into four groups depending on their specialities, namely, basic science and characterization, materials synthesis, fabrication and process, and application. Total number of researchers are 243, consisting of 53 pricipal researchers (professors and associate professors included), 63 senior researchers (assistant professors), 66 researchers, and 61 technicians and graduate students.

In the followings, achievements and current activities of the project are briefly described. In materials preparation Y-system as well as Bi- and Tl- systems have been synthesized. YBaCuO and BiSrCaCuO single crystals, typical size of 4x5x0.1 mm^3 were grown by several universities and reserach institutes. Thin films were prepared by various methods at many universities and institutes. Most of the physical properties, electrical, magnetic, thermoelectric, IR, Raman and photoacoustic spectroscopy were studied. Structural studies were also carried out by many groups. Some efforts of producing wire using Ag-sheathed YBaCuO and Bi-system and rapid solidification process was also tried.

Footnote #2

KAIST: Korea Advanced Institute of Science and Technology
KRICT: Korea Research Institute of Chemical Technology
KIMM : Korea Institute of Machinery and Metals
KERI : Korea Electric Technology Research Institute
ETRI : Electronics and Telecommunications Research Institute
KAERI: Korea Atomic Energy Research Institute
RIST : Research Institute of Industrial Science and Technology

It should be mentioned that there are several other active groups, besides the above mentioned groups, working on high Tc superocnductivity. They are universities such as Sogang U., Aju U., Hanyang U., and industy laboratories such as Samsung, Hyun Dai, Gold Star, Ssangyong, Daewoo, and others.

IV. High Tc Superconductivity Research in KAERI

Since Sept. 1987, research on high Tc superconductivity started in KAERI. Research activities in KAERI are classified into two directions; material development and development of application technology. The goals on superconductivity research in KAERI are as follows;

Long-term goals

- Fabrication of superconductive wire with $Jc>10^5 A/cm^2$
- Preparation of thin film with $Jc>10^6 A/cm^2$
- Fabrication of superconducting bearing.

Near-term goals

- Fundamental research on materials development
- Establishment of manufacturing techniques of thin film and bulk (Melt process, CVD etc)
- Investigation of the effects of neutron irradiation on superconductivity

The achievements of superconductivity research which have been carried out in KAERI since Oct. 1987 are summarized as follows;

- Effects of Fe doping on the phase transition in YBaCuO System[4]
- Pb-doped BiSrCaCuO superconductor was synthesized(May 1988)[5]
- Enhancement of Jc by repelletizing and resintering in Pb-doped BiSrCaCuO superconductor (July 1988)
- Formation process of high Tc phase in Pb-doped BiSrCaCuO superconductor [6]

- Effects of Pb doping on the lattice parameters of high Tc and low Tc phases[7]
- Effects of Sr/Ca ratio on the structure of superconducting phases in Pb-doped BiSrCaCuO superconductor [8]
- Effects of neutron irradiation on superconducting behavior[9]
- Effects of compaction pressure on texturization in Pb-doped BiSrCaCuO superconductor [10]
- Pellet expansion in Pb-doped BiSrCaCuO superconductor [11]
- Preparation of high magnetization bulk YBaCuO superconductor[12]
- Preparation of YBaCuO thick film by peritectic reaction[13]
- Preparation of high Jc YBaCuO superconducting thin film by CVD(Jc > 10^5 A/cm^2) [14,15]
- Fabrication of superconducting tape(Jc=1000A/cm^2)(Sept. 1989)

It may be valueable to point out that 105K Pb-doped BiSrCaCuO, superconducting tape, high Jc YBaCuO thin film, and high magnetization YBaCuO bulk superconductor were firstly fabricated by KAERI in Korea. It is also noticeable that some important results on high Tc superconductivity were issued out; a enhancement of Jc by repelletizing, pellet expansion, lattice parameter increase of superconducting phase by Pb doping, neutron irradiation effect on Tc, and effect of Sr/Ca ratio on the formation of superconducting phases in Pb-doped BiSrCaCuO system were investigated. Based on the above results, the superconductivity research in KAERI is now on turning point from basic research toward application.

V. Conclusion

Research on high temperature superconductivity in Korea is very active and more researchers are getting interested in this exciting field. It is encouraging that the government also has keen interest in the development of superconductivity technology. It is a rare opportunity for a developing country to start research science and technology at about the same time, along with the most advanced countries in the world. There are many problems, of course. More research fund is absolutely nescessary. And active participation of industy laboratories and centralizaition of the research effort of the government, by establishing of consortia and reserach centers, respectively, will be necessary to carry out goal oriented research successfully.

references

1. Bednorz and Muller, Z. Phys. B64, 189 (1986)
2. C.W. Chu et al. Phys. Rev. Lett. (1987)
3. The Dong-A Ilbo, March 19 (1987).
4. C. J. Kim, D. Y. Won, et al., " The orthorhombic-to-tetragonal phase transition in YBa(Cu, Fe)O ",Jour. Mat. Sci. 25, 2165 -2169 (1990)
5. H. G. Lee, D. Y. Won, et al., "Study on the stabilization of the high T phase in BiSrCaCuO system" Appl. Phys. Lett. 54 (4), 23 Jan. (1989)
6. C. J. Kim, D. Y. Won, et al., " The Formation of the High-Tc Phase in Pb-Doped Bi-Sr-Ca-Cu-O System" Japan. Jour. Appl. Phys. VoL. 28, No. 1, PP. L45 ~ L48, Jan. (1989)
7. C. K. Rhee, D. Y. Won, et al., "Effects of Pb Content on the Formation of the High-T Phase in the (Bi, Pb)-Sr-Ca-Cu-O System" Jap. Jour. Appl. Phy. VoL. 28, No. 7, PP. L1137 ~ L1139, July, (1989)
8. H. G. Lee, D. Y. Won, et al., "Structure and Superconducting Properties of the BiPbSrCaCuO System (0 < x < 4)" Jap. Jour. Appl. Phys. VoL. 28, No. 7, PP. L1151 ~ L1153, July, (1989)
9. Y. H. Herr, D. Y. Won, et al., " Effect of Neutron Irradiation on T of Pb-Doped BiSrCaCuO Superconductor" Japan. Jour. Appl. Phys. Vol. 28, No. 9, PP. L 1561 ~ L 1563, Sep. (1989)
10. C. J. Kim, D. Y. Won, et al., "Texture formation in PbBiSrCaCuO bulk superconductor" Jour. Mat. Sci. Lett. 9, 774 - 775 (1990)

11. C. J. Kim, D. Y. Won, et al,. " Expansion of BiSrCaCuO Pellets" Mat. Sci. Eng., B3, 501 -505 (1989)

12. C. J. Kim, D. Y. Won, et al,. " Preparation of YBaCuO Thick Film on (001) MgO Single Crystal by the Peritectic Reaction", submitted to 3rd International Symposium on superconductivity, Sendai, Japan. Nov. (1990)

13. H. G. Lee, D. Y. Won, et al,. "Formation of YBaCuO films on various substrates by Chemical Vapor Deposition", submitted to Jour. Mat. Sci.(1990)

14. H. G. Lee, D. Y. Won, et al,. "Pressure Dependence of the Formation of YBaCuO films by Chemical Vapor Deposition", submitted to Jap. Jour. Appl. Phy. (1990)

ISOTOPE SUBSTITUTION AS A METHOD OF HTSC STUDY

V. Ozhogin

Kurchatov Institute of Atomic Energy
Moscow 123182, USSR

The isotope substitution as a method of solid state studying is especially fruitful for HTSC because of a great variety of elements incorporated there. Generally speaking, an isotope has four features, not one: mass, spin, cross section for neutron scattering (σ_s) and neutron absorption (σ_a) and Moessbauer levels. Any of these features may be used for HTSC study.

One believes that an isotope mass shift of the critical temperature T_c is the "experimentum crucius" for the ascertainment of HTSC mechanism. However this is true only for monoelement superconductors (SC). The situation is more complicated for polyelement SCs and is discussed here on the base of various experiments including the Kurchatov Institute Isotope Group's (KIIG) study of the T_c shift for YBCO with the isotope substitution over three elements and of the T_c shift for HTSC Tl-compound with ^{18}O substitution for ^{16}O.

Nuclear magnetic and quadrupole resonances can also be related to the class of isotope effects. The relevant nuclei are ^{17}O, ^{43}Ca, $^{63;65}Cu$, $^{135;137}Ba$, ^{139}La, ^{183}W, $^{205;203}Tl$, ^{209}Bi, etc. The resonance parameters, e.g. resonance frequency shift, line shape and relaxation rates, provide a lot of information for the phenomenon understanding. KIIG concentrates itself on rare isotopes in ordinary HTSC and on ordinary isotopes in rare HTSC. The group exhibits here its recent results on ^{17}O in Tl(2212), on ^{43}Ca in Bi(2212), Tl(2212), $(LaCa)_2CuO_4$ and on ^{135}Ba in Y(123).

As to neutron diffraction experiments the isotope dependence of scattering intensity is also useful for HTSC study. KIIG presents the data from phonon spectra as determined by the neutron diffraction in ceramics YBCO with ^{65}Cu (σ_s=0.248 barn/a.m.u.) and with natural Cu (σ_s=0.124 barn/a.m.u.). The subtraction of the curves permits to identify the spectrum region relevant to Cu-atom vibrations and then also to oxygen atoms.

Very curious but complicated as well is an experiment proposed which uses the enormous difference between cross sections for neutron absorption in ^{157}Gd (242000 barns) and ^{160}Gd (0 barn) nuclei as ingredients of GdBCO. The thermal neutron irradiation of a sample with ^{157}Gd can create a quite unusual artificial solid state with unexpected SC properties.

The conclusions are as follows.

1. The isotope substitution is a powerful means for HTSC study.
2. The presence of an isotope shift of T_c is not a necessary evidence of electron-phonon mechanism of pairing.
3. The presence of an isotope shift of T_c is not a sufficient evidence of electron-phonon mechanism of pairing.
4. The magnitude of an isotope shift of T_c is an important test for any microtheory of HTSC.
5. It is highly desirable to invent and to measure the mass-isotope effects at $T \ll T_c \approx 100K$ (for example, $H_c(0) \sim M^{-\alpha}$, etc.) because all microtheoretical ideas were being formed in the assumption that only "zero" phonons are important.
6. The present day experimental situation with $T_c(M)$ is far from simplicity, especially in the vicinity of the structure instability.
7. Putting up and solving the physical questions is possible with the use of the difference in $\sigma^{(n)}_{scat}$ (or $\sigma^{(n)}_{abs}$) for various isotopes in HTSC compounds.
8. It is essential to overcome the difficulties in NMR experiments with ^{209}Bi for deeper comparison of Cu-O and Bi-O HTSC systems.
9. NMR, NQR and ENDOR with nuclei of "nonbasic" elements are promising:

«new nucleus -- new information»

10. One of the possible ways of increasing T_c for Tl-HTSC consists in avoiding Tl-defective positions which can induce Local Magnetic Moments on neighboring Cu atoms.

3. SPACE APPLICATIONS

SPACE APPLICATIONS FOR HIGH TEMPERATURE SUPERCONDUCTIVITY TECHNOLOGY - BRIEF REVIEW

Kumar Krishen, Ph.D.
NASA Lyndon B. Johnson Space Center
Houston, Texas 77058

INTRODUCTION

High temperature superconductivity (HTS) materials and devices are expected to provide revolutionary space applications which include: high-current power transmission; microwave, infrared, and optical sensors; signal processors; submillimeter wave components and systems; ultra stable space clocks; electromagnetic launch systems; magnetic replicas; and accelerometers and position sensors for flight operations. The HTS is expected to impact NASA's lunar outpost, Mars exploration, mission to Earth, and planetary/science exploration programs providing enabling and cost-effective technology. The operation of HTS devices and materials near and above liquid nitrogen (77K) can significantly reduce the cryogenic refrigeration burden associated with their use in space. In sensor applications, radiative and mechanical coolers can be used to cool the focal plane between 65K and 90K. The developments will mean size, weight and power savings, in addition to enabling or enhancing performance. The Naval Research Laboratory (NRL) plans to deploy a variety of passive microwave and millimeter wave components and devices in 1992 to study their performance in space. This NRL high temperature superconductivity space experiment (HTSSE) will drastically reduce the time required for insertion of HTS technology in space systems.

MILLIMETERWAVE AND MICROWAVE SYSTEMS

Microwave components and devices find extensive use in space data processing, communications, navigation, radar/tracking, and active and passive sensing systems [1]. Devices and components fabricated in the recent past include microstrip resonators, low-noise oscillators, band-pass filters, band reject filters, antennae, radar front ends, transmission/delay lines, couplers, high-frequency electronic circuits, superconductor/normal-metal/superconductor (SNS) and superconductor/insulator/superconductor (SIS) devices, and coplanar waveguides [2,3,4,5].

The SIS and SNS devices for use as mixers and local oscillators at frequencies up to terahertz and possibly a range of 60K to 80K have been fabricated at NASA's Jet Propulsion Laboratory (JPL). At the NASA Lewis Research Center, an electronic circuit operating at 33 to 37 GHz has been fabricated. Superconductor Technologies, Inc. (STI) has developed a 2.7 GHz thallium YBCO microstrip resonator, with peak power of 100 W, having a substantial improvement in Q compared to the gold resonator. The STI's HTS microstrip 5-pole Chebyshev filter shows a 9 db improvement with approximately three times bandwidth and a sharp cutoff. Millimeter wave mixers and antennae have been developed by several organizations. Lockheed developed superconducting high frequency antenna radiating elements [6] have shown great reductions in the size compared to a conventional antenna for the same radiating power. For array antennae, HTS microstrip feeds provide more than 20 db gain over copper microstrip for numbers of elements exceeding 200. The HTS will provide higher gain, greater efficiency, larger practical array antennas with the incorporation of feed elements, feedlines, and radiofrequency front-end filters. Beam steering with HTS phase shifters is also achievable.

SPACEBORNE/PLANET SURFACE SENSORS

The HTS applications in spaceborne and planet surface (such as Moon and Mars-based) sensing are numerous. Applications in cryocoolers include current leads, magnetic bearings, vibration isolation, and flux pump refrigerators. Studies show HTS has great potential for integrated optoelectronics. One application pursued by many organizations is in the infrared bolometers in wavelength regime longer than 20 μm. A revolutionary HTS device for magnetometer-based systems used in magnetic sensing and imaging is the superconducting quantum interference device (SQUID) [6]. The SQUID is more than one thousand times sensitive in measuring magnetic fields than any other magnetometer [7]. The SQUID can also be used in analog-to-digital conversion with enormous sensitivity/dynamic range and reduce the power consumption to a thousand times less than equivalent semiconducting circuits. Applications of HTS also are being pursued for SIS mixers for heterodyne detection, SNS tunnel junctions, and integrated focal plane array infrared sensors. The HTS current leads for cryocoolers have the potential to reduce cryogen boiloff by 30 to 40 percent and offer large bandwidth, dispersionless signal transmission capability [8]. A Lockheed-developed superconducting thermal isolator device will be tested in space on the NRL HTSSE payload scheduled for launch in 1992. It is expected to provide a 50 percent reduction in heat conduction. There is strong interest in both active and passive magnetic bearings for many space applications, including cryoturbopumps and optical data storage disks. The main advantages of these bearings are improved reliability, lifetime and vibration isolation.

A conspicuous gap in the sensing for astronomy arises in the region of 30 GHz to several terahertz. The HTS electronics with high sensitivity and wide operating frequency offers great potential in this region and is being developed through several programs [9].

OTHER APPLICATIONS

The use of HTS in developing magnetic attraction/repulsion for robotic, rescue, and docking applications is currently being studied. The HTS magnetic subsystems provide unique advantages in size and efficiency over conventional systems. To this end, magnetic replicas capable of trapping magnetic fields of at least 30 000G in bulk material are being developed through NASA participation [10].

The HTS electromagnetic launches can provide substantial applications for Earth launching as well as launches from the lunar and martian surfaces. The HTS can be used in energy storage, switching, and launch/payload coils. In addition, satellites in transition can be accelerated by using orbital energy storage/transfer satellites using HTS magnetic systems. In this scenario, satellites can be accelerated from low Earth orbit to geosynchronous orbit and beyond [1].

REFERENCES

[1] Krishen, Kumar, and Ignatiev, Alex, Future Superconductivity Applications in Space - A Review, Burnham and Kane (Editors) World Congress on Superconductivity, World Scientific Publishing Company, 1988.

[2] Hammond, Robert B.: HTSC Microwave Product Development at Superconductor Technologies, Inc. Presented at the Second World Congress on Superconductivity, Houston, Texas, September 1990.

[3] Riley, Landis A., et. al.: A Low Pass CPW Microwave Filter for the High Temperature Superconductor Space Experiment. Presented at the Second World Congress on Superconductivity, Houston, Texas, September 1990.

[4] Heinen, Vernon O.; Bhasin, Kul B,; and Long, K. J.: Emerging Applications of High-Temperature Superconductors for Space Applications. Presented at the Second World Congress on Superconductivity, Houston, Texas, September 1990.

[5] Nisenoff, M., et. al.: High Temperature Superconductivity Space Experiment (HTSSE): Passive Millimeter Wave Devices. Presented at the Second World Congress on Superconductivity, Houston, Texas, September 1990.

[6] Horizons, Lockheed Internal Communications, Roy A. Blay (Editor), Issue 28, Calabasas, CA, January 1990.

[7] Decher, Rudolf; Peters, Palmer; Sisk, Charles; and Urban, Eugene W.: Superconductivity Applications in Space Flight Science Experiments. Presented at the Second World Congress on Superconductivity, Houston, Texas, September 1990.

[8] Powers, C., and Arsenovic, P.: Electrical, Thermal, and Mechanical Characterization of HTcS Current Leads. Presented at the Second Congress on Superconductivity, Houston, Texas, September 1990.

[9] Supercurrents, The Superconductivity Magazine, Don Forbes (Editor), Vol. 8; Belmont, CA, March 1989.

[10] Chen, In-Gann; Liu, Jay; and Weinstein, Ray: Magnet Replicas. Presented at the Second World Congress on Superconductivity, Houston, Texas, September 1990.

A LOW PASS CPW MICROWAVE FILTER FOR THE NRL HIGH TEMPERATURE SUPERCONDUCTIVITY SPACE EXPERIMENT

A. Landis Riley, Brian D. Hunt, Wilbert Chew, Louis Bajuk, Marc C. Foote, Daniel L. Rascoe, Thomas W. Cooley

Jet Propulsion Laboratory
California Institute of Technology
Pasadena, CA 91109

ABSTRACT

A low pass coplanar waveguide microwave filter with a 0-9.5 GHz passband has been designed, fabricated, and tested at the Jet Propulsion Laboratory (JPL). The filter uses $YBa_2Cu_3O_7$ (YBCO) film produced at JPL on $LaAlO_3$ substrates. The filter exhibited a loss of about 0.5 dB at 9.5 GHz at 77 K compared to a loss of about 2.2 dB for a corresponding copper filter at 77 K. The coplanar waveguide structure used takes full advantage of superconductor deposited on a single side of the substrate. Five packaged filters of this design were delivered to the Naval Research Laboratory (NRL) as candidates for flight in the High Temperature Superconductivity Space Experiment.

I. INTRODUCTION

The goal of this work was to design and fabricate a high critical temperature (high T_c) superconductor low pass microwave filter for space applications for delivery to the Naval Research Laboratory (NRL) High Temperature Superconductivity Space Experiment (HTSSE). The HTSSE is implemented by NRL to demonstrate the feasibility of operating high T_c systems in the space environment through actual space flight experience. To provide a NASA unit for this experiment, we have used $YBa_2Cu_3O_7$ (YBCO) films fabricated at JPL. The device chosen for development was a low pass microwave filter using a coplanar waveguide (CPW) structure on a $LaAlO_3$ substrate.

Low pass filters are common microwave system components and can demonstrate the performance improvement made possible by exploiting superconductivity. The CPW structure chosen can accommodate microwave circuitry with a single conductor layer on a substrate. This alleviates the need to coat both sides of a substrate with superconductor.

II. DESIGN

CPW resonators using YBCO have previously demonstrated losses lower than for copper [1],[2] but building a filter with desired performance requires more design information. Despite the practical advantages of CPW, fewer design tools are available for CPW than for printed circuit structures such as stripline or microstrip, and design formulas are incompletely verified. Therefore, an empirical design approach was taken to develop the design as quickly as possible.

Test structures using normal metal on $LaAlO_3$ with CPW discontinuities (steps between CPW transmission line sections of different characteristic impedances) were measured to obtain approximate models of the discontinuities, and also to determine that the permittivity of $LaAlO_3$ is approximately 25. Comparing the measured frequency response of a prototype normal metal film filter to that of a YBCO film filter of the same design revealed the amount of frequency shift attributable to the internal inductance [3],[4] (including the superconductor kinetic inductance) in YBCO. The extra inductance could be modeled as a decreased phase velocity and increased impedance in each CPW transmission line segment in conventional microwave computer-aided design (CAD) software.

Using these models, the final filter design shown in Fig. 1 was made. The filter consists of tapered microstrip to coplanar waveguide transitions at the input and output of the circuit on the $LaAlO_3$ substrate, input and output coplanar transmission lines with 50 ohm characteristic impedances, and alternating sections of low and high characteristic impedance transmission line.

Hermetic microwave packages suitable for these materials in the space environment were also developed. The substrates were soldered to the Kovar packages using indium solder. Wiltron "K connectors™" were used to provide hermetic input and output coaxial connections. The packaged filter before weld-attachment of the lid is shown in Fig. 2.

III. RESULTS

The performance of 4 YBCO film filters at 77 K after processing and packaging is compared to that of a corresponding copper film filter at 77 K and at 297 K in Fig. 3. Losses up to the edge of the $LaAlO_3$ substrate were calibrated out. At the center of the measurement passband of 8.5 to 10.5 GHz to be used by NRL, the best YBCO filter has an insertion loss of about 0.5 dB, compared to about 2.2 dB for the cooled copper filter. The YBCO filters are superior to copper throughout the passband. Some variability in the frequency response of the YBCO filters can be seen. The operating temperature of 77 K is near the critical temperature T_c, which varies from 83 K to 88 K for these filters. The penetration depth is a strong function of temperature near the critical temperature.[3],[4] With different critical temperatures, the penetration depth would vary widely and give varying amounts of internal inductance, which would produce varying frequency responses. The filters shown had acceptable amounts of variation; the cut-off frequency remained within the measurement band for temperatures 70-80 K. The variation for one of the filters is shown in Fig. 4; this measurement included the loss of coaxial cables within the refrigerator, unlike the fixed temperature calibrated measurements in Fig. 3. Fig. 5 shows the effect of input power on transmission of the same filter, at -12 and -2 dBm compared to -22 dBm input power.

The packaged filters were subjected to wire bond pull tests (0.0007 in. diameter gold wire passed 2 gram pull tests), thermal cycling (immersed at least 5 times in liquid nitrogen, and cooled in a vacuum chamber once to 15 K), and vibration and shock levels specified by NRL to simulate launch conditions. RF performance was measured before and after these stresses, and found negligibly changed. Five packaged filters were delivered to NRL as candidates for a satellite launch projected for 1992.

IV. CONCLUSION

The development of these filters demonstrated that coplanar waveguide is a practical microwave structure for YBCO on $LaAlO_3$. A CPW microwave filter using currently available YBCO, after all the processing and assembly steps required to make a space-qualifiable package, can provide performance superior to a similar copper film filter. Better performance can be expected as YBCO films improve.

Acknowledgments

We thank J. Bautista for providing a cryogenic refrigerator for microwave measurements and use of a wire-bonder, and S. Chavez, C. Cruzan, and J. Rice for assembly and wire-bonding expertise. This research, performed at the Center for Space Microelectronics Technology, Jet Propulsion Laboratory, California Institute of Technology, was sponsored by the National Aeronautics and Space Administration (NASA), Office of Aeronautics and Exploration Technology, by the Defense Advanced Research Projects Agency, and by the Strategic Defense Initiative Organization, Innovative Science and Technology Office.

REFERENCES

[1] Valenzuela, A. A. and Russer, P., "High Q coplanar transmission line resonator of $YBa_2Cu_3O_{7-x}$ on MgO," Appl. Phys. Lett., 55, 1029-1031, (1989).

[2] Valenzuela, A. A., Daalmans, B., and Roas, B., "High-Q coplanar transmission line resonator of $YBa_2Cu_3O_{7-x}$ on $LaAlO_3$," Electron. Lett., 25, 1435-1436, (1989).

[3] Van Duzer, T. and Turner, C. W., Principles of Superconductive Devices and Circuits. New York: Elsevier, 1981.

[4] Bhasin, K. B., Chorey, C. M., Warner, J. D., Romanofsky, R. R., Heinen, V. O., Kong, K. S., Lee, H. Y., and Itoh, T., "Performance and modeling of superconducting ring resonators at millimeter-wave frequencies," 1990 IEEE MTT-S International Microwave Symposium Digest, 269-272 (1990).

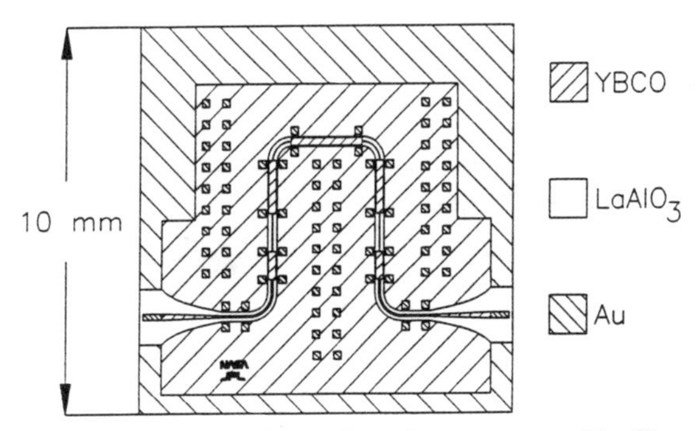

Fig. 1. Top surface of coplanar waveguide filter.

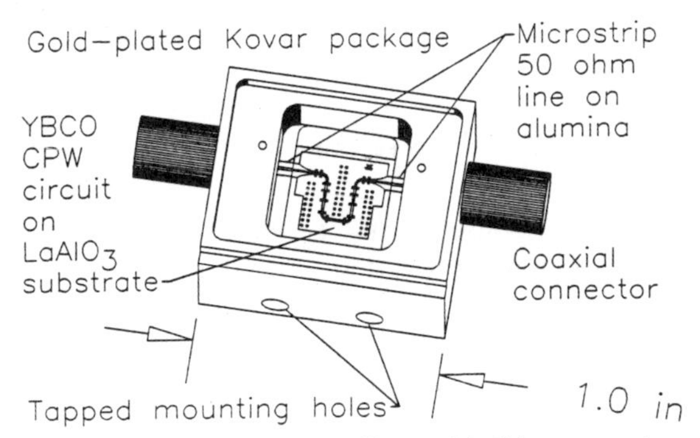

Fig. 2. Packaged filter with lid removed.

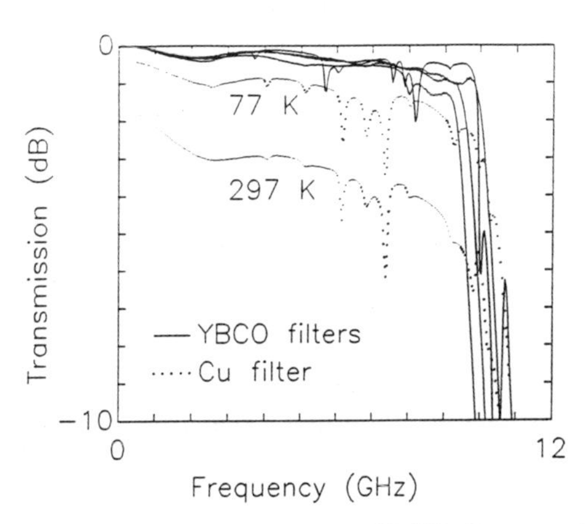

Fig. 3. Measured insertion losses of 4 YBCO filters and copper filter at 77 K, and copper filter at 297 K.

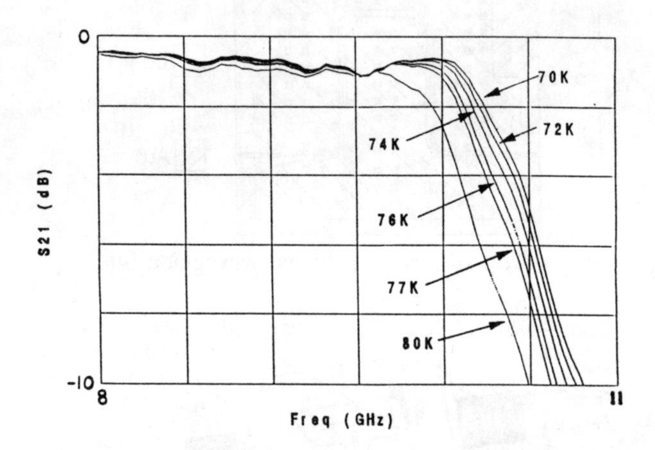

Fig. 4. Spectral response of filter with temperature as a parameter, measured inside refrigerator, including cables inside refrigerator.

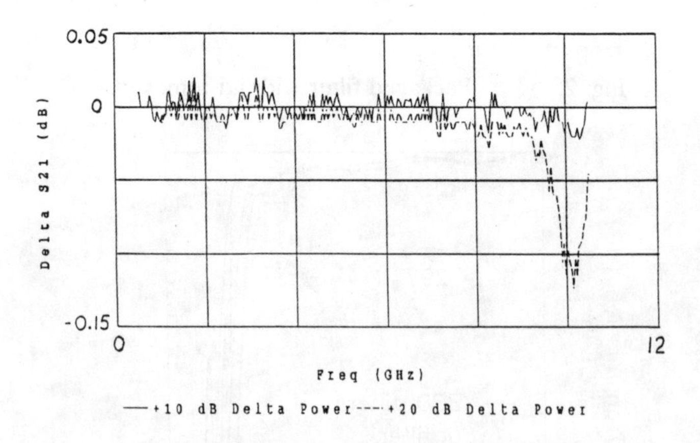

Fig. 5. Change of spectral response with change in input drive level, referenced to -22 dBm drive level.

Emerging Applications of High-Temperature Superconductors for Space Communications

Vernon O. Heinen
Kul B. Bhasin
Kenwyn J. Long
National Aeronautics and Space Administration
Lewis Research Center
Cleveland, OH 44135

INTRODUCTION

Space application of superconducting technology has previously been limited by the requirement of cooling to near liquid helium temperatures. The discovery of HTS materials with transition temperatures above 77 K along with the natural cooling properties of space suggest that space operations may lead the way in the applications of high temperature superconductivity. Proposed space missions require longevity of communications system components, high input power levels, and high speed digital logic devices. The complexity of these missions calls for a high data bandwidth capacity. To ensure adequate efficiencies, the microwave surface resistance of the component materials must be reduced to as low a level as possible. Incorporation of high critical temperature superconducting (HTS) thin films into some of these communications system components may provide a means of meeting these requirements as well as offering enhanced efficiency and reductions in size and weight.

The use of high T_c superconductors in a microwave system requires development of thin films on microwave substrates which can be patterned into desired microwave circuits such as filters, phase shifters, ring resonators and delay lines. The superconducting thin films for microwave circuits need to be deposited on low-dielectric-constant and low-loss substrates, have smooth morphology, high critical temperature T_c, high critical current density J_c, and low surface resistance R_s. Furthermore, films on the substrates must be evaluated in devices such as microstrip or ring resonator circuits to determine the quality factor, Q, and various loss factors before appropriate microwave circuit applications can be developed.

STATUS OF HIGH-T_C SUPERCONDUCTOR PROPERTIES

To obtain high quality YBaCuO films on suitable substrates, the lattice constants of the substrate must be matched to those of the film, and there must be no detrimental chemical reactions between the substrate and the film. In addition, the film composition must be as close as possible to the desired stoichiometry. Many of the physical and chemical deposition techniques used to obtain high quality films require post-annealing at high temperatures. This high-temperature anneal causes undesirable chemical interactions at the film-substrate interface, making it unsuitable for microwave applications[1]. To circumvent this problem, an in situ annealing procedure, which allows lower growth temperatures, has been used to grow epitaxial films using a laser ablation technique[2].

The best laser-ablated film had a T_c of 89.8 K immediately after deposition, as determined by a standard four-point resistance measurement. From x-ray diffraction data, the film was determined to be c-axis aligned. In Table I, we list the performance of $YBa_2Cu_3O_7$ thin films on various microwave substrates along with physical properties of these substrates. As can be seen, the value of J_c was greater than 10^6 A/cm^2 at 77 K.

Surface resistance characterization of superconducting films offers valuable information on the film quality for microwave surface applications. Currently, surface resistance values are obtained by cavity[3,4] and stripline measurements[5]. Correlation between material properties (i.e., T_c, dc conductivity above T_c, and penetration depth) and surface resistance are still not well understood for new high T_c superconducting films. Conductivity is a complex quantity, $\sigma = \sigma_1 + j\sigma_2$. For $\sigma_2 \gg \sigma_1$ one can obtain the surface resistance of superconducting film

$$R_s = 0.5\sigma_1 \mu / \sigma_2^{3/2} \tag{1}$$

where σ_2 is related to the penetration depth λ by

$$\sigma_2 = 1/\omega\mu\lambda^2 \tag{2}$$

To obtain superconducting film surface resistance values that are lower than those of a normal metal, the smallest values of σ_1 and λ are desired. Miranda, et al.[6] have measured microwave transmission in a waveguide for superconducting films. From the transmission data, using the two-fluid models, σ_1 and λ have been obtained. The surface resistance for films deposited on $LaAlO_3$ was calculated. In figure 1, which is adopted from reference 7, we show how the quadratic variation f^2 of the surface resistance varies with frequency for laser-ablated YBaCuO films on microwave substrates.

The surface resistance is several orders of magnitude lower than that of copper. Surface resistance, penetration depth, and microwave conductivity measurements provide valuable information on the quality of these films for microwave circuits.

Microstrip resonators patterned from thin films on microwave substrates allow direct measurement of microstrip losses. We have fabricated microstrip ring resonators operating at 35 GHz from laser-ablated YBaCuO thin films deposited on lanthanum aluminum substrate[8]. Several groups have studied resonator circuits at lower frequencies[5,9-11]. The resonator circuits we fabricated were patterned by standard photolithography using negative photoresist and a "wet" chemical etchant. These resonators were characterized using a Hewlett-Packard 8510 Automatic Network Analyzer, operating in a WR-28 waveguide. Two features are apparent; (1) the coupling changes with temperature (the coupling coefficient increases with decreasing temperature) and (2) the resonant frequency shifts with temperature. This change is a consequence of the dependence of internal impedance of the strip on the varying normal and superconducting electron densities.

The best resonators measured to date have shown unloaded Q values ranging from 2500 to 1000 at 20 and 77 K, respectively. This corresponds to a surface resistance value of, at most, 15 mΩ at 77 K at 35 GHz, a value two to three times better than that of copper at the same temperature and frequency. The 33-37 GHz YBaCuO ring resonator circuit developed at LeRC is viewed as a precursor to frequency-selective filters and may have potential application in enhancing the efficiency of radiating antenna elements. Such HTS resonating circuits have a high Q as compared to equivalent normal metal circuits and may afford reduction in size and mass of electronically-steered millimeter wave antennas.

POTENTIAL APPLICATIONS

A. Passive Microwave Circuits

High T_c superconducting thin films have shown lower surface resistance than copper. Low conductor losses have been demonstrated for a high T_c superconducting ring resonator circuit. Low surface resistance and conductor losses are desirable in passive microwave circuits used in communication and radar systems since they offer increased bandwidth, reduction in loss and size, and provide low noise. A complete system analysis of the impact of high T_c superconducting microwave circuits remains to be undertaken. From a block diagram of a satellite transponder (fig. 2), we have considered several potential applications of HTS microwave circuits in satellite communications system components. Based on results obtained to date on the performance of superconducting microstrip resonator circuits with high Q values, one can easily project the application of superconducting passive circuits as low loss, high Q filters[13], high Q resonators, delay lines, power splitter combiners, and resonator stabilized oscillators.

HTS materials may be most effective in enhancing the efficiency of phased-array antennas when incorporated into interconnects and power dividers. Thin films of YBaCuO offer a reduction in resistive heating loss relative to gold and copper, and could conceivably be designed into microstrip lines. This frequency-dependent resistive heating loss effect is most significant at frequencies below Ka-band (~20-30 GHz). However, as frequency increases individual elements become smaller. The number of required elements increases, necessitating more complex interconnects, thereby making the overall savings from HTS materials more significant.

At submillimeter frequencies ($f \gtrsim 300$ GHz), RF ohmic losses in antenna elements of normal metals are on the order of a fraction of a dB. HTS antenna patch elements would thus yield only a slight advantage over normal metal components in this frequency range. However at millimeter-wave operation (30 GHz $\lesssim f < 300$ GHz), ohmic losses in antenna elements and interconnects are significant in normal metals, making benefits from HTS materials appreciable. The small antenna sizes required at these frequencies are more readily encapsulated in cryogenic envelopes than are microwave-frequency elements. The critical current density of HTS thin films should exceed the current densities in proposed space communication antenna systems, which is expected to be on the order of 10^5 A/cm^2 or less.

Granular YBaCuO and BiCaSrCuO detectors have been fabricated and tested at 77 K, between 24 and 110 GHz[12]. Superconducting microstrip lines were used as IF and/or video output lines. All detectors showed a dramatic increase in sensitivity below 77 K. The theoretical high-frequency limit (determined from the superconductor energy gap) for these detectors is in the low terahertz range. However, the experimental video response decreases by about one order of magnitude when the carrier frequency is increased from 100 to 300 GHz. This research determined that the properties of HTS ceramics, and thus the resulting detector performance, are highly dependent upon the substrate material and are most likely tied to the dielectric properties of the substrate.

In addition to these applications, extremely low loss phase shifters using superconducting switches are also feasible. In figure 3, we show a phase shifter which utilizes superconducting-normal-superconducting switches in place of FET/diode switches. The switches are fabricated from high T_c thin films of YBaCuO. The switches operate in the bolometric mode with the film held near its transition temperature. Radiation from a light source raises the temperature and consequently causes the film to become resistive. If the switches in the reference path are illuminated, they too will become resistive. The switches on the opposite side of the device are superconducting. Since each switch is positioned one quarter of a wavelength from the junction, the signal will be reflected from the delay path in phase. A similar phenomenon occurs at the output port. To achieve the desired phase shift, the opposite set of switches is illuminated. Figure 3 shows the predicted behavior for a 180° phase shifter, with an exceptionally narrow insertion loss envelope and excellent return loss.

Figure 4 illustrates an example of a hybrid semiconductor/superconductor device. It is possible that by combining the excellent low noise properties of GaAs devices with the low loss and low noise properties of superconducting transmission lines, one can achieve ultra-low-noise receivers for satellite communications applications. If these promising concepts of high T_c superconducting devices are actually brought to fruition, then one can conceive of their use in low loss, low noise superconducting phased array antennas in space communications systems such as shown in figure 5. HTS transmission lines can provide low loss feed networks for antenna arrays.

B. E-BEAM DEVICES

There are several areas for possible application of HTS material in electron beam devices. These include the slow-wave circuit, the cathode, and the focussing magnet. Superconducting slow-wave circuits are likely only in very limited applications due to the heating caused by interception of the electron beam by the slow-wave circuit. Another possible area of application of HTS material is in the cathode. Electron field emission from the cathode may be enhanced when the cathode surface is in the superconducting state. Research in this area is in its very early stages and not likely to be applied in the near future.

What is likely to be the first application of HTS material in e-beam devices is for use as the focusing magnet. At the present time most traveling wave tubes used in space use periodic permanent magnet (PPM) focusing to confine the electron beam because PPM focusing is light and requires no power. The disadvantage of it is that the attainable magnetic fields are less than that required for optimum performance of the tube. This results in reduced frequency stability, gain flatness, attainable output power, and overall efficiency. A larger solenoidal field can be used for confined flow focusing to overcome these problems resulting in a much improved tube. Because of the large number of potential applications of HTS magnets there are many groups working toward their development.

CONCLUSIONS

Microwave components have been fabricated from HTS materials which have superior performance over similar nonsuperconducting components. The development of HTS devices has large potential impact on the communication system for future space missions. There remains a significant amount of progress to be made in the development of materials and devices to the extent of applications of HTS technology to microwave systems. System studies must be performed to determine any overall advantage of HTS system along with the necessary coolers over conventional technology.

REFERENCES

1. Valco, G.J., et al.: Sequentially Evaporated Thin Y-Ba-Cu-O Superconducting Films on Microwave Substrates. NASA TM-102068, 1989.

2. Warner, J.D., et al.: Growth and Patterning of Laser Ablated Superconducting YBCO Films on $LaAlO_3$ Substrates. NASA TM-1-2436, 1989.

3. Klein, N., et al.: Millimeter Wave Surface Resistance of Epitaxially Grown $YBa_2Cu_3O_7$ Thin Films. Appl. Phys. Lett. vol. 54, no. 8, Feb. 20, 1989, pp. 757-759.

4. Miranda, F.A., et al.: Millimeter Wave Surface Resistance of Laser Ablated Y-Ba-Cu-O Superconducting Films. Accepted for publication in Appl. Phys. Lett., Sept. 1990.

5. Fathy, A., et al.: Microwave Properties and Modeling of High T_c Superconducting Thin Film Meander Line. IEEE MTT-S International Symposium (Dallas, TX, May 1990) Digest, 1990, pp. 859-862.

6. Miranda, F.A., et al.: Microwave Conductivity of Superconducting Bi-Sr-Ca-Cu-O Thin Films in the 26.5 to 40.0 GHz Frequency Range. Physica C., vol. 188, 1990, pp. 91-98.

7. Inam, A., et al.: Microwave Properties of Highly Oriented $YBa_2Cu_3O_7$ Thin Films. Appl. Phys. Lett., vol. 56, no. 12, Mar. 19, 1990, pp. 1178-1180.

8. Bhasin, K.B., et al.: Performance and Modeling of Superconducting Ring Resonators at Millimeter-wave Frequencies. IEEE MTT-S International Microwave Symposium (Dallas, TX, May 1990) Digest, 1990, pp. 269-272. (Also NASA TM-102526, 1990).

9. Valenzuela, A.A., and Russer, P.: High-Q Coplanar Transmission Line Resonators of YBCO on MgO. Appl. Phys. Lett., vol. 55, no. 10, Sept. 4, 1989, pp. 1029-1031.

10. McAroy, B.R., et al.: Superconducting Stripline Resonance Performance. IEEE Trans. Magn., vol. 25, no. 10, Oct. 1989, pp. 1104-1106.

11. Takemoto, J.H., et al.: Microstrip Ring Resonator Technique for Measuring Microwave Attenuation in High T_c Superconducting Thin Films. IEEE Trans. Microwave Theory Tech., vol. MTT-37, no. 10, Oct. 1989, pp. 1650-1652.

12. Konopka, J., et al.: Microwave Detectors Based on Granular High-T_c Thin Films. IEEE Transactions on Microwave Theory and Techniques, vol. 38, no. 2, Feb. 1990.

13. Fiedziuszko, S.J., Holme, S., and Heidmann, P.: Novel Filter Implementations Using HTS Materials. To be published.

TABLE I. - KEY PROPERTIES OF MICROWAVE SUBSTRATE MATERIALS

Material	Highest T_c achieved	Dielectric constant	Loss tangent	Lattice size (Å)
Magnesium oxide (MgO)	88	9.65	$4x10^{-4}$	4.178 (100)
Lanthanum aluminate ($LaAlO_3$)	90	22	$5.8x10^{-4}$	3.792 (110)
Lanthanum gallate ($LaGaO_3$)	88	27	$2x10^{-3}$	3.892 (110)
Sapphire (Al_2O_3)	71	9.4	$1x10^{-6}$	5.111 $(01\overline{1})$
Yttria stabilized zirconia (ZrO)	89	27	$6x10^{-4}$	3.8795 (100)
Silicon (Si)	—	12	$10x10^{-4}$	5.43 (100)
Gallium arsenide (GaAs)	—	13	$6x10^{-4}$	5.653 (100)

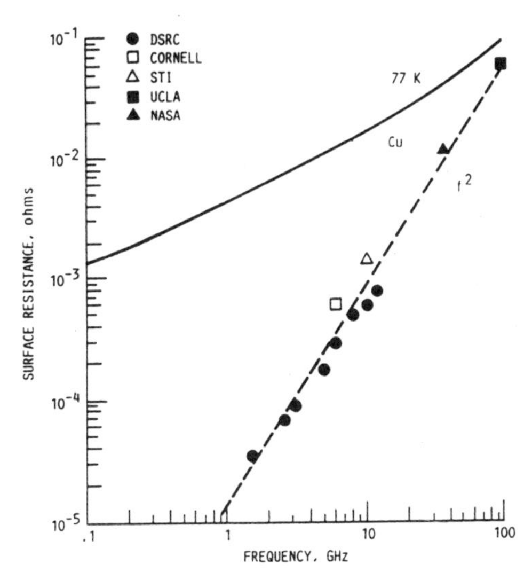

FIGURE 1. - SURFACE RESISTANCE OF LASER ABLATED Y-Ba-Cu-O FILMS ON $LaAlO_3$ SUBSTRATE VERSUS FREQUENCY. ADOPTED FROM APPLIED PHYSICAL LETTERS VOLUME 56, P.P. 1178-1180. NASA DATA OBTAINED BY MICROWAVE CONDUCTIVITY MEASUREMENTS.

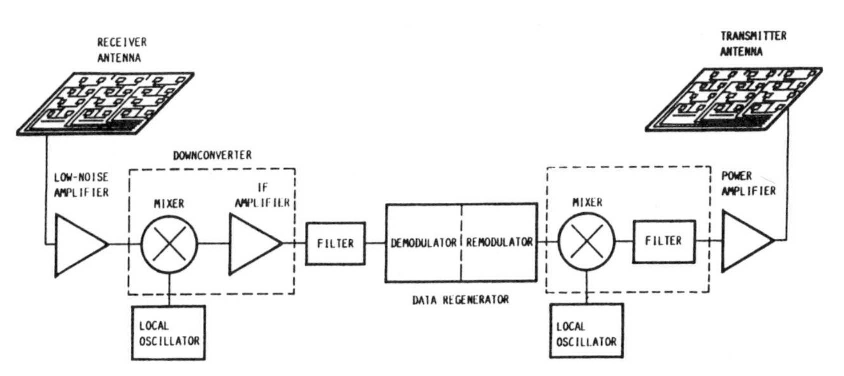

FIGURE 2. - BLOCK DIAGRAM OF A SATELLITE TRANSPONDER.

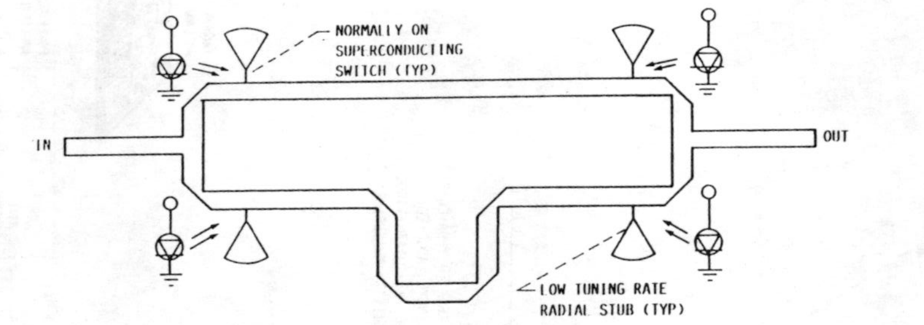

(a) OPTICALLY CONTROLLED HIGH-T_c SUPERCONDUCTING SWITCH-LINE PHASE SHIFTER.

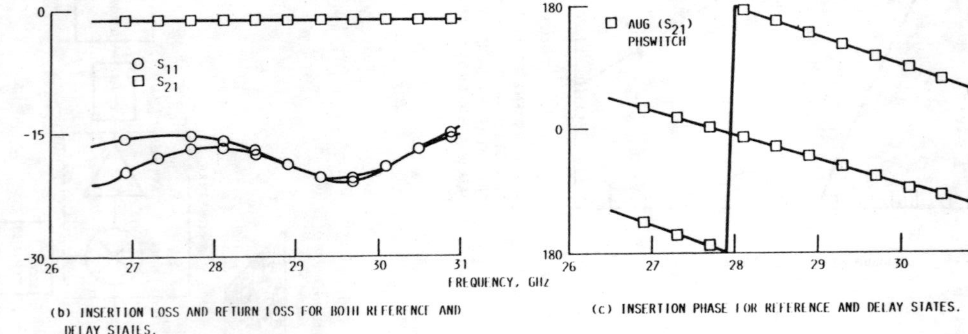

(b) INSERTION LOSS AND RETURN LOSS FOR BOTH REFERENCE AND DELAY STATES.

(c) INSERTION PHASE FOR REFERENCE AND DELAY STATES.

FIGURE 3.

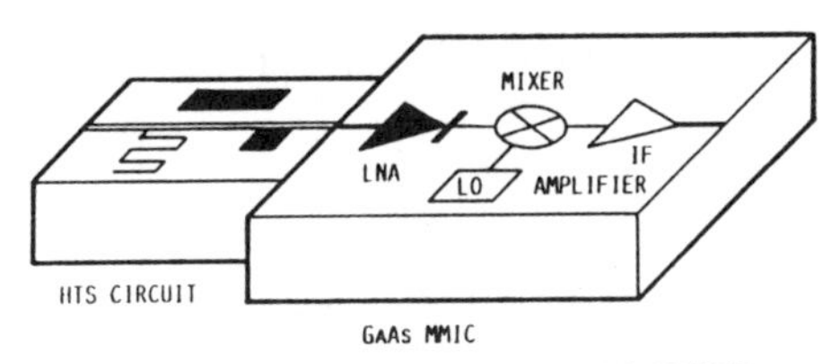

FIGURE 4. - SUPERCONDUCTING GaAs MMIC HYBRID RECEIVER.

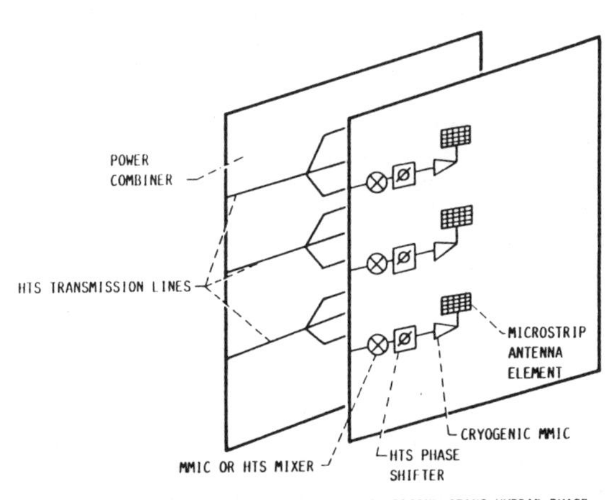

FIGURE 5. - CONCEPTUAL DIAGRAM OF A MMIC-SUPERCONDUCTING HYBRID PHASE ARRAY ANTENNA.

SUPERCONDUCTING MAGNETIC ENERGY STORAGE FOR FUTURE NASA MISSIONS: THE LUNAR ORBIT-TO-SURFACE MICROWAVE BEAM POWER SYSTEM

Karl A. Faymon
NASA - Lewis Research Center
Cleveland, Ohio 44135.

ABSTRACT:

An alternative to landing individual power systems (nuclear or otherwise) for planetary surface bases is to "beam power down" from an orbiting power platforms by means of microwave beams. In the case where the power platform orbit is non-synchronous, provision must be made for energy storage on the surface to power the station when it is in the eclipse portion (out of view) of the orbiting power platform. For reasonable transmission distances (non-synchronous orbits) for microwave beams, adverse duty cycles (long eclipse times and short beam times) result. The surface energy storage must be charged during the beam time. Adverse duty cycles will require high charge rates, possibly an order of magnitude greater (or more) than those seen in typical earth orbit missions.

Low energy densities and greatly reduced lifetimes may result from very high charging rates when conventional electrochemical storage systems are used with orbiting microwave beam power systems. Superconducting Magnetic Energy Storage (SMES) with its high efficiencies, independence upon duty cycle and charging rates and no apparent limitation on lifetime by number of charge-discharge cycles appears to be suited to this type of application. This paper explores the combination of microwave beamed power from orbit combined with SMES on the lunar surface from the viewpoint of minimizing the "total" power system mass delivered to the lunar vicinity.

INTRODUCTION:

NASA has previously carried out a study to determine the applicability of SMES for future NASA space missions [1]. This study indicates that the attributes of SMES are such that it has potential for specific NASA mission applications. NASA has studied a number of applications of microwave beam power [2-5]. The Second Beam Power Workshop [5] has concluded that microwave beam power is a viable technology option for many future NASA space missions. One particular application under consideration is that of beaming power from a planetary orbit to the surface thus replacing the surface power system. Only under very special circumstances is it feasible to transmit microwave power from a stationary (synchronous) orbit because of the great transmission distances involved. An option is to "beam power down" from a lower (than synchronous) orbit.

In many respects a beamed power system has many of the attributes of a solar based power system. During the time the solar based system is in the sunlight portion of its orbit, the system must collect enough solar power/energy to power the spacecraft and at the same time charge the storage system (on board the spacecraft) so that it can provide power to the spacecraft when it is the eclipse phase of the orbit. Similarly, an orbital beam power platform must "beam" down enough energy to the surface station to power the station and at the same time charge the surface energy storage system (while it is in view of the station) so that it can supply power to the station while the orbiting platform is out of sight of the surface station. This is shown schematically on Figure I.

Solar based power systems in low earth orbit have duty cycles (the ratio of eclipse time to beam time) of less than unity and this is the design condition for conventional electrochemical energy storage systems. As is evident from Figure I, the duty cycles for microwave beam power platforms are significantly different than those for low earth orbit solar based systems - as a rule these duty cycles are much greater than unity. This will have an adverse effect upon conventional electrochemical energy storage systems because of the high charging rates which are required for the energy storage system. To evaluate this effect (adverse duty cycles) a number of configurations - various orbits and station locations - have been studied to determine whether any one energy storage technology will show any advantage over another.

In particular, conventional electrochemical energy storage was evaluated in comparison with superconducting magnetic energy storage (SMES) for utilization in a lunar surface base power system which receives its power from a microwave beam power platform in polar orbit. SMES by virtue of its high magnetic field has high force levels which must be structurally constrained. By locating the SMES on the planetary surface it may be possible to constrain these forces by utilizing planetary surface features.

Surface stations located at 30 and 60 degrees north lunar latitudes were analyzed. Two orbiting microwave power platforms in orbits positioned 90 degrees relative to each other were assumed - in elliptical orbits with orbit apogee located over the north lunar pole. Orbit configurations studied included variations in orbit apogees from 1000 to 3000 kilometers altitude. As will be shown in this paper, a considerable mass savings advantage accrues to the use of SMES as the energy storage medium. Also, because of the various trade-offs which are exist, an "optimum" orbit (apogee altitude) exists. This "optimum" is a function of the station location (latitude) and is different for the SMES station as compared to the station which utilizes the conventional electrochemical storage. Elliptical (Molniya type) orbits were selected for this application.

A previous study has shown that such orbits result in less total system mass than circular orbits. This is indicated on Figure II.

THE LUNAR ORBIT TO SURFACE MICROWAVE BEAM POWER SYSTEM:

The elements, both the orbital power transmission system and the surface station receiving system are shown in Figure III. Each of these elements has supporting subelements and subsystems such as structure, thermal control, etc., The major elements of a microwave beam power system are listed in the following:

ELEMENTS OF AN ORBIT-TO-SURFACE MICROWAVE BEAM POWER SYSTEM

ORBITING POWER PLATFORM	SURFACE STATION
Power Source, (assumed to be nuclear).	Microwave Receiving Rectenna.
Power Management and Distribution System.	Power Management and Distribution System.
Orbiting Platform Loads.	Surface Station Loads.
RF Generator.	* Energy Storage System.
Spaceborne Microwave Power Transmitting Antenna.	

* It is the attributes of the surface station energy storage system which will be analyzed to determine which system will result in a total minimum mass (orbiting and surface) delivery to the lunar vicinity. The "rectenna" is the antenna which receives the high frequency microwave beam and converts it to direct current - which is the output of the rectenna.

The orbital geometry in Figure I shows the configuration for both the 30 and 60 degree north lunar latitude surface stations. Figure I shows the particular configuration where the orbit passes directly over the stations. The orbit in this case is assumed to be polar with apogee directly over the lunar north pole.

In actuality, the moon is a rotating body. As such the relationship between the orbiting power platform and the surface receiving station is constantly changing. The "look angles" from the orbiting platform to the ground station do not always remain in the orbit plane but will have a sideward component as the moon rotates beneath the platform. If we postulate two elliptical orbits at 90 degrees relative to each other, the surface stations at 30 and 60 degrees north lunar latitude will have a reasonable look angle geometry as the moon rotates. The surface station will always be within sight of one of the orbiting power platforms.

ENERGY STORAGE SYSTEM FOR THE MICROWAVE BEAM POWERED LUNAR SURFACE BASE:

Because of the high cost of delivering mass to the lunar vicinity and surface, it is desired to minimize the mass of the microwave beamed power system. Energy storage systems always loom as one of the most massive and voluminous subsystems of a power system which requires energy storage. Because of the "adverse duty cycle", the energy storage system stands out as a major contributor to both the landed mass and orbital mass of the microwave beam power system..

In [1] a survey was carried out to compare the attributes of the various energy storage technologies. As a result of this survey,(the results are presented in Table I)', and a cursory analysis of the various storage technologies, it was decided to carry out a more detailed analysis of SMES in comparison to the conventional electrochemical energy storage systems. (The references in the right hand column of Table I refer to the references in [1].)

A detailed computer model of the surface receiving station was constructed [6]. All elements and subelements, (subsystem efficiencies, thermal control radiators, etc.) were included in this model. An orbital geometry model was also constructed to calculate the duty cycles, beam times, eclipse times and transmission distances for both circular and elliptical orbits for arbitrary locations on the lunar surface (latitude). The spaceborne, or orbital power system was not modeled in as great a detail. Detailed information on some of the elements of the airborne power transmission system (i.e. the radio frequency generators, etc.) were not available. However, for the purposes of this study it was felt that this was not required.

A critical input to this analysis was that of the transmission efficiencies as a function of orbital and system parameters. This is shown graphically in Figure IV and was taken from [7] where it is described in detail. The transmission efficiency becomes one of the critical factors in determining the spaceborne power system requirements and mass. The model of the surface power system was used to calculate the power required at the receiving antenna to meet all station load and energy balances.

The mass of the spaceborne system was calculated by means of a transfer function relating the power output of the transmitting antenna to the mass of the system. The transfer function has the units of kg(power system mass)/kW(antenna output power). This was treated as a system parameter and the results were investigated over a range of this parameter. Within the bounds of this study it was felt that the various power technologies applicable to this application could be adequately represented by such a transfer function.

ANALYSIS OF THE ENERGY STORAGE OPTIONS FOR A MICROWAVE POWERED LUNAR STATION

Figure I shows the orbit and microwave beam related geometries for stations located at 30 and 60 degrees north lunar latitudes. Shown here is the geometry for elliptical orbits with apogee directly over the lunar north pole.
The moon is a rotating body, relative to these lunar orbits. As such the orbit geometries shown on Figure I are indicative only of a specific instant in the lunar rotation cycle. The relatives positions of the orbiting platform and the surface station are continually changing as the moon rotates beneath the orbit. If only a single platform in one orbit were assumed there would be times when the surface station is completely out of view during some orbits of the orbiting platform.

If a configuration utilizing two orbits is used where the orbit planes are normal to each other, it is possible to have the surface station in view of one or the other platforms at every orbit of one or the other platforms. The effect this would have upon the analysis carried out here would be to affect the microwave beam transmission efficiency since now there would be times when the incident microwave beam has a sideward look angle component as the moon rotates beneath the orbit. This is easily incorporated into data given in Figure IV which now shows only the in-plane variations as the platform passes directly over the station. It is assumed that the rectenna is "fixed" on the lunar surface - parallel to the surface.
Because of the rectenna size and dimensions, a considerable mass penalty would accrue if the antenna were to be steerable. The advantage of a steerable antenna would be that it could always be oriented to be normal to the incoming microwave power beam resulting in increased transmission efficiencies. For purpose of actual mission design, the "worst case" situation with respect to duty cycle must be used for mission design purposes. The "worst case", for design purpose, is that case where the orbiting platform has the greatest sideward look angle component to the ground station.

For purposes of this study, the orbit geometry shown on Figure I was used. A more precise geometry which includes the out-of-plane effects was not felt to be warranted and would only effect the final magnitudes of the results but would have little effect upon the conclusion as to the relative evaluations of SMES versus conventional electrochemical energy storage. The input data detail and accuracy of other components of the orbiting system were felt to warrant this conclusion and approach.

RESULTS OF THE ANALYSIS:

The results of the study are shown in tabular form on Table II. Here are listed the critical parameters which will allow a comparison of SMES and conventional electrochemical energy storage systems for the particular application addressed in this paper. The inputs to this analysis are given on Table III.

As is evident from the results presented in Table II, the total ground receiving station mass is reduced for the SMES over that of the electrochemical energy storage system. In addition the power required at the transmitting antenna is also less for the SMES than that required for the conventional storage system.

This behavior stems from two factors. The high round trip efficiency, (charging-discharging efficiency) of the SMES storage system results in less power required at the transmitting station (antenna). The SMES is relatively independent of the duty cycle and thus the energy density of that system does not incur a penalty at the larger duty cycles as does the more conventional electrochemical storage system. This is especially pronounced at the lower altitudes where the duty cycles are largest because of the orbit geometry. However, as the orbit altitude increases, the orbit period also increases and even though the duty cycle decreases to a more favorable value, the eclipse time increases in magnitude and thus requires more energy storage capability. This results in a increase in storage system mass as is shown on Table II. The total system mass also increases as the orbit altitude increases. There is an altitude at which the total system mass tends to minimize because of the various trade-offs which occur. This is explained in the following.

At the lower altitudes where the duty cycle is "extremely adverse", the rectenna mass tends to dominate. As the apogee altitude increases, the duty cycle decreases and in some interval, the rectenna mass decreases faster than the storage system mass increases due to the increase in eclipse time. After a certain point the decrease in duty cycle is such that the rectenna mass decreases more slowly and the storage system mass increases more rapidly, due to the increase in eclipse time. After this point the result is an increase in total system mass as altitude increases further. This behavior results from the non-linear aspects of the orbit characteristics.

The total system microwave beam power system mass is then the sum of the ground receiving station mass and the mass of the airborne power system:

Total Mass = Transmitter power X Transfer function + ground station mass.

Values of the transfer function, which has the units of kg/kW are in the ranges of 25 to 40 kilograms of orbital system mass per kilowatt of transmission power. The actual value of this parameter does not effect the <u>relative</u> results presented here.

CONCLUSIONS:

Preliminary indications are that SMES will show considerable advantages from the viewpoint of total system mass when incorporated into a Orbit-to-Surface Beam Power System. These advantages accrue through the increase in round trip charge-discharge efficiency of the SMES as compared to conventional electrochemical energy storage systems and the relative independence of the SMES upon high charging rates also contributes to reducing the mass of this system in comparison to electrochemical systems by alleviating the penalty in energy density that occurs in electrochemical systems which see high charging rates.

For the parameters used for this study, an "optimum altitude" exists for the lunar application for elliptical orbits for the orbiting power station. This comes about for the lower apogee altitudes because the duty cycle decreases with increasing altitude faster than the eclipse period increases at the lower apogee altitudes. Thus, the rectenna mass, which is highly dependent upon duty cycle decreases faster that the storage system mass (which is dependent upon the eclipse period magnitude) increases. At the higher apogee altitudes the reverse occurs.

The question arises as to whether or not it is possible to build SMES energy storage systems which are indicative of the values used in this study. Reference [8] indicates that energy densities on the order of present day electrochemical systems are possible. Reference [9] contains a preliminary design of a helium based SMES which approaches today's electrochemical system energy densities. Reference [10] contains a cursory analysis based on the Virial theorem which indicates that with advanced composite materials to constrain the large forces which result in SMES systems it is possible for SMES to be competitive with other energy storage options on an energy density basis.

Cooling of the SMES magnets (by cryogens ?) must also be addressed and this would result in some mass penalty. However, for a manned base application the cooling equipment could possibly be integrated with other base systems. Thus, with some advances in technology, SMES with energy densities such as those postulated here could become a reality. If High Temperature Superconducting Magnetic Energy Storage should become feasible, the energy densities postulated here could actually be exceeded. Because of the other desirable characteristics of SMES "cascading benefits" could result [1] and SMES could become the optimum storage option for applications such as were studied here.

ACKNOWLEDGEMENT:

I wish to extend my gratitude to Cadet Daniel J. Fonte Jr. of the US Air Force Academy who adapted a computer program and carried out the calculations for this work. Cadet Fonte was at the NASA-Lewis

Research Center during the summer of 1990 under the auspices of the USAF Cadet Summer Research Program.

REFERENCES:

1. Faymon, K.A., and Rudnick, S.J., "High Temperature Superconducting Magnetic Energy Storage for Future NASA Missions", Proc. of the 23rd IECEC, pages 511-14, Denver, CO. (August 1988).

2. Cull, R., and Faymon, K.A., "Orbit To Surface Beamed Power For Mars Bases Expansion", Proc. of the 24th IECEC, pages 479-84, Washington D.C. (August 1989).

3. Faymon, K.A., "Microwave Beam Powered Mars Airplane", Presented at the 25th IECEC, Reno, Nevada, (August 1990).

4. Fay, E.H., and Cull, R., "Lunar Orbiting Microwave Beam Power System", Presented at the 25th IECEC, Reno, Nevada (August 1990).

5. "Second Beamed Space Power Workshop", Joint Workshop Sponsored by the Langley Research Center and the Lewis Research Center, held at the Langley Research Center, Hampton, VA. February 28 - March 2, 1989.

6. Fonte, D., "Microwave Space Beam Power Analysis - Lunar Mission", Study performed at the Lewis Research Center under the auspices of the US Air Force Academy Summer Research Program, May - June 1990.

7. Hoffert, M., Miller, G., Heilweil, B., Ziegler, W., and Kadiramanglan, M., "Microwave and Particle Beam Sources and Propagation", Proc. of the SPIE held at Los Angeles, CA. (January 1988).

8. Leung, E.M.W., Hilal, M.A., Parmer, J.F., and Peck, S.D., "Lightweight Magnet for Space Applications", IEEE Trans. on Magnetics, Vol. Mag-23
No.2 (March 1987).

9. Eyessa, Y.M., Boom, R.W., and McIntosh, G.e., "Superconductive Energy Storage for Space Applications", Applied Superconductivity Center, University of Wisconsin, Madison, WI (July 1982).

10. Faymon, K.A., "Some Thoughts on Energy Storage for Beamed Power", Informal presentation to NASA-LeRC/DOE-ANL Joint Superconductivity Team, Lewis Research Center, (March 1989).

TABLE I. ENERGY STORAGE TECHNOLOGY ALTERNATIVES FOR PHOTOVOLTAIC SPACE POWER SYSTEMS
LOW EARTH ORBIT APPLICATIONS: LESS THAN 1 MW POWER SYSTEMS

Energy Storage Technology	W-Hrs/kg SOA 1989	W-Hrs/kg Advanced 2000+	Round Trip Efficiency	Cycling Flexibility	Rate of Discharge	Cascading Benefits	Estimated Life	Ref's.
Batteries:								
NiCd	10-20	NA	75%	Low	Low	9	5 years	12,14,22
NiH2	50	80	80%	Medium	Medium	6	5 years	12,14,15, 22
NaS	NA	100	80%	Low	Medium	5	10 years	21,22
Lithium	NA	80	80%	?	?	8	10 years	22
Fuel Cells:								
H2-O2	30	180	60%	High	Medium	2	10 years	12,16,17, 18,22
Capacitors: High Voltage Systems	10	20	90%	High	Very High	7	Long	21
Flywheels: Composite Materials (Man Rated)	5	20-30	80%	High	High	3	Long	13,19,20
Superconducting Magnetic Storage:								
Conventional Superconducting Magnetic Storage	30	85	95%	Very High	Very High	1	Indef.	2,3,21
HTSC Magnetic Storage	NA	125	95%	Very High	Very High	1	Indef.	Estimate

LERC-ANL.90

TABLE II.

LUNAR SURFACE STATION - NORTH LATITUDES: 30˚ & 60˚
185.2 km PERIAPSIS - 135˚ BEAM ANGLE

APOAPSIS ALTITUDE: km		60˚ EC	60˚ SMES	30˚ EC	30˚ SMES
1000	DUTY CYCLE	3.630211	3.630211	4.734034	4.734034
	STOR. SYST. MASS	3615.07	2704.824	4010.88	2829.556
	TOTAL MASS	6058.297	4657.828	6866.716	5058.561
	TRANS. POWER	1140.932	973.117	1284.186	1003.232
1500	DUTY CYCLE	2.53804	2.53804	3.572604	3.572604
	STOR. SYST. MASS	3647.31	2860.066	4092.895	3083.59
	TOTAL MASS	5724.671	4567.847	6564.436	5060.718
	TRANS. POWER	1130.443	1020.611	1375.92	1042.194
2000	DUTY CYCLE	1.983477	1.983477	2.947007	2.947007
	STOR. SYST. MASS	3782.76	3022.021	4313.927	3350.557
	TOTAL MASS	5686.256	4613.602	6596.157	5205.025
	TRANS. POWER	1360.424	1352.352	1376.452	1261.331
2500	DUTY CYCLE	1.65511	1.65511	2.551793	2.551793
	STOR. SYST. MASS	3960.012	3194.269	4589.699	3627.713
	TOTAL MASS	5770.264	4724.679	6765.947	5415.544
	TRANS. POWER	1782.108	1675.727	1850.028	1612.576
3000	DUTY CYCLE	1.442872	1.442872	2.277564	2.277564
	STOR. SYST. MASS	4164.347	3379.312	4894.873	3913.524
	TOTAL MASS	5923.308	4877.557	7008.964	5664.466
	TRANS. POWER	1936.272	1911.254	2253.119	2073.068
	NORTH LATITUDE	60˚		30˚	

POWER: kW
MASS: kg

EC: ELECTROCHEMICAL STORAGE   SMES: SUPERCONDUCTING MAGNETIC ENERGY STORAGE

TABLE III.

Lunar Surface Stations - North Latitudes: 30˚ and 60˚

Molniya Type Orbits - Apogee Over North Pole

Periapsis: 185.2 kM (100 NM) - Beam Angle: 135 Degrees

Input Data: Common to Electrochemical and SMES

PMAD Radiator: 450˚ K
Storage Radiator: 450˚ K
Rectenna Power Density: 2 kW / m^2
Rectenna Specific Weight: 2 kG / kW
Radiator View Factors: 0.5
Antenna: 35 meter diameter
Microwave Frequency: 300 GHZ
Station Power
  Beam Time: 100 kW
  Eclipse Time: 100 kW
Gravitational Constant of Moon: 1.62 m / sec^2
Radius of Moon: 1739 kM
Advanced Radiator Technologies

Specific Input: SMES

Energy Dependent Specific Weight: 100 watt-hrs / kG
Power Dependent Specific Weight: 5 kG / kW
Round Trip Efficiency: 97 %
Discharge Efficiency: 98.5 %
PMAD Efficiency: 95 %

Specific Input: Electrochemical Storage

Energy Dependent Specific Weight: 100 watt-hrs / kG
Power Dependent Specific Weight: 8 kG / kW
Round Trip Efficiency: 70 %
Discharge Efficiency: 85 %
PMAD Efficiency: 93 %

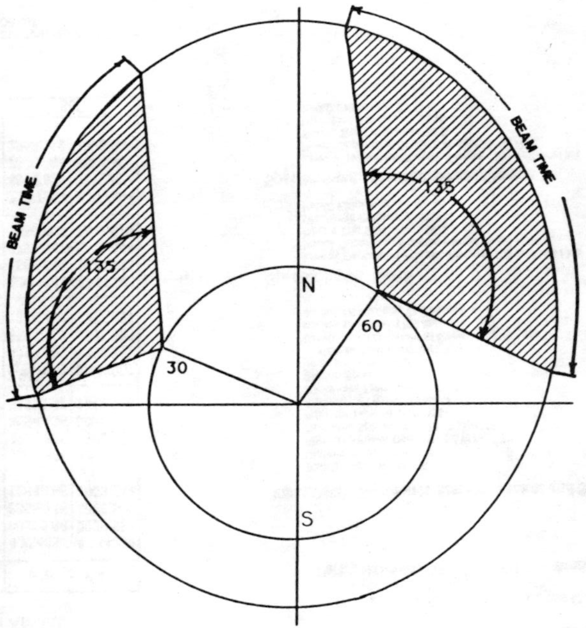

FIGURE I. Polar orbits for microwave power beaming to lunar stations located at 30° and 60° north latitudes.

WHY ELLIPTICAL ORBITS FOR ORBIT-TO-SURFACE MICROWAVE BEAMED POWER ?

COMPARISON OF CIRCULAR AND ELLIPTICAL LUNAR ORBITS WITH SAME TRANSMISSION EFFICIENCIES

- **ELLIPTICAL (MOLNIYA TYPE) ORBITS HAVE SHORTER PERIODS**
  - LESS ADVERSE DUTY CYCLES : LESS POWER REQUIRED AT TRANSMITTER
  - SHORTER ECLIPSE TIMES : REDUCED ENERGY STORAGE CAPACITY REQUIRED

ORBIT	ALTITUDE	APOGEE ALTITUDE	ORBIT PERIOD	DUTY CYCLE	BEAM TIME	ECLIPSE TIME
CIRCULAR	LOW	1500 KM	4.59 HRS	3.77	0.97 HRS	3.63 HRS
ELLIPTICAL	LOW	1500 KM	3.27 HRS	3.57	0.72 HRS	2.56 HRS
CIRCULAR	HIGH	3000 KM	8.14 HRS	2.77	2.16 HRS	5.98 HRS
ELLIPTICAL	HIGH	3000 KM	4.79 HRS	2.28	1.46 HRS	3.33 HRS

POLAR ORBITS : PERIGEE FOR ELLIPTICAL ORBIT -- 182.5 KM
SURFACE STATION AT 30° N LATITUDE : BEAM WIDTH 135.°

FIGURE II.

SES90.1

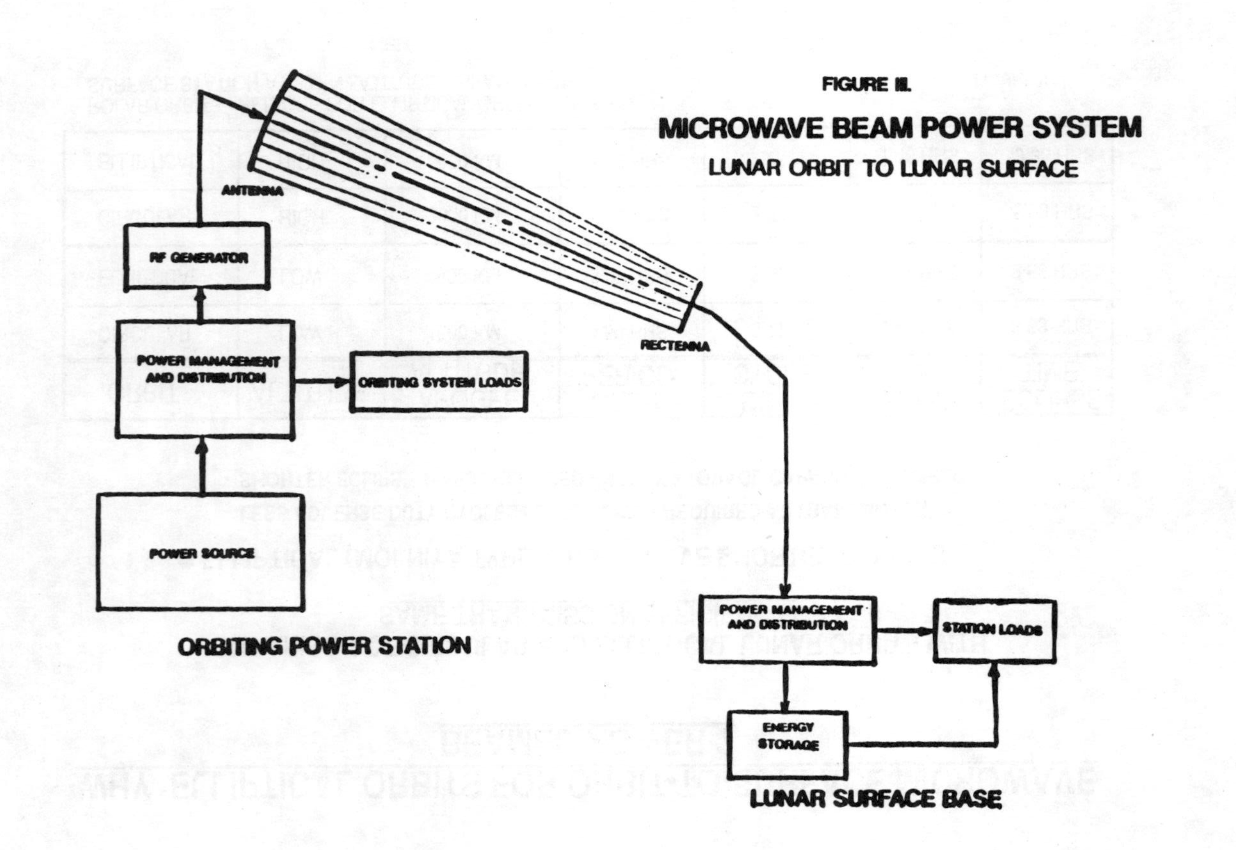

FIGURE III.

MICROWAVE BEAM POWER SYSTEM

LUNAR ORBIT TO LUNAR SURFACE

FIGURE IV.

MICROWAVE BEAM TRANSMISSION EFFICIENCY

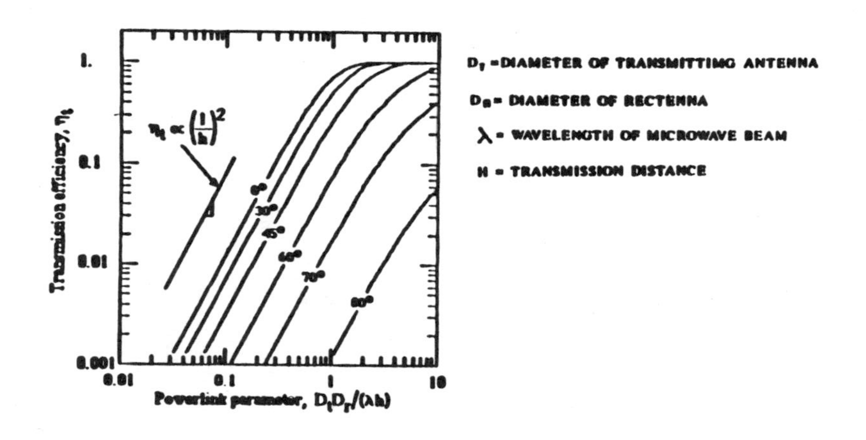

BASED ON NYU MICROWAVE BEAM MODEL

SUPERCONDUCTIVITY APPLICATIONS IN SPACE FLIGHT SCIENCE EXPERIMENTS

Rudolf Decher, Palmer Peters, Charles Sisk, and Eugene Urban
Space Science Laboratory
NASA Marshall Space Flight Center
Huntsville, Alabama 35812 U.S.A.

ABSTRACT

Superconductivity plays a major role in the development of gravitational physics space flight experiments. These experiments require measurements of extremely small relativistic effects with utmost precision. Their implementation is made possible only by employing the unique properties of superconductors and low temperature physics. Superconductivity applications in three space flight experiments presently under development in the laboratory will be discussed, including the Gravitational Probe B, the Satellite Test of the Equivalence Principle, and the Superconducting Gravity Gradiometer.

INTRODUCTION

Gravitational physics experiments represent a special category of space flight experiments because of their unprecedented requirements in measurement sensitivity and accuracy. The objectives of these experiments include the measurement of relativistic gravity effects to test gravitational theories, especially Einstein's General Relativity Theory, and to test fundamental principles of physics. Only by utilizing the properties of superconductivity and superconducting devices does it become possible to meet the difficult instrument requirements in accuracy and sensitivity. Three experi-

ments which use superconductivity at liquid helium temperature will be discussed: The Gravitational Probe B (GP-B), the Satellite Test of the Equivalence Principle (STEP), and the Superconducting Gravity Gradiometer (SGG). These experiments have been under development in the laboratory for several years with funding provided by NASA. The purpose of this discussion is to show applications of superconductivity in space flight experiments. In the limited space available it is not possible to explain in detail the scientific objectives and instrument design of the experiments.

Each of the three experiments will be enclosed in a liquid helium dewar. The low temperature provided by the liquid helium (superfluid helium) is not only necessary to obtain superconductivity, but is needed also to meet other experiment requirements. The very high mechanical stability, thermal uniformity, and low electrical noise in sensors and circuits which can be obtained are necessary to achieve experiment objectives.

Various phenomena of superconductivity are employed in the implementation of these experiments. SQUIDS (Superconducting Quantum Interference Devices) serve as sensitive magnetometers and position detectors. The Meissner effect is used to obtain magnetic suspension of test masses and to provide shielding against external magnetic fields. The London moment of a spinning superconductor is used to obtain a gyroscope spin axis readout. The property of magnetic flux conservation in superconducting circuits is employed in all of the above applications.

GRAVITY PROBE B

Gravity Probe B (GP-B) is a satellite experiment to perform a new test of Einstein's General Relativity Theory, as well as other theories of gravity, by measuring the relativistic precession of superconducting gyroscopes in Earth orbit.[1,2] The relativistic precession of an orbiting gyroscope was derived from Einstein's Theory by L. I. Schiff and a space flight experiment was proposed by W. H. Fairbank and R.

Cannon, all at Stanford University. The experiment is being developed by a team at Stanford University with participation from Lockheed Missiles and Space Company, Inc. The principal investigator is C.W.F. Everitt. For several years Marshall Space Flight Center developed in-house technology for GP-B in collaboration with Stanford University including superconducting thin film coatings and circuits and gyro rotor fabrication techniques.

Gravity Probe B will contain four unique gyroscopes to measure two relativistic effects. It will fly in a 650 km circular, polar orbit. With the gyro spin axes in the orbit plane and parallel to the line-of-sight to the reference star Rigel (Figure 1), the relativistic precessions predicted by Einstein's General Relativity Theory are as follows. The geodetic effect will cause the gyro spin axis to precess by 6.6 arcsec in the plane of the orbit during 1 year. The motional or gravitomagnetic effect causes a spin axis drift of 0.042 arcsec per year in the plane of the celestial equator. The spin axis precession will be measured relative to the fixed star Rigel by a star tracking telescope which is part of the experiment package. The GP-B instrument is enclosed in a superfluid helium dewar which represents the major portion of the spacecraft. The operational lifetime of the experiment in orbit is 1 to 2 years. The goal of the experiment is to achieve a precision of 0.001 arcsec/year in the measurement of the relativistic precessions. To measure these very small changes in the spin axis orientation requires extremely stable and sensitive gyroscopes and the elimination or reduction to very low levels of all disturbing forces. Years of research have established an account of all conceivable error sources, and ingenious solutions were developed to make these measurements feasible. Superconductivity is utilized in very novel ways to obtain a gyroscope spin axis readout and to achieve shielding against external magnetic fields. The gyroscope rotor consists of a solid quartz sphere of 1.5 inch diameter and spherical to better than 1 μ inch, coated with a very uniform niobium thin film. The rotor is suspended inside the quartz gyro housing by electrical fields. The superconducting niobium film has two functions: it provides a conducting surface for the electrical suspension of the rotor, and it is part of the readout system which measures the rotor spin axis orientation. The concept of the gyro readout is illustrated in Figure 2. The spinning superconducting surface of the gyro rotor generates a magnetic moment (London moment) which is always aligned along the spin axis. The rotor spins inside a superconducting loop located on the gyro housing. The readout loop is coupled to a SQUID magnetometer. As the spin axis precesses, the magnetic flux through the gyro readout loop changes. This flux change, which is detected by the SQUID magnetometer, is a measure of the spin axis precession. In reality the pickup loop is made of multiple turns of a thin film superconductor for optimum coupling to the SQUID.

The magnetic flux through the readout loop generated by the London moment of the spinning rotor is at maximum when the spin axis is at a right angle to the plane of the loop. In this case the magnetic field has a magnitude of 1.14×10^{-7} w (Gauss) where w is the angular velocity of the rotor. With the intended spin rate of 170 Hz the field of the London moment becomes 1.2×10^{-8} Tesla which is the readout signal of the gyroscope. The accuracy and resolution of the readout is limited by several factors, mainly by SQUID noise and interfering magnetic fields.

The basic signal to be measured is a very small and very slowly changing dc type signal which is buried in the 1/f noise of the SQUID. However, by rolling the spacecraft slowly (10 minutes per revolution) about the line of sight to the reference star, the gyro precession signal appears as a sinusoidal signal at the roll rate of the spacecraft. This is a tremendous advantage in overcoming the readout noise problem because narrow band signal processing techniques can be applied. Laboratory experiments with a 200 MHz RF SQUID have shown that a resolution of 1 milliarcsec can be obtained with an integration time of 70 hours. Using dc SQUIDs which have a better noise characteristic than RF SQUIDs it may be possible to reduce the integration time required for 1 milliarcsec resolution to perhaps 2 hours. To achieve these high resolutions in the readout system, temperature variations of the SQUID must be kept below

10^{-4} K for the mission duration, which is one of several design requirements for the experiment.

Another critical design requirement for the GP-B experiment is magnetic shielding of undesired dc and ac magnetic fields to extremely low levels at the gyroscope location to avoid corruption of the readout signal. The magnetic field from trapped flux in the gyro rotor must be below 10^{-7} Gauss because of the dynamic range limit of the SQUID. DC background magnetic fields can be caused by flux trapped in superconductors resulting from ambient magnetic fields as the experiment is cooled below the transition temperature. The Earth's magnetic field (approximately 0.5 Gauss) penetrates into the dewar through the window for the star tracking telescope. Because of the spacecraft roll, an ac Earth magnetic field appears at the gyro readout loop. This ac field has to be attenuated by a factor of 10^{13} to achieve the specified readout resolution.

Three different types of shields are employed to obtain the required magnetic shielding. First, the experiment cavity inside the dewar is surrounded by a cylindrical MU-metal shield. Second, inside the MU-metal shield, there is a superconducting magnetic shield formed by a cylindrical sheet of superconducting foil (lead bag) open at one end for insertion of the GP-B instrument and tightly closed at the other end. The space inside the lead bag has a background field of less than 10^{-7} Gauss. The GP-B instrument will be inserted and cooled down in this low field environment to minimize trapped flux. A unique technique developed by Cabrera at Stanford University is employed to achieve these low field levels. A long cylindrical lead bag foil, open at one end, is folded such that it has a very small inside volume. The lead bag is cooled through the transition temperature inside the dewar in an evacuated probe. Once superconducting the folded foil is expanded mechanically to obtain the shape of a tube. Because of the Meissner effect, new flux cannot enter the inside of the superconducting lead bag, and the original flux which existed inside the folded bag stays constant. The magnetic field inside the expanded bag is now reduced by an amount corresponding to the increase in area since the field is the flux per unit area. This principle is illustrated in Figure 3. By inserting another folded bag into the expanded first bag, the process can be repeated to obtain further reduction of the inside magnetic field.

This process, sketched in Figure 4, is repeated several times to obtain the required low magnetic field inside the dewar. In Figure 4a the first bag has been expanded and is filled with liquid helium. The vacuum probe containing the second folded bag is inserted into the dewar (b), the bottom of the probe is opened, and liquid helium rises inside the probe cooling the lead bag to below the transition temperature (c). The first bag is removed after its bottom has been cut open, (d) and (e), the vacuum probe is removed, and the bag is expanded providing the new shield with reduced inside field, (f) and (a). Each subsequent bag expansion takes place in a lower field. With five cycles a magnetic field of less than 10^{-7} Gauss is achieved. The shield foil has to be cooled very slowly to reduce thermoelectric currents in the foil caused by thermal gradients which lead to trapped flux. This process requires approximately 20 hours per cycle.

The combined MU-metal and lead bag shield reduce the dc magnetic field in the experiment cavity to less than 10^{-7} Gauss, and the transverse ac (Earth magnetic) field is reduced by a factor of approximately 10^{9}. The third level of shielding results from each gyroscope being enclosed in a cylindrical superconducting shield. Additional shielding is provided by the components of the readout system and gyrorotor generating a total ac shielding factor of 10^{13}.

A laboratory model of the GP-B instrument is undergoing testing at Stanford University. Hardware for a flight test of a GP-B gyroscope on the Space Shuttle is presently being designed.

SATELLITE TEST OF THE EQUIVALENCE PRINCIPLE (STEP)

The goal of this experiment is a test of the Weak Equivalence Principle, sometimes expressed as equivalence of inertial and gravitational mass, which states that all bodies fall with the same acceleration in the same gravity field independent of their composition. This principle is the foundation of Einstein's General Relativity Theory. Any violation of it, no matter how small, would require corrections of basic concepts of physics. There are conceivable violations of the Equivalance Principle related to particle physics which would show up only at much higher experiment accuracies than have been achieved so far (few parts in 10^{11}). A cryogenic space experiment to test the Weak Equivalence Principle with an accuracy of 1 part in 10^{17} was proposed by Everitt and Worden of Stanford University. The large (10^6) increase in accuracy is made possible by performing the experiment in Earth orbit and by using superconductivity and cryogenic technology.[3,4] Development of the experiment apparatus is in progress at Stanford University, and the STEP mission is presently subject to a joint NASA/ESA study.

The concept of the experiment is shown in Figure 5. Two cylindrical, coaxial test masses made of different material are in free fall in Earth orbit inside a satellite. If there is a violation of the Equivalence Principle, the two test masses would experience slightly different accelerations and would move axially relative to each other. The differential acceleration is measured by an extremely sensitive superconducting differential accelerometer (position sensor). The sensitive axis of the accelerometer has a fixed orientation in inertial space as indicated in Figure 5. As the experiment orbits the Earth, the differential acceleration changes direction and varies with the period of orbital revolution. The whole experiment is enclosed in a liquid helium dewar to provide the necessary cryogenic environment for the superconducting circuits and for mechanical and thermal stability.

The principle of the accelerometer is shown in Figure 6 (ground-based version). The test masses are cylindrical in shape and are arranged in a coaxial geometry (one inside the other) so their centers of mass coincide. Each test mass is suspended in a superconducting magnetic bearing which constrains radial motion of the masses but permits free motion along their common axis which is the sensitive axis of the accelerometer. The position of the test masses along the sensitive axis is measured by sense coils connected to a SQUID detector. The test masses are isolated from each other by a superconducting magnetic shield, and the complete accelerometer is enclosed in another superconducting magnetic shield. The STEP experiment may use three accelerometers to permit comparison of several test masses made of different materials.

The principle of the superconducting magnetic bearing is illustrated in Figure 7 for one test mass. The bearing is made of superconducting wire attached to a cylindrical support structure. Adjacent wires carry currents of opposite direction to eliminate any net magnetic moment. The suspension of the superconducting test mass in the bearing is the result of the Meissner effect. Non-superconducting test masses will be coated with a superconducting film to achieve magnetic suspension. In the low-gravity environment of space, only a small magnetic force (i.e., small super currents) will be necessary to keep the test masses in position. External disturbance forces (e.g., atmospheric pressure) acting on the spacecraft are compensated by a drag-free control system which reduces accelerations experienced by the instrument to a level of 10^{-9} cm sec^{-2} or less. The magnetic bearing provides radial motion constraints of the test masses. The critical requirement for the bearing design is to provide free motion along the common axis of the test masses with minimal axial forces. Along the direction of the sensitive axis, force variations in space and time must be less than 1 dyne cm^{-1} and 10^{-13} dyne sec^{-1}, respectively, for kilogram-size test masses. The magnetic bearing developed at Stanford University can meet these requirements.

The acceleration of the test masses is sensed by a SQUID position detector which measures differential and common mode acceleration of the test masses. Figure 8 illustrates the principle of operation of the position detector. M1 and M2 are the test masses; L1, L2 and L4, L5 are the corresponding position sensing coils. The complete circuit is superconducting and the total flux trapped in the circuits (caused by persistent currents in the various loops) will remain constant. If, for example, M1 is displaced, the inductances L1 and L2 will change (M1 acting like a tuning slug) and consequently the currents through the circuit will change, because the flux in a superconducting circuit will remain constant. The resulting change in current through L3, which depends on the difference in position of the test masses, is detected by a SQUID magnetometer. The overall arrangement is such that the SQUID coupled to L3 senses only differential acceleration while the second SQUID coupled through a transformer detects common mode acceleration of the test masses. The position detector has a displacement sensitivity of 10^{-13} cm. The sensitivity is 10^{-14} and 10^{-9} cm sec^{-2} Hz$^{-1/2}$ for differential acceleration and for common mode acceleration, respectively.

The general configuration of the STEP satellite as envisioned at this time is shown in Figure 9. Three accelerometers are enclosed in a liquid helium dewar. The whole instrument inside the dewar is surrounded by a lead bag superconducting magnetic shield. Some of the spacecraft systems are indicated in the sketch. The instrument operates at a temperature just below the lambda point (2.2 K) of liquid helium.

SUPERCONDUCTING GRAVITY GRADIOMETER

A very sensitive gravity gradiometer employing superconducting technology was proposed by H. J. Paik of the University of Maryland. NASA intends to fly the Superconducting Gravity Gradiometer (SGG) in an Earth-orbiting satellite to map the gravity field of the Earth with higher accuracy than has been possible in the past.[5,6] The goal of the experiment is a sensitivity of 10^{-4}E Hz$^{-1/2}$. E (Eotvos), the unit of measure of

gravity gradient, equals 10^{-9} sec^{-2}. For example, at the Earth's surface the gravity gradient has a value of 3100 E. The instrument is a three-axis gradiometer containing three single-axis gradiometers arranged in an orthogonal geometry and integrated with a six-axis superconducting accelerometer (SSA). With a different mission this instrument could provide a very accurate test of the inverse square law of gravity for distances from 10 to 10^4 km.

The concept of the single-axis SSA is shown in Figure 10. The gravity gradient is determined by measuring the differential acceleration of the two proof masses (m_1, m_2) which are separated by a distance of 16 cm. The niobium proof masses are supported by mechanical springs permitting motion along the common axis of the masses which is the sensitive axis of the gradiometer. Acceleration will cause motion, i.e., displacement of the masses which is detected by superconducting coils (L5, L6) coupled to a SQUID. Displacement of the proof mass modulates the inductance of the coils through which persistent currents I_{d1} and I_{d2} are flowing. The flux in the closed loop circuit must remain constant, and any change in coil inductance will result in a current change through the SQUID coil Ls which is detected by the SQUID. The current changes caused by a common acceleration of the two proof masses can be cancelled out by proper adjustment of the current ratio I_{d1}/I_{d2}. The gradiometer, therefore, will be sensitive to differential accelerations only. The common mode acceleration is detected by an identical superconducting circuit (not shown) with one of the persistent currents reversed.

The proof masses are constrained in position along the sensitive axis by magnetic suspension generated by persistent currents through coils L3 and L4. Superconducting magnetic suspension is employed to generate a "negative spring" which compensates for the stiffness of the mechanical spring providing essentially free motion of the masses along the sensitive axis.

At the center of the three-axis SGG is a superconducting accelerometer (SSA) which measures linear and angular acceleration of the instrument against a levitated niobium proof mass in all six degrees of freedom. The shape of the proof mass and the arrangement of levitation and sensing coils are shown in Figure 11. Acceleration is sensed by 6 inductance bridges with a total of 24 "pancake" coils coupled to SQUIDs. A second set of 24 coils is used for levitation and feedback. The accelerometer uses force rebalance with the proof masses kept in position by a feedback system and the feedback signal is the readout. The expected sensitivity for linear acceleration is 10^{-13} g $Hz^{-1/2}$ and for angular accelerations is 10^{-11} rad sec^{-2} $Hz^{-1/2}$ The SSA fits into a 10 cm cube. The SSA is located inside the SGG at the origin of the three gradiometer axes.

The complete instrument assembly weighs about 40 kg and fits into a spherical volume of 30 cm diameter. The SGG is enclosed in a superfluid helium dewar. The spacecraft orbit altitude will be typically 200 km with a mission lifetime of 6 months. A mission and spacecraft study was completed in 1989. A three-axis laboratory model of the SGG has been assembled and is undergoing tests.

REFERENCES

[1]Parkinson, W., Everitt, C.W.F. and Turneaure, P., "The Gravity-Probe-B Relativity Gyroscope Experiment: An Update on Progress," Advances in the Astronautical Sciences, edited by R. D. Culp and T. J. Kelly, paper number AAS 87-011, p.141 (1987).

[2]SPIE Proceedings 619 (SPIE, Bellingham, WA) (1986). (several papers on GP-B).

[3]Worden, P. W., "Cryogenic Equivalence Principle Experiment: Discussion and General Status," Proceedings of Third Marcel Grossmann Meeting on General Relativity, Shanghai (1982).

[4]Worden, P. W., "Measurement of Small Forces with Superconducting Magnetic Bearings," Prec. Engineering 4, 139 (1982).

[5]Moody, M. V., Chan, H. A. and Paik, H. J., "Superconducting Gravity Gradiometer for Space and Terrestrial Applications," J. Appl. Phys. 60(12) (1986).

[6]Superconducting Gravity Gradiometer Mission, Study Team Technical Report, edited by S. H. Morgan and H. J. Paik, NASA Technical Memorandum 4091 (1988).

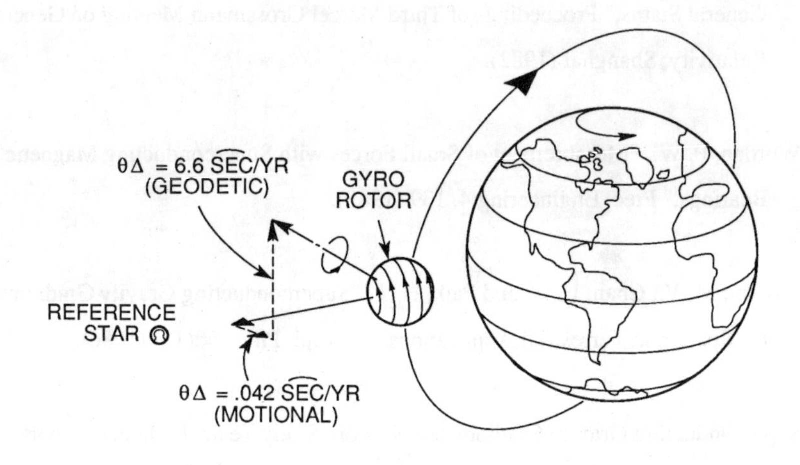

Figure 1. Relativistic Gyro Precession

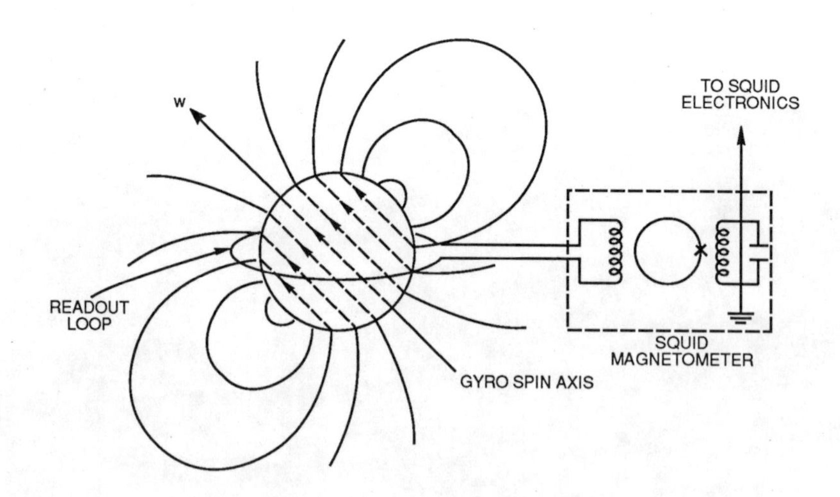

Figure 2. Gyroscope Readout

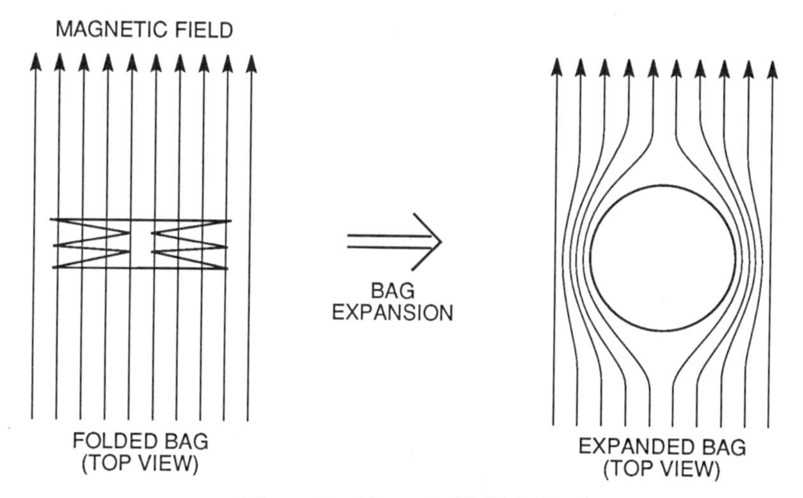

Figure 3. Magnetic Field Reduction

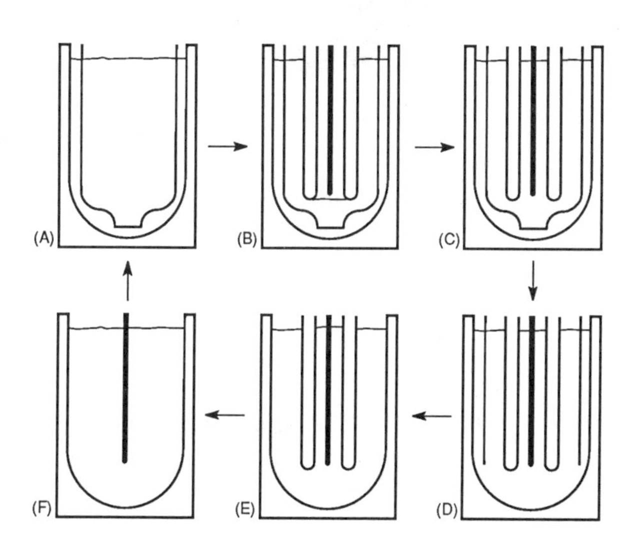

Figure 4. Lead Bag Expansion Cycle

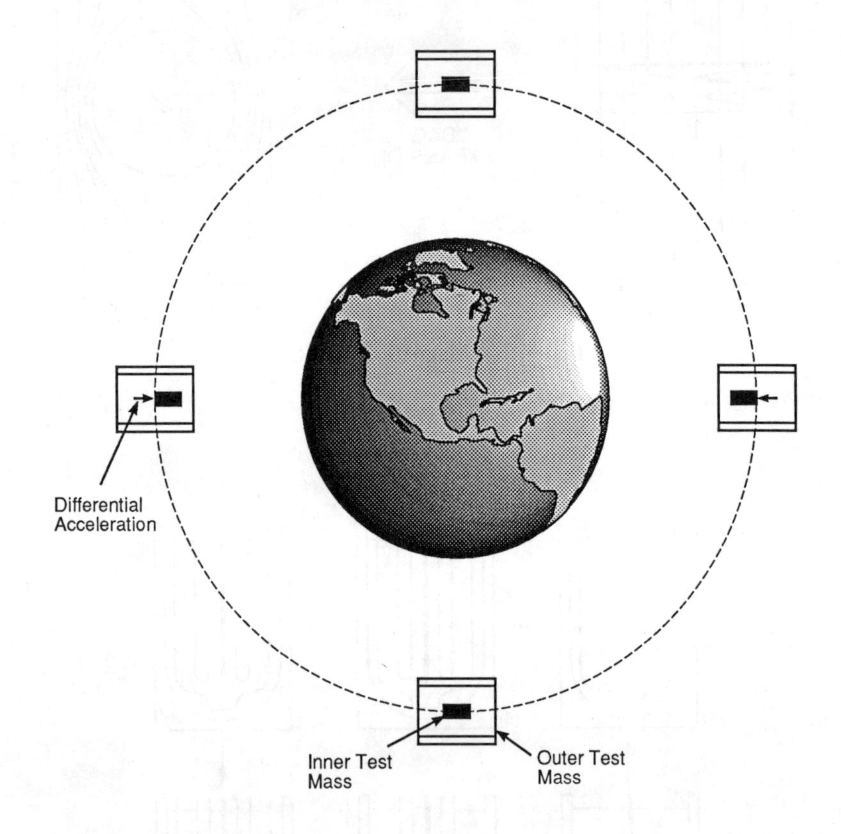

Figure 5. STEP Experiment Concept

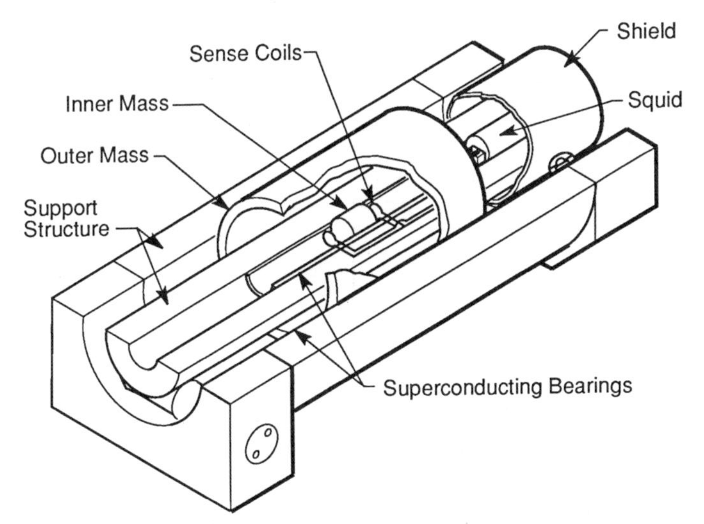

Figure 6. Differential Accelerometer

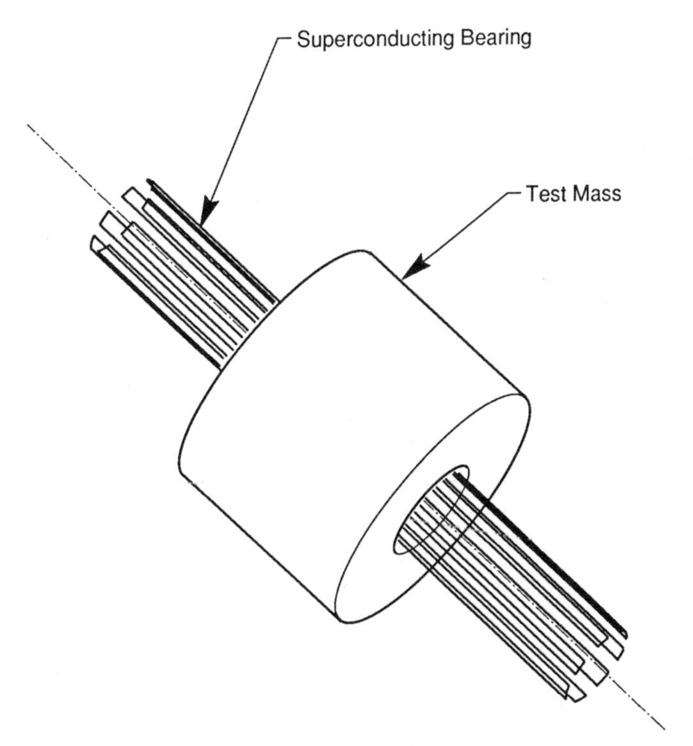

Figure 7. Superconducting Magnetic Bearing

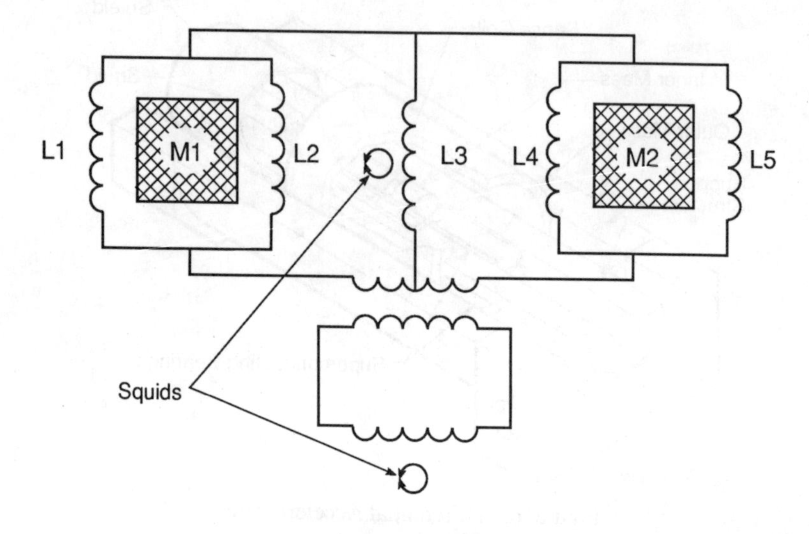

Figure 8. Position Sensing Circuit

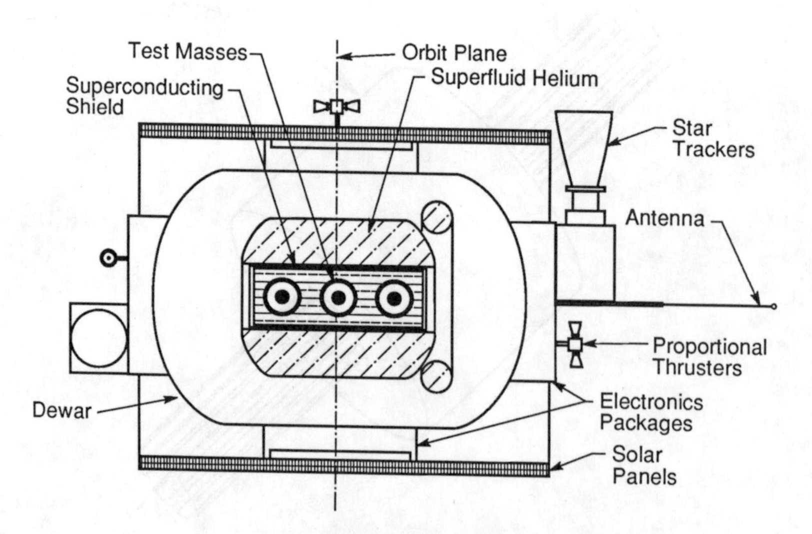

Figure 9. STEP Satellite Configuration

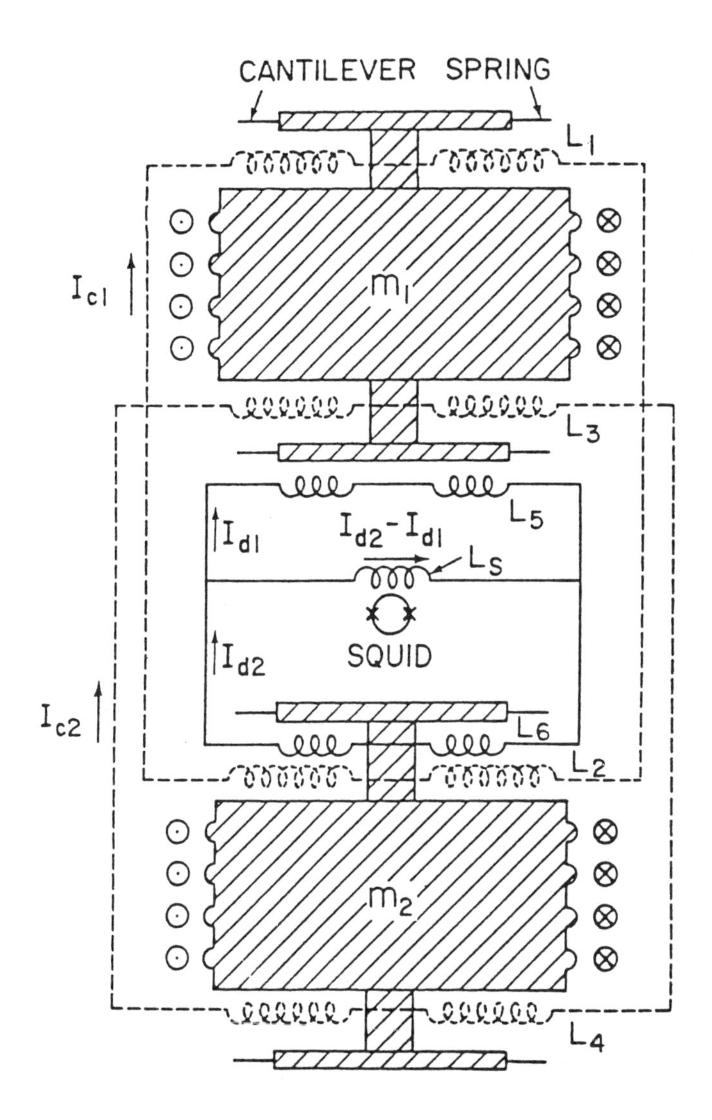

Figure 10. Single-Axis Superconducting Gravity Gradiometer

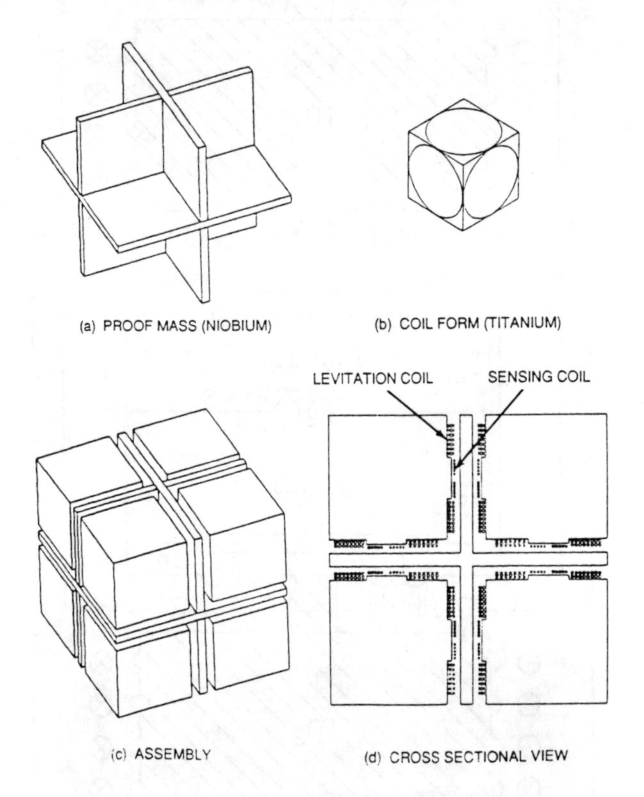

Figure 11. Six-Axis Superconducting Accelerometer

Progress in the development of High Tc superconductor IR bolometers

B. Lakew
STX Corporation, Lanham, MD 20706

J. Brasunas
NASA/Goddard Space Flight Center, Greenbelt, MD 20771

ABSTRACT

In our previous reports we have described the construction and performance of Mark I,the first high Tc superconducting (HTS), transition-edge, composite IR bolometer(1). We report here the improvements that have been made since with Mark II and Mark III. The thickness of the HTS thin film substrate(LaAlO3) was reduced from 500 μm to 75 μm , and the HTS/substrate's other dimensions were reduced by a factor of 40. A 2.5 μm thick Silicon IR absorber substrate was used instead of 25 μm. The time constant of the detector is now 150 ms down from 32 s. The detectivity D* is now $1.5x10^8 cmHz^{1/2}W^{-1}$ at 5 Hz up from $5.5x10^6$ at 2 Hz. 1/f excess noise was observed and was 20 times higher than the expected thermal fluctuation noise.

INTRODUCTION

The Cosmic Background Explorer(COBE) satellite launched in November 1989 used very sensitive liquid helium cooled infrared detectors with detectivities D* of the order of 10^{14} $cmHz^{1/2}/W$. A year after the launch and a very successful mission with many outstanding discoveries about the early universe, as anticipated COBE has run out of cryogen.

For missions to the outer planets the duration of the journey, which can be up to 7 years, and weight limitations make it impossible to

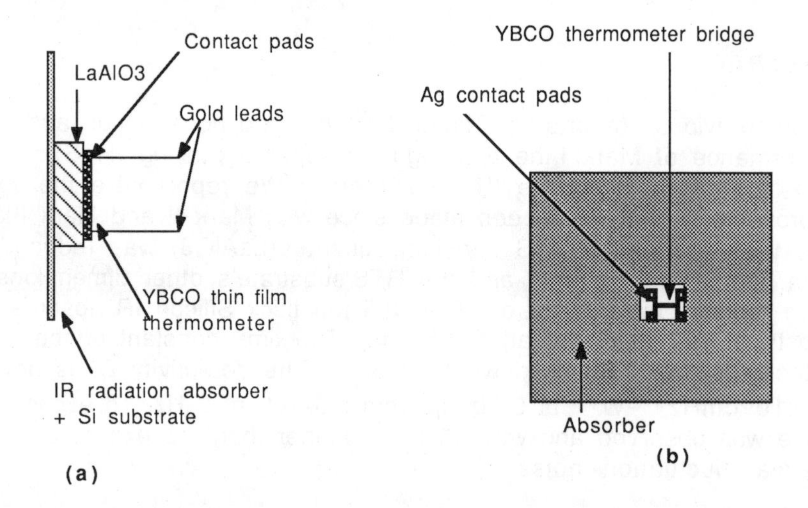

Fig. 1 Outline of the Composite Infrared spectrometer.

carry cryogens to cool infrared sensors in the focal plane of instruments. The Voyager interferometers, which have obtained extremely valuable data from the outer planets, used relatively insensitive room temperature thermal type infrared(IR) detectors ($D^* \sim 10^9$).

The Composite Infrared Spectrometer (CIRS), a proposed infrared experiment on the Cassini orbiter science payload, will use two thermopile detectors at the focal planes of its mid and far IR Fourier transform spectrometers (15 to 50 μm and 30 to 1000 μm respectively). Below 15 μm quantum detectors will be used.

Cassini is a proposed post-Voyager mission to Saturn and Titan. It is a NASA/ESA joint effort and will be launched in 1996. The CIRS thermopiles are near room temperature thermal detectors and will have a $D^* \sim 3x10^9$ and a time constant of 25 ms. They will operate at the instrument temperature of about 170 K.

Radiative and mechanical coolers can now be used to cool the focal plane to a range of temperatures between 65 and 90 K.

The advantages of operating at lower temperatures are :

(i) a reduction of the significant noise sources (Johnson noise and thermal fluctuation noise, which is the limiting noise source) with the decrease in temperature,
(ii) reduction of the time constant of the detector . If the thermal conductance (G) of the detector system is unchanged then cooling reduces the heat capacity(C) . The time constant (tau = C/G) is therefore reduced which also means that the response speed of the detector is increased.

Because there are no thermal IR detectors optimized for operation between these temperatures currently available, our goal has been to develop improved performance composite thermal IR sensors (Fig 1) operating near 80 K. The type of detector being developed is a transition-edge bolometer utilizing the sharp resistance change at Tc of a high Tc superconducting YBaCuO thin film. The composite bolometer uses separate elements for radiation absorption, temperature measurement, and thermal isolation.

The calculated near phonon limited performance of the detector around 80 K ($D^{*}\sim10^{10}$) makes it theoretically better than the thermopile detectors proposed to fly on Cassini.

The improved performance of the high Tc bolometer would enhance significantly the determination of hydrocarbons and nitriles known to be present in the atmosphere of Titan. A greater array of trace molecules(i.e. CO, HCN...) could also be observed over a wide range of altitudes.

BOLOMETER CONSTRUCTION IMPROVEMENTS

The hoped for characteristics of the improved bolometer are summarized in Table 1.

Table 1. High Tc Composite Bolometer Characteristics

-Operating temperature	70 to 90 K
-IR absorber area	~ 1x1 mm
-Wavelength coverage	15 to1,000 μm
-Time constant	10 to 30 ms
-Detectivity	> 3×10^{9} $cmHz^{1/2}/W$

Since Mark I our efforts have centered around the reduction of the heat capacity, the time constant and the 1/f noise of the detector. Using a simple technique the thickness of the HTS substrate(LaAlo3) was reduced from 500 μm to 70 μm . The other dimensions of the HTS/substrate was reduced about 40 times.

The paterned YBaCuO thin films on $SrTiO_3$ or $LaAlO_3$ were made by The National Institute of Standards and Technology in Boulder, CO. For Mark II and III thin films were deposited by laser ablation. The films were then integrated into the composite bolometer and tested at Goddard Space Flight Center in Greenbelt, MD.

Table 2 shows the relative dimensions of Mark I, Mark II and Mark III.

Table 2. High Tc composite bolometers comparative dimensions.

	Mark I	Mark II	Mark III
IR absorber	$2.5 \times 2.8 mm^2$ x 25μm	$3.5 \times 1.7 mm^2$ x 2.5 μm	$1 \times 1.6 mm^2$ x 2.5 μm
HTS substrate	$1.8 \times 2.7 mm^2$ x 500μm ($SrTiO_3$)	$350 \times 350 \mu m^2$ x500μm ($LaAlO_3$)	$350 \times 350 \mu m^2$ x75μm ($LaAlO_3$)
YBaCuO thin film	20 μm wide 0.2μm thick meander	20 μm wide 0.2 μm thick meander	10 μm wide 0.2 μm thick bridge

The length of the gold leads which make electrical contact to the High Tc film thermometer was reduced from about 2.5 cm to about 3.5 mm thus increasing the thermal conductance of the detector system and reducing the contribution of the gold leads to the heat capacity.

DETECTIVITY AND TIME CONSTANT

The performances of the bolometers are summarized in table 3.

Table 3. Improvements in detector performance.

	Mark I	Mark II	Mark III
Time constant	32 s	750 ms	150 ms
Detectivity D*	$5x10^6$ at 2 Hz	$3x10^7$ at 1 Hz	$1.5x10^8$ at 5 Hz

NOISE

All our bolometers have shown substantial 1/f excess noise at the transition temperature. At 10 Hz their noise level is about $1\text{-}2x10^{-7}V/Hz^{1/2}$ per volt of dc bias and is higher than the noise calculated from our C, G and expected thermodynamic fluctuations. Other groups have observed similar excess 1/f noise(2).
Excess noise less than $10^{-7}V/Hz^{1/2}$ per volt of dc bias has also been obtained with laser ablated YBaCuO thin film on top of a SrTiO3 buffer layer on a sapphire sustrate(3).
The main causes of excess noise in our detectors could be film quality and/or equilibrium temperature fluctuations due to poor thermal coupling between the HTS film and the substrate.

FUTURE IMPROVEMENTS

We have made much progress in reducing the heat capacity of our high Tc composite bolometers. Mark IV will have an even thinner LaAlO3 HTS substrate. We anticipate the time constant to be near 50 ms and the D* near $1x10^9$. More effort will be put to reduce the 1/f excess noise by using HTS thin films with high Jc and better thermal contact to the substrate.

ACKNOWLEDGEMENTS

We are pleased to acknowledge the assistance of Carol Sappington of NASA/GSFC code 724.
This work is currently supported by the Planetary Definition and Development Program, and OAST Code R, both at NASA Headquarters.

REFERENCES

J.C. Brasunas, S.H. Moseley, B. Lakew, R. Ono, D. McDonald, J Beall, and J. Sauvageau, "Construction and performance of a high Tc superconductor composite bolometer", J. Appl. Phys.,**66**(9), 4551 (1989)

R. Black, L. Turner, A. Mogro-Campero, T. McGee, and A. Robinson, " Thermal fluctuation and 1/f noise in oriented and unoriented YBaCuO films", Appl. Phys. Lett., **55**(21), 2233(1989)

S. Verghese, P. Richards, K. Char and S. Sachtjen, " Fabrication of a high Tc superconducting bolometer", SPIE vol. **1292** , 137(1990)

ELECTRICAL, MECHANICAL, AND THERMAL CHARACTERIZATION OF HIGH T_c SUPERCONDUCTING CURRENT LEADS

C. Powers, P. Arsenovic, G. Oh
Goddard Space Flight Center
Materials Branch, Code 313
Greenbelt, Md. 20771

ABSTRACT

High T_C superconducting (HTSC) electrical current leads and ground straps will be used on future NASA payloads with sensitive instruments operating at cryogenic temperatures. These may include Cassini, EOS, and Space Station, to name several examples. The use of these new leads will greatly decrease electrical noise by providing better grounding. An increase in the lifetime of accompanying cryogenic cooling systems may also result.

HTSC current leads from several vendors have been subjected to a program of electrical, mechanical, and thermal testing. Preliminary results from x-ray analysis, critical current density, electrical resistance, effective Young's modulus, tensile strength, thermal conductivity, and thermal diffusivity measurements are reported. It is expected that this testing program will result in the certification of HTSC electrical current leads in future NASA missions.

INTRODUCTION

A program has been initiated in the Materials Branch at the Goddard Space Flight Center to develop methods for characterizing candidate HTSC material for possible space flight use. Part of this program has been directed at measuring various physical quantities of HTSC electrical current leads that may be used in cryogenic instruments. The advantages of using current leads constructed from HTSC material over conventional current leads are higher electrical conductivity and lower thermal conductivity. The disadvantages of using HTSC material are its mechanical properties. Since all HTSC materials are ceramics, they have a low ductility and have a widely varying strength due to inherent defects typical of ceramics.

In this paper, preliminary measurements of electrical resistivity, critical current density, effective Young's modulus, tensile strength, thermal conductivity, thermal diffusivity, and x-ray analysis of various $YBa_2Cu_3O_{7-x}$ specimens are reported. A simple four-point contact technique is used for both the electrical resistivity, ρ, and critical current density, J_C, measurements. Mechanical measurements are made by a standard technique using an

Instron tensile tester. Thermal conductivity, *k*, and thermal diffusivity, **D**, measurements are made using an implementation of Angstrom's temperature-wave method.[1] This method to measure thermal diffusivity uses a heater (Peltier junction) attached to one end of a specimen rod to send a periodic heat wave down the rod. Two thermometers are mounted along the specimen to follow the wave. The phase difference between the temperature oscillations at the two thermometers and the ratio of the amplitudes of these oscillations are used to calculate thermal diffusivity. By calibrating the heater to determine the amplitude of the heat wave, the thermal diffusivity measurements can be converted to thermal conductivity.

SAMPLE DESCRIPTION

Three candidate HTSC current leads have been subjected to electrical, mechanical, and thermal testing. These current leads are identified as current leads A, B, and C for this paper. All specimens are made of $YBa_2Cu_3O_{7-x}$. Other candidate HTSC current leads will be tested as specimens become available.

Current lead A, which contains 15% silver by weight, has a cylindrical shape with a diameter of 0.188 cm, a typical length of 3.772 cm, and a density of 5.43 gm/cm^3. Current lead B is made from a pre-shapable tape which was cut to size and then baked. This lead has a rectangular shape with a height of 0.072 cm, a width of 0.496 cm, a typical length of 2.887 cm, and a density of 3.09 gm/cm^3. Current lead C was made from powder pressed in a die with a pressure of 69 MPa. This lead has a rectangular shape with a height of 0.184 cm, a width of 0.193 cm, a typical length of 2.436 cm, and a density of 4.76 gm/cm^3.

EXPERIMENTAL METHOD

Electrical measurements

Electrical resistivity measurements are made using a standard 4-point contact technique. A Keithley 580 micro-ohmmeter is used to make these measurements. The ohmmeter uses a source current of 100 mA. These measurements are made in a Janis 8CNDT cryostat at various temperatures between 68 and 300 K.

A 4-point contact technique is also used to make critical current density measurements. Low resistance contacts (to prevent local heating) to the HTSC specimens are made using a mixture of silver and epoxy. Current is supplied through a specimen by a Sorensen programmable voltage supply and a Kepco 15-20M amplifier. The current is slowly increased until the voltage drop across the specimen begins to increase. A Keithley 197 multimeter is used to measure current and a Fluke 8505A multimeter is used to measure voltage. These measurements are made in the cryostat at various temperatures between 68 and 90 K.

Mechanical measurements

Mechanical testing of HTSC current leads to determine Young's modulus and tensile strength is done using a standard Instron tensile tester. Samples are tested using a 20,000 lb. load cell, which applies increasing tension to each specimen at a fixed rate. Specimens are held in place by grips at each end. A crosshead speed of 0.05 inches/minute and a chart recorder speed of 2 inches/minute are used. The chart recorder plots applied load vs. crosshead displacement. All tests are conducted at room temperature. Yield stress is calculated by dividing the maximum load on the specimen by it's cross sectional area. Young's modulus is calculated from the slope of the linear region of the specimen's stress - strain curve.

Thermal measurements

In this implementation of Angstrom's temperature-wave method to measure thermal conductivity and diffusivity , a Peltier junction is used to generate a periodic heat wave and two chromel-alumel thermocouples made of 0.0076 cm diameter wire are used to measure the temperature along the specimen being tested (Figure 1). One thermocouple is connected to the specimen 0.635 cm from the Peltier junction, and the other is 1.905 cm from the Peltier junction. These thermocouples are connected to reference thermocouples which are connected to an aluminum disk used for temperature control. The reference thermocouples are used to eliminate any voltage on the specimen thermocouples caused by operating at cryogenic temperatures. The Peltier junction is also mounted on the aluminum disk which is connected to the high-conductivity copper sample mounting rod supplied with the cryostat. The sample mounting rod is used to control the absolute temperature at which a measurement is made. Presently, measurements are made in vacuum (0.027 Pa) between 125 and 300 K. The Peltier junction is bonded to the aluminum disk and the specimen using STYCAST 2850FT epoxy manufactured by Emerson & Cuming. All thermocouples are also bonded using STYCAST 2850FT.

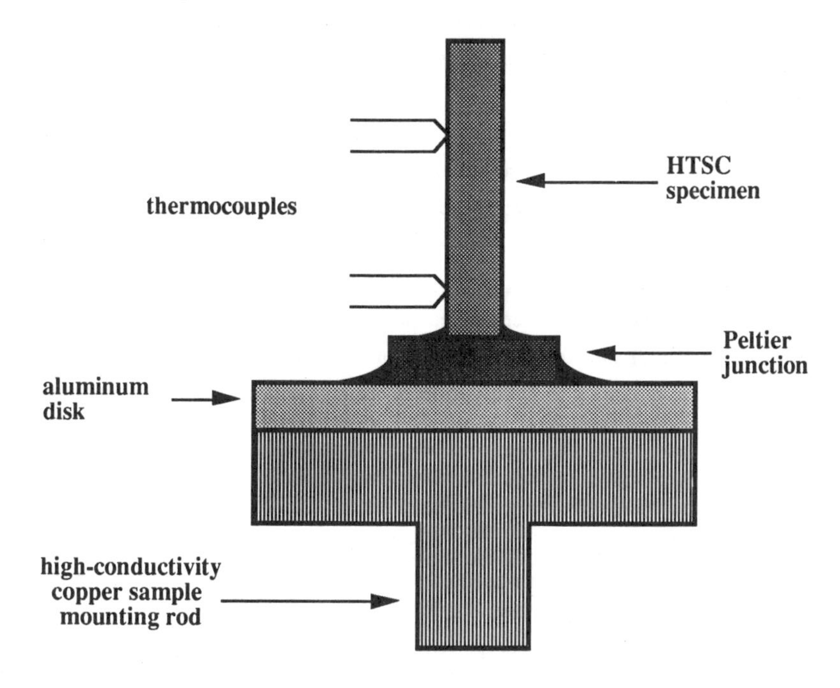

Figure 1. Basic apparatus for measuring thermal conductivity and diffusivity.

A Hewlett-Packard 3314A function generator and a Kepco 50-2M amplifier are used to produce an oscillating current through the Peltier junction. A Keithley 197 digital multimeter is used to measure the current through the Peltier junction. A Fluke 8505A digital multimeter and a Keithley 196 system digital multimeter are used to measure the voltage across each thermocouple.

Once the temperature of the aluminum block and test specimen have reached equilibrium, a periodic current is supplied to the Peltier junction to generate a thermal wave through the specimen rod. When the temperature oscillations at each thermocouple reach steady state, a BASIC program is initiated on a Fluke 1722A instrument controller to read and record the voltage across each thermocouple and the current through the Peltier junction. For these experiments the frequency of oscillation of the thermal wave ranges between 0.01 and 0.05 Hz, so the voltage across each thermocouple is sampled once a second giving between 20 to 100 measurements per cycle.

Prior to making the above mentioned thermal measurements, an experiment was performed to measure the thermoelectric power, Q, versus temperature for the Peltier junctions used in these experiments (MELCOR FC 0.7-8-05L Frigichip). This experiment was done to calculate the Peltier coefficient, Π, for these Peltier junctions, which is used to calculate thermal conductivity. The thermoelectric power was measured by mounting a thermocouple on both sides of three Peltier junctions which were then connected to an aluminum disk for mounting in the cryostat. A resistor (heater) was also mounted on one side of the Peltier junctions. These Peltier junctions were purchased as a lot of ten, with three being used to determine the Peltier coefficient for the lot and the other seven to make thermal conductivity and diffusivity measurements on the HTSC specimens.

The Kepco 50-2M amplifier was used to supply a current through the resistor, which caused heat to flow through the Peltier junctions. This heat flow caused a temperature gradient and a voltage drop across the Peltier junctions which was measured using the thermocouples and three multimeters. A Keithley 740 scanning thermometer was used to measure the temperature at each thermocouple. Two Keithley 197 digital multimeters and one Fluke 8505A digital multimeter were used to measure the voltage drop across each Peltier junction. The Fluke instrument controller was used to record the measurements from the multimeters and thermometer. These measurements were made in the cryostat at various temperatures between 68 and 300 K.

X-ray measurements

X-ray analysis is performed using a Scintag x-ray diffractometer controlled by a DEC MicroVAX computer. A time average method of data collection is used, where multiple scans on the sample being analyzed are obtained. The intensity data at every angular position for each individual scan are added together. At the end of the data collection, each intensity is divided by the number of scans to arrive at the correct count rate. After the raw data are collected, a background intensity correction is performed, followed by a lattice parameter calculation with a standard data reduction program.

X-ray analysis generates several results. First, it provides a good measure of the purity of the specimen, by the positions and relative intensities of the diffraction peaks. Diffraction scans are compared with a standard reference pattern obtained from the National Institute of Science and Technology (NIST). A second use of XRD is estimating the oxygen content of $YBa_2Cu_3O_{7-x}$. This material exhibits superconducting properties at cryogenic temperatures only for a small range of oxygen content. As the oxygen level is decreased from O_7 to O_6, the material undergoes a phase transition from orthorhombic to tetragonal and is no longer superconducting. The positions of the x-ray diffraction peaks undergo a small but measurable shift with changes in oxygen content. This corresponds to small changes in the lattice parameters and hence the unit cell volume of the samples involved. Using a previously developed correlation curve of oxygen level vs. unit cell volume, the oxygen content of a specimen can be determined by x-ray diffraction.[2] It can then be seen if this oxygen level is in the correct range for the sample to show superconducting properties.

DATA ANALYSIS

Electrical data

Electrical resistivity is calculated from the voltage and current measurements mentioned previously, using the following equation,

$$\rho = \frac{VA}{Il} \qquad (1)$$

where V is the voltage drop across the specimen, l is the distance between the voltage contacts to the specimen, I is the current through the specimen, and A is the cross sectional area of the specimen. The electrical resistivity vs. temperature for all three current leads is shown in Figure 2.

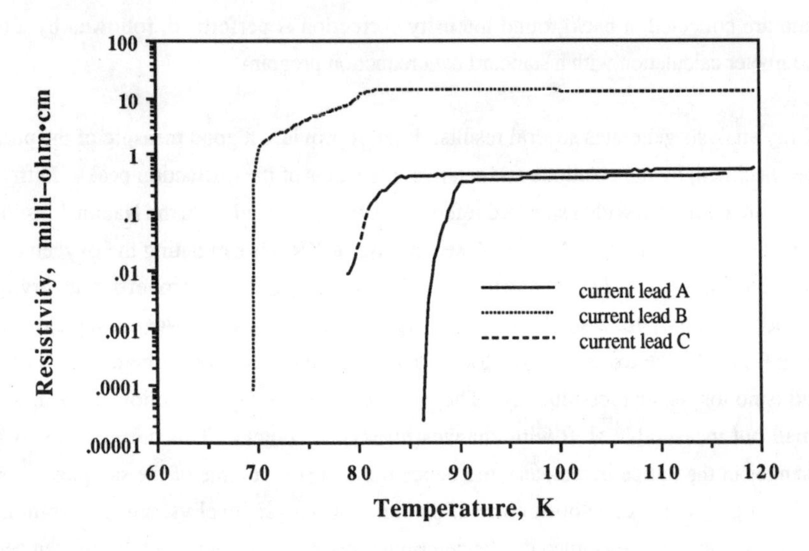

Figure 2. Electrical resistivity vs. temperature for two of the current leads tested.

The critical current was calculated using a first order curve fit to the voltage "lift-off" data mentioned previously. The critical current density was calculated by dividing the critical current by the cross-sectional area of the test specimen. The critical current density vs. temperature can be seen in Figure 3. Both the electrical resistivity and critical current density have an uncertainty of about 10% due to the size of the voltage contacts to a specimen.

Current lead B has two critical current densities, the first critical point partially destroys the superconducting state in the specimen and the second point completely destroys it. These two critical points are probably due to the existence of two superconducting phases in current lead B.

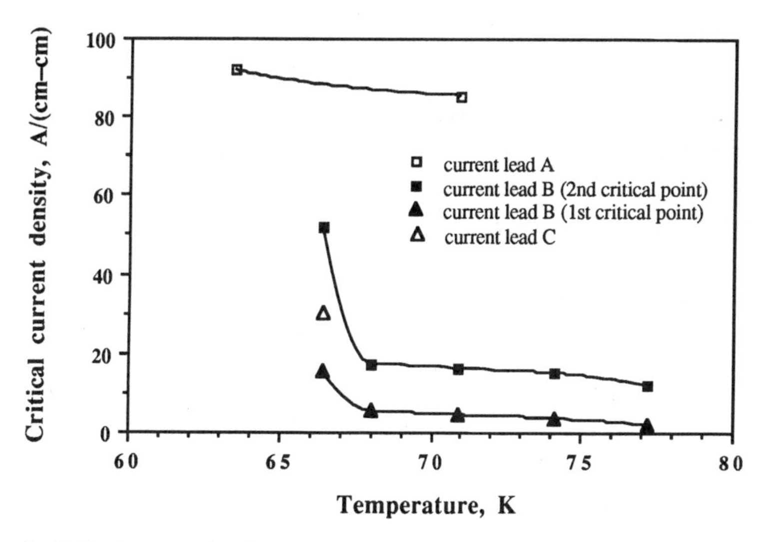

Figure 3. Critical current density vs. temperature for the three current leads tested.

Mechanical data

The results of the mechanical measurements show significant scatter, as is typical of the strengths and modulii of ceramics. The following table shows the average Young's modulus, E, and the average tensile strengths for the three current leads.

Table 1. Young's modulus and tensile strength for the three HTSC current leads.

Current lead	Average E (GPa)	Average Strength (MPa)
A	90 ± 10	20 ± 10
B	70 ± 12	7 ± 4
C	77 ± 8	10 ± 5

Thermal data

The thermal diffusivity of a specimen is calculated from the thermocouple voltage measurements mentioned previously using a curve fitting routine to determine the magnitude and phase of the temperature oscillations at each thermocouple. These values are then used in the following equation to calculate diffusivity,

$$\mathbf{D} = \frac{\pi f L^2}{\phi \ln(a_0/a_1)} \tag{2}$$

where f is the frequency of oscillation, L is the distance between the thermocouples, ϕ is phase different between the thermocouples, a_0 is the amplitude of the temperature oscillation at the thermocouple closest to the Peltier junction, and a_1 is the amplitude of the temperature oscillation at the second thermocouple. This equation assumes that the amplitude of the heat wave has decayed to a negligible value at the end of the specimen farthest from the Peltier junction. The thermal diffusivity vs. temperature for the three current leads is shown in Figure 4. Due to equipment problems, there is only one data point for current lead C.

For negligible lateral heat losses to the surrounding environment or thermocouples the phase difference between the thermocouples is approximately equal to the logarithm of the amplitude ratio, and the following equation can be used to calculate diffusivity,

$$\mathbf{D} = \frac{\pi f L^2}{\phi^2} = \frac{\pi f L^2}{\ln(a_0/a_1)^2} \,. \tag{3}$$

In these experiments, the value of diffusivity calculated using only the phase difference is always about 10% larger than the values calculated using the amplitude ratio, indicating some lateral heat loss. Modified diffusivity calculations which account for losses due to the thermocouples only predict the value calculated from the phase difference to be about 2% larger than the value from the amplitude ratio. This other 8% difference is characteristic of a lateral heat loss, which presently has not been identified.

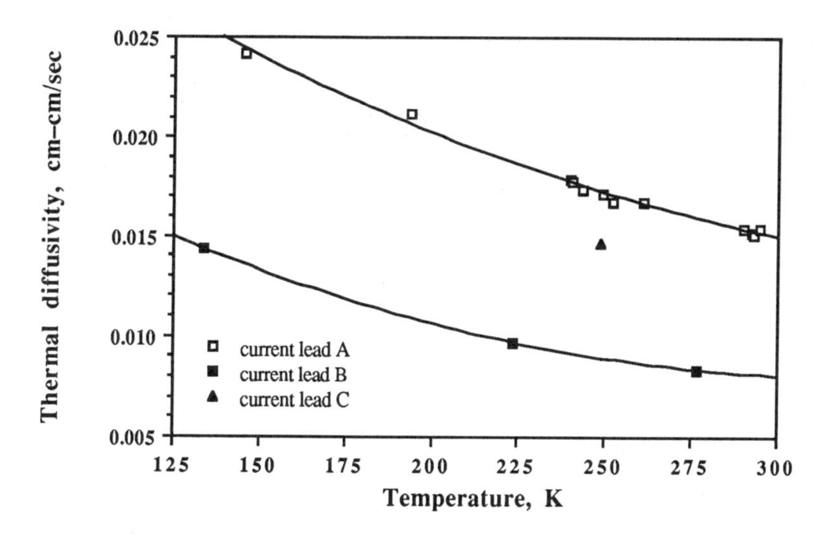

Figure 4. Thermal diffusivity vs. temperature for the three current leads tested.

Thermal conductivity is calculated from the amplitude of the drive current through the Peltier junction, the amplitude of the temperature oscillations at the interface between the Peltier junction and test specimen, and the diffusivity values calculated previously using the following equation,

$$k = \frac{I_o \Pi \sqrt{\mathbf{D}}}{T_o A \sqrt{2\pi f}} \tag{4}$$

where I_o is the amplitude of the current through the Peltier junction, Π is the Peltier coefficient, T_o is amplitude of the temperature oscillations at the interface, and A is the cross sectional area of the specimen.[3] The thermal conductivity vs. temperature for all three current leads can be seen in Figure 5. T_o is calculated using the thermocouple voltage measurements that were used to calculate diffusivity by fitting them to the following equation for the temperature along a specimen as a function distance and time,

$$T(x,t) = T_o e^{-\mathbf{k}x} \sin(2\pi f t - \mathbf{k}x) \tag{5}$$

where x is the distance from the Peltier junction along the test specimen, t is time, and **k** is defined as $\sqrt{\pi f/\mathbf{D}}$.

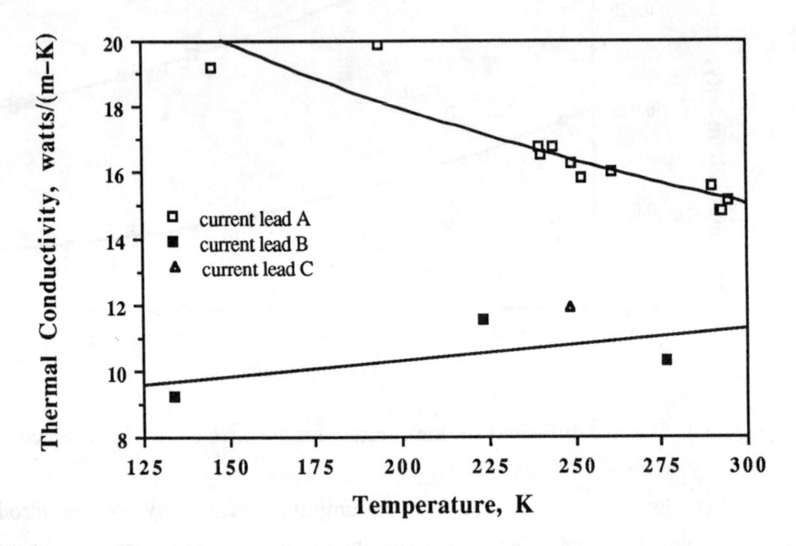

Figure 5. Thermal conductivity vs. temperature for the three current leads tested.

The Peltier coefficient, Π, for the Peltier junctions used in these measurements was calculated from thermoelectric power measurements using the following equation

$$\Pi = TQ \tag{6}$$

where T is absolute temperature, and Q is the thermoelectric power mentioned previously.[4] Thermoelectric power is defined as the voltage drop across a Peltier junction divided by the temperature gradient across the junction. The Peltier coefficient for the junctions used in these measurements is shown in Figure 6 as a function of temperature. The Peltier coefficient for the three junctions has a standard deviation of 10%. The calculated values of the Peltier coefficient vs. temperature were fit to a fifth order polynomial equation to be used in equation (4).

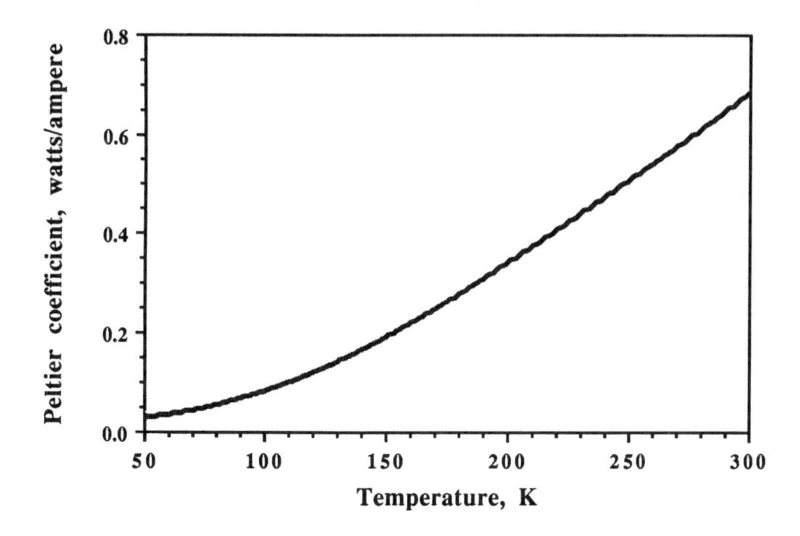

Figure 6. Peltier coefficient vs. temperature for the Peltier junctions used in these experiments.

The thermal diffusivity and conductivity of a cylindrically shaped fused SiO_2 specimen (from Corning Glass Works) with a diameter of 0.307 cm and a length of 4.075 cm was measured using this technique to test the accuracy of the technique. The measured values agree with values reported in the TPRC Data Series to within the uncertainty of these reported values.[5,6] Previous work showed that this technique has an accuracy of ± 6% for thermal diffusivity measurements.[7] The thermal conductivity measurements have an additional 10% uncertainty due variations in the Peltier coefficient.

X-ray analysis

All of the samples analyzed by x-ray diffraction were shown to be reasonably pure and to have an oxygen content in the correct range (6.5 to 7) to be superconducting at cryogenic temperatures. It should be emphasized that the changes in unit cell dimensions observed from the slight peak shifts in the x-ray pattern are very small, and considerable care must be taken when examining the material to ensure reliable results. The following table shows the calculated unit cell volume and oxygen level for the three current leads.

Table 2. Unit cell volume and oxygen level for the three HTSC current leads.

Current lead	Unit Cell Vol. $(Angstroms)^3$	Oxygen Level
A	173.14	7.0
B	173.30	6.9
C	173.35	6.9

RESULTS & CONCLUSIONS

Since the electrical resistivity of all three current leads approaches zero below 77 K (the expected upper operating temperature for current lead applications), any of the three would perform better as an electrical conductor than copper or manganin, which are present current lead materials. Current lead A would be a more favorable material than B or C because its transition temperature is well above 77 K, and because it has a higher critical current density.

As has been the problem with HTSC materials since their discovery, the mechanical properties of all three current leads are typical of ceramics. While Young's modulus is 60 to 80% of the value reported for manganin and copper, the tensile strength of these current leads is only 2.5 to 6.5% of the value reported for manganin and copper. Current lead A, which contains 15% silver, has a higher Young's modulus than the other two current leads and a tensile strength that is two to three times larger.

The thermal conductivity of all three current leads is several orders of magnitude lower than copper, and lower than manganin as well. The thermal conductivity of current lead A is only slightly less than manganin, while the thermal conductivity of B and C is about half of manganin over the temperature range that measurements were made.

Presently, the current lead that contains 15% silver appears to be the best candidate for cryogenic applications. This material has the best electrical properties (highest transition temperature and largest critical current density), and the best mechanical properties (highest Young's modulus and tensile strength). While this current lead has the highest thermal conductivity among the three, it still has a thermal conductivity that is lower than copper and manganin.

More testing of HTSC current leads is presently under way to improve the measuring techniques employed, and to increase the amount of data available on HTSC current leads. Future plans include a complete survey of available current leads and improvements in the mechanical and thermal testing. Implementing non-destructive mechanical testing should improve the mechanical testing.[8] The use of infra-red imaging and better techniques to calibrate Peltier junctions should produce more accurate thermal measurements.

ACKNOWLEDGEMENTS

The authors would like to thank Dr. J. Hull of Argonne National Laboratory, Ms. S. Wise and Mr. M. Hooker of Clemson University, and Mr. J. Colony of NASA - Goddard Space Flight Center for supplying test samples to make this work possible. The authors appreciate the efforts of Mr. C. Clatterbuck for his help in connecting the thermocouples and Peltier junction to the specimen rods and for his silver/epoxy mixture. The authors would also like to thank Mr. C. Taylor for making the thermocouples used in these tests and Mr. S. Pagano for his help with the cryostat and his efforts in constructing the test apparatus. A thorough and rigorous analysis of the experimental data was made possible through many conversations with Dr. H. Leidecker, with whom most aspects of thermal wave propagation were discussed. Dr. H. Leidecker also first suggested using a Peltier junction to generate a heat wave.

REFERENCES

1. Touloukian, Y.S., Powell, R.W., Ho, C.Y., and Nicolaou, M.C., Thermal Diffusivity -- Vol. 10 of Thermophysical Properties of Matter -- The TPRC Data Series, pp. 28a-37a, IFI/Plenum Data Corp., New York, 1973.

2. Arsenovic, P., and Flom, Y.," X-ray Characterization of HTSC Yttrium-Barium-Copper-Oxide", 1989 GSFC Research and Technology Report, Greenbelt, Md., 1990.

3. Howling, D.H., Mendoza, E., and Zimmerman, J.E., "Preliminary Experiments on the Temperature-Wave Method of Measuring Specific Heats of Metals at Low Temperatures", Proc. Roy. Soc., London, A229, pp. 86-109, 1955.

4. Ashcroft, N.W., and Mermin, N.D., Solid State Physics, pp. 256-258, Saunders College, Philadelphia, Pa., 1976.

5. Touloukian, Y.S., Powell, R.W., Ho, C.Y., and Nicolaou, M.C., Thermal Diffusivity -- Vol. 10 of Thermophysical Properties of Matter -- The TPRC Data Series, pp. 399-400, IFI/Plenum Data Corp., New York, 1973.

6. Touloukian, Y.S., Powell, R.W., Ho, C.Y., and Nicolaou, M.C., Thermal Conductivity-Nonmetallic Solids -- Vol. 2 of Thermophysical Properties of Matter -- The TPRC Data Series, pp. 183-189, IFI/Plenum Data Corp., New York, 1973.

7. Powers, C., "A Technique to Measure the Thermal Diffusivity of High T_c Superconductors", Proceedings to the AMSAHTS '90 Conference, in press.

8. Jiang, H., Arsenovic, P., Eby, R. K., Liu, J. M., and Adams, W. W., "Nonlinear Elasticity of Strong Fibers", Polymer Preprints, Japan, 36, 1987.

MAGNETIC REPLICAS COMPOSED OF HIGH T_C SUPERCONDUCTORS

Roy Weinstein, In-Gann Chen, and Jia Liu
Institute for Beam Particle Dynamics and
Texas Center for Superconductivity
University of Houston, Houston, Texas 77204-5506

ABSTRACT

Work on the development of magnet replicas, based upon the Incomplete Meissner Effect in type II superconductors, is reported.

When imperfections, or pinning centers, exist in a superconductor (SC), the SC can trap field rather than expel it. SCs, with trapped field, behave like a metallic permanent magnet, except that they copy, essentially precisely, the fields used to activate them.

The resulting magnets, which are being developed from high T_c materials, have several very desirable features:

- They can trap and thereby copy, precise fields of the magnets used to activate them, in imprecise bulk superconductor. They are therefore replicas.
- They are light. For example an iron particle beam electromagnet for steering, weighing three tons, can be replaced by a replica weighing 200 lbs. They are even lighter than permanent magnets (88% of weight).
- They use no power except for replacement of liquid nitrogen.
- The cryostat requirements are simple.
- They hold promise of very high field strength.
- As a result of the replica property the same magnet replica can be a dipole in one application and a quadrupole or sextupole in the next, thus lowering on board inventory.
- They are less expensive than ordinary magnets.
- No wire is needed, thus bypassing a major problem in high T_c superconductors.

Very low weight, low inventory need due to interchangability, possibility of ambient temperatures below T_c, and power savings make them particularly interesting for aerospace applications. They are also very suitable for particle beam magnets, large generators or motors, and industrial separators.

To date we have produced Y-123 samples capable, when fabricated as a cylinder, of trapping fields of 15,000 Gauss. Recent data indicate that trapped fields of 60,000 Gauss are achievable at 77K, and we expect to achievc half of this value within a year. Even higher trapped fields are possible at lower temperatures.

Several phenomenological laws have been discovered for magnet replicas. Two of these help to stabilize replicas against the well known Giant Creep Effect (loss of field).

We have performed initial steps of fabrication of larger magnets from small samples. We have also operated a small motor on a magnet replica.

We desire to develop samples capable of trapping 60,000 Gauss, and to develop methods of production of large volume samples, so that we may proceed to prototyping.

I. INTRODUCTION

An R and D program to develop practical magnet replicas based upon high T_c superconductors has been in progress for about two years.

Usually superconductors (SC's) expel a magnetic field (Meissner Effect). However, type II SC's may trap the field. The effect is usually referred to as the Incomplete Meissner Effect. If small imperfections are present in the SC, the field can be "expelled" into the imperfection region rather than outside the superconductor. When the external field is turned off, the collapse of such internal flux lines is prevented by induced persistent super-currents within the SC (Faraday's Law).

The possibility of magnets based upon such a trapped field was studied in low T_c SC's by a group at Stanford University and the Electrical Power Research Institute [1].

Magnets based upon the Incomplete Meissner Effect in high T_c materials have several promising advantages:

- Replication. If there are many trapping centers per unit volume, the captured field, for example inside a cylinder, is an essentially perfect replica of the activating field. Thus complex fields of precisely machined magnets can be copied in imprecise bulk SC.

- Weight. The resulting magnet replicas are light. For example, an iron particle-beam electromagnet for beam steering, weighing three-tons, can be replaced by a replica composed of $YBa_2Cu_3O_7$ (Y-123 material) weighing 200 lbs. Even when used simply as permanent magnets, the weight of a replica is 88% that of a SmCo ferromagnetic permanent magnet.

- Power. No power is needed for magnet replicas, once activated, except for replacement of liquid nitrogen (LN).

- Cryostats. Because LN is used, cryostat requirements are greatly simplified.

- High Field. Recent experimental data indicates that magnet replicas of at least 60,000 Gauss are achievable. This is about three-times the fields attainable with permanent ferromagnets. As we shall describe below, materials for 15,000 Gauss magnets have already been developed. The only upper limit known at the present time to replica fields is about 300,000 Gauss.

- Low Inventory. For beam steering magnet applications, the same magnet replica can be a quadrupole in one application, and a dipole or sextupole in the next.

- Low Capital Cost. Replica cost estimates typically run about 10% to 30% of the cost of the parent magnet.

- No Wire Needed. Magnet replicas do not require wire, only bulk material. They can be made even if the wire problem in Y-123 material proves insurmountable.

The low weight, low inventory, power savings, and possibility of ambient temperatures below T_c make magnet replicas particularly attractive as a candidate for space applications. Other attractive uses are for accelerator external beam magnets, motor and generator magnets, medical MRI, and industrial particle separators.

II. PROGRESS IN INCREASING TRAPPED FIELD

We began our work with sintered samples in which 0.3 Gauss was trapped. Table 1 shows our progress to date. Prior to spring of 1990, a steady improvement of materials resulted in an increase in trapped field by a factor of 20,000, from 0.3 to 6,000 Gauss[2]. At that time the best materials we produced were melt-textured samples of $YBa_2Cu_3O_7$ following the method of Salama, et.al.[3].

More recently we have increased the number of imperfections (pinning centers) in the SC by proton bombardment. An experiment was done at the Harvard Cyclotron. Small tiles of sintered and melt-textured material were exposed to a beam of 160 MeV protons. The melt-textured samples which were used in this bombardment were capable, before exposure, of fabrication to produce a 3,000 Gauss magnet. Each typical small tile could itself capture about 300 Gauss. After exposure to about 10^{16} protons/cm^2 the retained fields were increased by as much as a factor of five. (See Fig. 1.) Single tiles retained as much as 1,500 Gauss. With the bombarded materials it would be possible to fabricate a 15,000 Gauss magnet replica.

If one defines the ratio of the retained magnetic field after bombardment to that before bombardment to be R, the data is reasonably well represented by[(1), 4]

$$R = 1 + (R_{max} - 1)(1 - e^{-F/F_0}), \qquad \text{(Eq. 1)}$$

---

(1) The introduction of new pinning centers appears to occur at a markedly lower value of F than does loss of superconducting properties. It seems clear, however, that at high F, R will begin to decline so that Eq. 1 must be considered to be approximate.

where F is the fluence of protons in units of number protons/cm^2, and $F_0 \approx 7 \times 10^{15}$ proton/cm^2.

Table 1

Progress of Magnetic Field Trapped within Y-123 HTS at 77K at IBPD/TCSUH

Date	Field Trapped on Disk (Gauss)	Field Extrapolated to Cylinder (Gauss)	Comment
12/88	–	0.3*	Poor Y-123 scintered ceramic
2/89	2	10.0*	Good Y-123 scintered ceramic
4/89	4.5	18.0	Optimum thickness Y-123 scintered ceramic
5/89	10	50	Y-123/Ag composite, scintered
6/89	80	700	Y-123/Ag composite, partially melt grown, sub-mm size grain
8/89	380	4,000	Melt texture Y-123, sub-cm size grain, off ab plane
9/89	600	6,000	Melt texture Y-123, sub-cm size grain, ab plane
10/89	1,280	6,000	Same as 9/89, with limited fabrication

* Cylindrical sample tested. (Entries under cylinders, having no *, are calculations based on Vortex current and Persistent current model including non-linear magnet field effect.

A partial fabrication of a mini-magnet, from several bombarded tiles, was performed. The predicted and observed values of the trapped field in this mini-magnet were about 4,300 Gauss. Fig. 2 shows an activation curve for this mini-magnet. What is shown is the trapped field, B_T, as a function of the activating field, B_0.

In our early research[2] we discovered that B_T and B_0 are related by

$$B_T = B_{T,max}\,(1 - e^{-B_0/B_{T,max}}) \qquad \text{(Eq. 2)}$$

where $B_{T,max}$ is the maximum trappable field in the sample used. This equation appears to be a basic law since it has held with great accuracy for all of the materials we have tried. We will comment on its basic origin in the near future.

Since the time of the bombardment we have developed materials which can retain 1,000 Gauss in a single tile without irradiation. It will be interesting to see if bombardment also improves these improved samples by additional large factors.

III. STABILITY

We learned in our early research, as did others, that the field trapped in a magnet replica decreases with time according to the law

$$B_T(t) = B_T(0) - c \log t. \qquad \text{(Eq. 3)}$$

We experimented with a wide variety of Y-123 material including sintered, silver mixed sintered, and melt-textured, and found that $C/B_T(0)$ was almost identical for all of these materials[2] if they were activated to $B_{T,max}$. If t, in Eq. 3, is in minutes, our observations were consistent with

$$C/B_{T,max}(0) \sim 0.043 \pm 0.009. \qquad \text{(Eq. 4)}$$

Eqs. 3 and 4 indicate that a magnet replica can be stabilized to the order of a few percent per year by simply "aging" it, for example, for a couple of months before it is used.

Another method of stabilization was found [2]. Eq. 4 holds if $B_T(0) = B_{T,max}$. We also discovered

$$C/B_T(0) \propto (B_T(0)/B_{T,max})^2 \qquad \text{(Eq. 5)}$$

i.e., if the material is initially activated to $B_T < B_{T,max}$, the stability of the trapped field increases quadratically with $B_T(0)/B_{T,max}$.

In this regard we have also found that for pinning centers created by proton bombardment, $C/B_{T,max}$ is about 10% larger than for other materials.

IV. MODELS OF CURRENT FLOW, AND MEASUREMENT OF J_C

We have accurately measured the trapped fields surrounding sample tiles, and have modeled the current flow in the material in an attempt to theoretically calculate the fields. It is found that the observed field can be very well reproduced by a current model with two components.[5] First, we assume a surface current density, J_s (Amp/cm), on the sample. This same assumption, called an Amperean current, successfully models the fields around, for example, a SmCo permanent magnet. In Y-123 material, however, a second current component is needed. We assume for this a uniform volume current, J_V (Amp/cm^2), flowing in concentric circles in the sample (see insert of Fig. 3a). We have developed analytical and numerical fitting programs. The input parameters are the diameter and thickness of the tiles. A fit to the field is required, and best values of J_S and J_V are deduced. The quality of the fits obtained are excellent. Fig. 3 shows the fitting program applied to a sample of melt-textured Y-123 material. Fig. 3a shows the field on the axis of the sample, as the distance, Z, from the sample is varied. Fig. 3b shows the field near the surface of the sample (x,y plane) as x and y are varied.

The presence of the J_V term makes the trapped field around a tile of Y-123 material more peaked than for a permanent ferromagnet of the same aspect ratio. This effect is shown in Figs. 4a and b.

Once J_S and J_V are known, the magnetic moment (in e.m.u.) of the sample can readily be calculated[5]

$$m = \pi a^2 t/10 (J_V a + J_S) \quad \text{(Eq. 6)}$$

where a and t are radius and thickness of the SC sample respectively.

We find the moment, calculated in this way, to be the same as that found experimentally by the use of Vibrating Sample Magnetometer (VSM) or SQUID devices.

VSM and SQUID measurements are a common approach to obtaining a value of the critical current in the SC material, J_c. The magnetic moment is interpreted by the Bean model[6] to yield J_c. In the same way, measurements of B(x,y,z) around a magnetized sample yield the magnetic moment and from this, J_c. As an example of the J_c's in our samples, we find that a melt-textured proton irradiated sample, capable of making a replica tile of 1,300 Gauss, had $J_c \approx 26{,}000\ A/cm^2$.

The method described here, developed as an offshoot of our magnet replica work, is applicable to any size sample. It works better, in fact, for large samples, in which sample size is larger than Hall probe size. It appears to be more suitable to industrial production of large Y-123 monoliths than are VSM or SQUID.

V. PROTOTYPES

We are anxious to move on to the production of prototypes. We believe that trapped fields of at least 30,000 Gauss will be achieved within a year. While we do not intend to cease developing materials, we do believe that it is time to produce prototypes of useful magnets. This requires considerable construction budgets which we have not yet been able to attain.

In the interim, we have focused on small devices. The mini-magnet characterized in Fig. 2 was discussed previously. We have also taken a small electric motor and excised its magnet. The magnet was replaced with a mini-magnet of about 1,000 Gauss. We then operated the hybrid motor, drawing 3.5 Watts. As expected all worked well. Enough ideas were generated in this exercise so that a 1 kw motor using a magnet replica will soon be tried. While the first experiment produced only a toy, this line of experimentation is capable of serious consequences. Large motors consume about 8%, of the total power used, in powering their own magnets. The efficiency of such motors (and their torque) can be significantly improved by using magnet replicas, and in large motors the LN requirement is tolerable.

VI. CONCLUSION

Magnet replicas, based upon the incomplete Meissner effect, are theoretically capable of fields possibly as high as 300,000 Gauss. Materials have been developed, using proton bombardment, which presently allow fabrication of replicas of 15,000 Gauss. Thirty-thousand Gauss appears probable within a year, and the collection of all present experimental evidence indicates that 60,000 Gauss will soon be achievable.

Laws governing the activation, decay, and fabrication of magnetic replicas have been found. The effects of replica decay can be controlled by aging, or reduced by activating to less than the maximum trappable field.

A by-product of this work has been the development of a method for measuring J_c in large sized bulk material. This method measures values of J_c in agreement with those measured on small samples by VSM or SQUID.

A small motor has been run using a magnet replica in place of the motor's original magnet. More serious prototyping is now possible for particle beam magnets, motor and generator magnets, and industrial separators.

VII. ACKNOWLEDGEMENTS

We wish to thank Prof. Pei Hor, Dr. Y. Tao, and Mr. J. Bechtold of the University of Houston Physics Department for X-ray analysis and VSM measurements. This work is supported in part by grants from the Texas Center for

Superconductivity at the University of Houston, the Electric Power Research Institute, and the NASA Johnson Space Center.

VIII. REFERENCES

1. Rabinowitz, M., et.al., "Very Incomplete Meissner Effect", Lett. al Nuovo Cimento 7, 1 (1973); Appl. Phys. Lett. 22, 599 (1973); Appl. Phys. Lett. 30, 607 (1977); IEEE Magn. 11, 548 (1975).
2. Weinstein, R., Chen, I.G., Liu, J., and Parks, D., "Persistent Magnetic Fields Trapped In High T_c Superconductor", Appl. Phys. Lett. 56, 1475 (1990).
3. Salama, K., Selvamanickam, V., Gao, L., and Sun, K., "High Current Density in Bulk $YBa_2Cu_3O_7$ Superconductior", Appl. Phys. Lett. 54, 2352 (1989).
4. Weinstein, R., Chen, I.G., Liu, J., "Permanent Magnets of Y-123 Material Bombarded with 160 MeV Protons", to be submitted to Appl. Phys. Lett.
5. Chen, I.G., Liu J., Weinstein, R.,"Trapped Magnetic Field and Related Transport Critical Current Density in $YBa_2Cu_3O_7$ Superconductors", submitted to Physics C.
6. Bean, C.P.,"Magnetization of Hard Superconductor", Phys. Rev. Letters 8, 250 (1962).

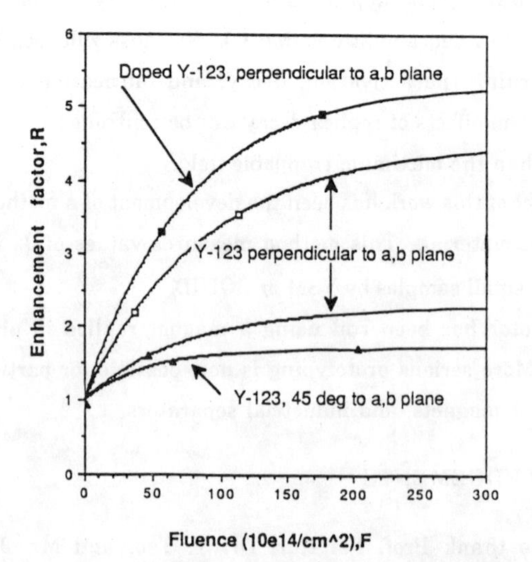

Fig. 1 Enhancement of field trapping capability of melt-textured Y-123 samples by 160 MeV proton bombardment

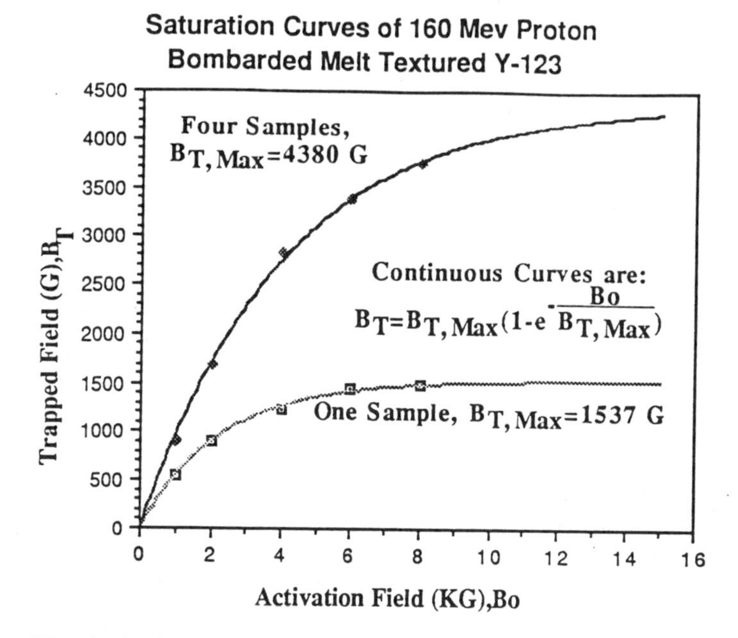

Fig. 2. Activation curve of one sample, and a mini-magnet composed of four samples

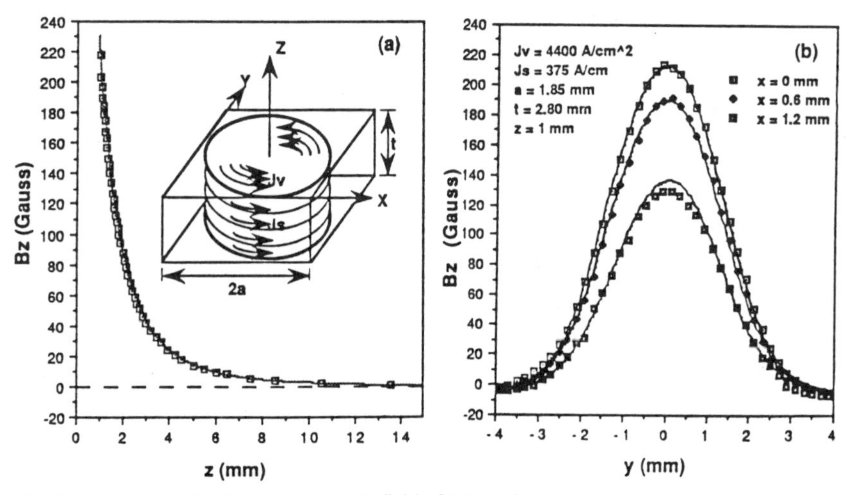

Fig. 3. Measured and calculated magnetic field of (a) varying z, and (b) varying x and y. Insert shows the schematic illustration of Js+Jv model.

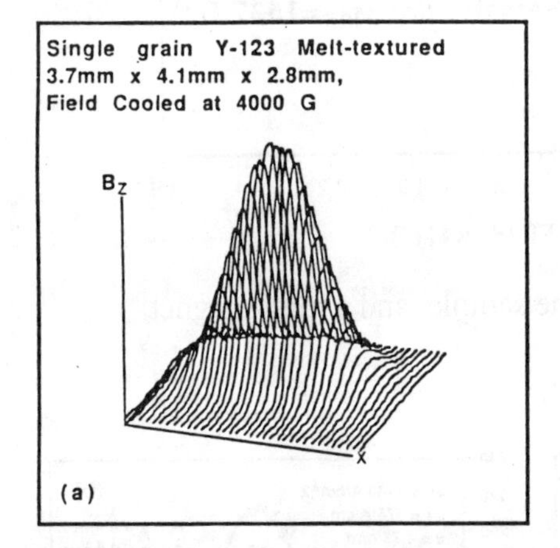

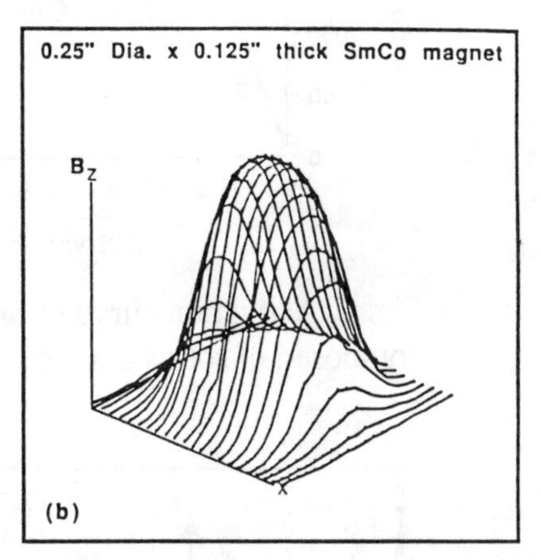

Fig. 4 The spatial distributions of (a) field trapped in a textured Y-123 sample and (b) in a SmCo permanent magnet.

4. CHARACTERIZATIONS

Morphology of High-J_c CVD-Film and Crystal Structures of Simple Superconductors

Tsuyoshi Kajitani, Takeo Oku, Kenji Hiraga, Hisanori Yamane,
Toshio Hirai, Syoichi Hosoya, Tsuguo Fukuda, Katsuyoshi Oh-Ishi,
Satoru Nakajima*, Masae Kikuchi, Yasuhiko Syono, Kazuo Watanabe,
Norio Kobayashi and Yoshio Muto

Institute for Materials Research, Tohoku University,
Sendai 980, Japan

* CASIO Computer Co. Ltd., Ishikawa-cho, Hachioji 192, Japan

Abstract

A TEM study on the high-J_c ($6.5 \times 10^4 A/cm^2$ at 77.3K and 27T ($B \perp c$)) CVD-YBC film deposited on the $SrTiO_3$ (100) substrate has revealed that the film consists of small domains, typically 20~50nmϕ × 10nm in size and preferentially oriented perpendicular to c-axis. The domains are separated by faulted regions where extra layers of about 20nm in diameter are surrounded by deformed lattices. Rectangular precipitates, about 20~50nm × 10nm in size, are frequently seen in the high-resolution transmission electron microscope images taken with the beam parallel to [010].

X-ray and neutron diffraction studies on the superconducting and non-superconducting Nd-Ce-Cu-O, Pr-Ce-Cu-O, Tl-Ba-Cu-O and La-Sr-Cu-O (powders and single-crystals) were carried out to investigate structural changes in the Cu-O layers and/or the copper/apical-oxygen chemical bonds. Three-dimensional representations of the electron density distribution in the Cu-O layers and at the apical-oxygen site are obtained by the Fourier- and differential-Fourier-syntheses techniques. The anisotropic shapes of oxygen atoms located in the Cu-O layer in the Nd-Ce-Cu-O, Pr-Ce-Cu-O and La-Sr-Cu-O superconducting cuprates are noted.

The copper/apical-oxygen distances are more or less longer in the superconductors relative to the non-superconductors which are isomorphous to the superconductors except in their oxygen contents.

I. Introduction

A remarkable advance has been made in the preparation of high-J_c YBC film with a J_c=6.5 × 10^4 A/cm^2 at 77.3K and 27T (B⊥c) by the present authors, Yamane et al. [1] and Watanabe et al. [2,3]. This J_c is higher than that of other high-J_c films and wires synthesized by Satchell et al. [4], Heine et al. [5], Sato et al. [6] and Uno et al. [7].

In the previous TEM observations done in the film surface normal direction by Watanabe et al. [2], thin precipitates, conceivably the flux-pinning-centers, were found. There is large anisotropy in the J_c, which becomes very high when the magnetic field, B, is applied perpendicular to the c-axis but becomes low when the B is applied parallel to the c-axis. Because of this anisotropy, it is conceivable that the flux-pinning-centers may be plate-like regions precipitated parallel to the ab-planes. The aim of the present HR-TEM work was to search for the flux-pinning-centers in samples, cut and polished perpendicular or parallel to the film surface.

Recently, a close relationship between the T_c of the superconducting cuprates and the location of the copper/apical-oxygen distances in a CuO_6 octahedron or in a CuO_5 pyramid has been reported by Murayama et al. [8] and Kaldis et al. [9]. They suggested that the T_c becomes higher when the copper/apical-oxygen distance becomes shorter. However, Nelmes et al. [10] observed an elongation of the copper/apical-oxygen distance in $La_{1.85}Sr_{0.15}CuO_{4-y}$ by the neutron diffraction at high pressure, where the T_c becomes higher about 5 K than at 1 atm. Kwei et al. [11] also reported only a slight decrease of the copper/apical-oxygen distance in the cooled $YBa_2Cu_3O_7$ with decreasing temperature.

In this study, the copper/apical-oxygen distance is measured carefully in Tl-Ba-Cu-O and La-Sr-Cu-O. In addition to this, crystal structural changes, particularly in the Cu-O layers, both at the superconducting and the non-superconducting states were elucidated. Both the electron-doping-type cuprates, Nd-Ce-Cu-O and Pr-Ce-Cu-O, and the hole-doping-type cuprates, Tl-Ba-Cu-O and La-Sr-Cu-O were studied.

II. Experimental Procedures

2.1 HR-TEM Study

The CVD-YBC films were deposited on $SrTiO_3$ (100) substrates. β -diketonate complexes i.e. 2,2,6,6-tetramethyl-3,5-heptanedionate chelated Y^{3+}, Ba^{2+} and Cu^{2+} respectively, were used as the three vapor sources. The films were deposited in a mixed gas atmosphere of $Ar/O_2 = 3/1$ at 10 Torr on the substrate which was heated at about 1173K close to the melting point of $YBa_2Cu_3O_{7-y}$. The post-annealing was done in the O_2 atmosphere at 1 atm. To view the cross-section of the CVD-film, two pieces of cut substrate and film were pasted together, film to film, then cut again perpendicular to the film and polished in an Ar ion sputtering mill.

HR-TEM observations were made on a 400KV electron microscope, JEM-4000EX.

2.2 X-ray and neutron diffraction

Single-crystal samples of Nd_2CuO_{4-y}, $Nd_{1.83}Ce_{0.17}CuO_{4-y}$, Pr_2CuO_{4-y} and $Pr_{1.85}Ce_{0.15}CuO_{4-y}$ were grown from sintered powder rods in an infrared-heated floating zone furnace. The travelling-solvent-floating-zone (TSFZ) technique [12] was applied with the use of CuO-flux. The zone-travelling-rate was controlled at 0.5mm/h.

Single crystal samples of $La_{1.92}Sr_{0.08}CuO_{4-y}$ and $Tl_2Ba_2CuO_{6-y}$ were grown by the conventional CuO flux method in a platinum crucible or in a gold tube, respectively.

Powder samples of each cuprate were also prepared by the standard solid-state reaction technique. X-ray powder diffraction measurements (CuKα radiation) were carried out to determine the lattice parameter temperature dependency in the range from 10K to 1200K. At high temperatures above ambient, the powder samples were placed in air or in He atmosphere. In the partial reduction in the oxygen content occurred in the range 350K to 1100K. X-ray 4-circle diffraction measurements ,with monochromatized MoKα radiation, were done on the single crystals, typically $0.1 \times 0.2 \times 0.02$ mm^3 in size. Single-crystal neutron diffraction studies, with monochromatized neutron ,λ =1.0 Å, were done on Pr_2CuO_{4-y} and $La_{1.92}Sr_{0.08}CuO_{4-y}$, which were grown sufficiently large in order to compare the structural parameters with those determined from the X-ray data.

III. Experimental results

3.1 HR-TEM study

Photographs 1 and 2 show high-resolution images obtained with the incident beam parallel to [010]. Relatively large grains about 30nm × 20nm in size are observed in the photograph 1. Those grains or precipitates may not correspond to the disk-shaped precipitates observed previously by Watanabe et al. [2] and can not be identified as CuO because of the differences in the lattice constants.

Photograph 2 shows complicated high-resolution image which was frequently observed. The complexity is conceivably due to many faulted regions formed parallel to ab-planes. The faulted regions consist of extra-layers of about 20nm in diameter and deformed regions adjacent to them. Similarly faulted regions in the YBC sintered-powders have been reported by Hiraga et al.[13]. Undeformed YBC-domains, interleaved by the faulted regions, are about 20~ 50 nmϕ x 10 nm in size. The electron diffraction pattern taken with the incident beam parallel to [010] did not change when the specimen was tilted by 5° , which is appreciably wider than the usual crystals. Such morphology indicates that the sample was grown quickly at a temperature just below the melting point of $YBa_2Cu_3O_{7-y}$ where the homogeneous nucleation process followed by the preferential crystal growth in the ab-planes occurs.

Photograph 3 shows a low magnification TEM image obtained with the incident electron beam parallel to [001]. Round precipitates are shown and they are covered with Moiré-fringes, indicating that the precipitates are semicoherent with the matrix and have very similar crystal lattice. These precipitates might not be the ones that Watanabe et al. [2] observed previously.

3.2 X-ray and neutron diffraction study on $Nd_{2-x}Ce_xCuO_{4-y}$ and $Pr_{2-x}Ce_xCuO_{4-y}$

Figure 1, the Nd_2CuO_{4-y}-Ce_2CuO_{4-y} pseudo-binary phase diagram obtained previously[14], shows the antiferromagnetic/super-conducting and metallic/non-metallic phase boundaries, the latter being defined by the resistivity minimum points. In the vicinity of the resistivity minimum points, the lattice constant temperature dependency curves show anomalous breakes (arrows in Fig.2). Similar changes are observed in the Pr-Ce-Cu-O system, but the anomalous points are found at lower temperatures.

Thus the end-members Nd_2CuO_{4-y} and Pr_2CuO_{4-y}, are in the non-metallic or semiconductor regime, but $Nd_{2-x}Ce_xCuO_{4-y}$ and $Pr_{2-x}Ce_xCuO_{4-y}$ with $x > 0.13$ are in the metallic regime at room temperature. The partly reduced samples, quenched from temperatures of more than 1200K becomes more metallic i.e. electric conductive.

Figure 3 , an ORTEP drawing of Nd_2CuO_{4-y} with $y \sim 0$ shows the thermal ellipsoids determined from X-ray single-crystal diffraction data, which were reported separately [14]. Attention is focused on structural changes due to the Ce^{4+} substitution for Nd^{3+} or Pr^{3+} and to the partial reduction of the oxygen content. An interesting structural alteration was found in the Cu-O layer, where the thermal ellipsoids of the O_1 atoms located at 4c(0 1/2 0) site are vertically elongated ,as shown in Fig.3. This elongation is much greater than the experimental error. By using Fourier- and differential-Fourier-syntheses-techniques, electron density distributions in the Cu-O layer were derived, and these exhibited a similar elongation of O_1 atom. Figures 4 (a-d) show three dimensional representations of the electron density distribution in Nd_2CuO_{4-y}, (Figs. 4(a) and (b)), and $Nd_{1.83}Ce_{0.17}CuO_{4-y}$, (Figs. 4(c) and (d)). Figures. 4(a) and 4(c) show the Cu-O layer in the oxygen saturated samples which were annealed at 670K in air. The oxygen contents are reduced in the samples shown in Figures.4(b) and 4(d). Partial reduction of the oxygen content was achieved by annealing the Nd_2CuO_{4-y} and $Nd_{1.83}Ce_{0.17}CuO_{4-y}$ in air and in Ar gas at 1373K, respectively followed by quenching in liquid N_2. These diagrams demonstrate that the O_1 atom is elongated vertically in the oxygen-saturated Nd_2CuO_{4-y} in Fig. 4(a) but the same atom is elongated horizontally toward copper sites in Fig. 4(c). On the other hand, partial reduction in the oxygen content in $Nd_{1.83}Ce_{0.17}CuO_{4-y}$ leads to an additional vertical elongation of O_1 (see Figs. 4(c) and (d)). Similar changes are observed in the Pr-Ce-Cu-O system. Figure 5 shows the electron density in the Cu-O layer in the reduced i.e. ,annealed in Ar gas and then quenched, $Pr_{1.85}Ce_{0.15}CuO_{4-y}$ in which the shape change is more moderate compared to the Nd-Ce-Cu-O system. Neutron diffraction measurements were also performed for a Pr_2CuO_{4-y} single-crystal in the oxygen saturated state, and those confirm the vertical elongation of O_1 atom. This means that the elongated shape of O_1 atom , deduced from the X-ray data, is due to the vibration or local displacement of the atom, since neutrons are mainly scattered by nuclei, in contrast to X-rays, which are scattered by electron clouds. However, the nature, static or dynamic, of the movement is not yet determined. It may also be pointed out that the shape of the oxygen atoms in the Cu-O layer are changed somewhat differently by the Ce^{4+} substitution and by the partial reduction of the oxygen content. Since partial reduction of the oxygen content is necessary to improve the electric conductivity and to undergo the superconducting transition in the Ce_{4+}-doped T'-phase, the vibration or local displacement of oxygen atoms shown in the reduced samples may be quite important. On the other hand, the horizontal expansion of oxygen atoms by the Ce^{4+} substitution may be interpreted simply in terms of the bond length effect, since the bond length between the monovalant copper, Cu^{+}, and oxygen is shorter, 1.84Å, than that of divalent copper, Cu^{2+}, and oxygen, 1.95Å. Decreasing the copper valency from Cu^{2+} to Cu^{+1}, by the Ce^{4+} substitution, a fluctuation in the copper/oxygen distances occurs.

3.3 X-ray diffraction study on $Tl_2Ba_2CuO_{6-y}$

This crystal undergoes a superconducting transition at 85K only if it is annealed at about 870K in Ar or N_2 gas for one or two hours and then quenched in liquid N_2 [15,16]. The superconducting transition does not occur in a well oxidized specimen which is considered to be too much hole-doped for the superconductivity to occur.

We have focused our attention to both the copper/apical-oxygen distance and the vibration or local displacement of the oxygen in the Cu-O layer in the three $Tl_2Ba_2CuO_{6-y}$ samples, one in the as-grown state and two in differently reduced states. The single-crystals were annealed at 810K in Ar and then cooled in a furnace or annealed at 870K in N_2 and then quenched in liquid N_2. The T_c of these samples ,65K and 85K, were measured on separate but identically reduced sintered pellets.

Figure 6 shows three-dimensional representation of the differential-Fourier-syntheses obtained from the single-crystal X-ray Bragg intensity at room temperature. The picture shows the electron density distribution which is only due to oxygen atoms. The electron densities of oxygen atoms O_1 at 4c(0 1/2 0) site and O_2 at 4e(0 0 z) with z=0.114~ 0.121 are shown. Copper atoms are located at the center of the oxygen octahedra. Fig. 6 shows two interesting structural changes. The copper/apical-oxygen distance appears to be increased from 2.645(14)Å to 2.815(8)Å by the partial reduction ,while T_c increases from 0K to 85K. The O_1 atom looks nearly spherical in the non-superconducting sample (T_c=0K) but becomes disk-like shape in the T_c=85K sample. The thermal factor, u_{ij}, also shows similar anisotropy. The O_1 atom is vibrating in the directions perpendicular to the Cu-O chemical bonds in the high-T_c sample, and shows some similarity to Nd-Ce-Cu-O and Pr-Ce-Cu-O systems.

In addition to these shape changes, the site occupation probability at the Tl-site at 4e(0 0 z) with z=0.202 decreased about 5% in the T_c=85K sample. The missing Tl atoms are located at an impurity site, 4e(0 0 z) with z=0.230. This is probably due to a partial oxygen loss from the Tl-O double layers in the T_c=85K sample.

3.4 X-ray and neutron diffraction study on $La_{1.92}Sr_{0.08}CuO_{4-y}$

The end member crystal, La_2CuO_{4-y}, undergoes a tetragonal/orthorhombic structural transition at about 500K. Birgeneau et al. [17] found a soft TA phonon mode at 1/2 1/2 0 in the tetragonal lattice, which corresponds to the shearing deformation of the ab-planes. The CuO_6 octahedra rotate by about 4° in the tetragonal to orthorhombic transition. Thus, there are two different structural alterations which may have a relationship with the superconductivity in the La-Sr-Cu-O system.

Some pieces of the as-grown $La_{1.92}Sr_{0.08}CuO_{4-y}$ single-crystals were annealed at 873K or 1373K in air and then quenched in air or in liquid N_2 to reduce the oxygen content, respectively. The quenched crystals becomes less hole-doped than at the as-grown state and the lattice constants are also changed. X-ray single-crystal diffraction measurements were done at room temperature for three different samples i.e. as-grown sample and samples quenched from 873K and 1373K. The copper/apical-oxygen distance decreases from 2.393(4)Å in the as-grown sample to 2.388(9)Å and 2.376(7)Å in the samples quenched from 873K and 1373K, respectively, and these values are shorter than the distance in the more hole-doped sample $La_{1.85}Sr_{0.15}CuO_{4-y}$, 2,414(8) Å[18]. Thus the copper/apical-oxygen distance increases by hole-doping in the La-Sr-Cu-O system, at least at room temperature.

The neutron single-crystal diffraction data taken for the as-grown $La_{1.92}Sr_{0.08}CuO_{4-y}$ indicates that the copper/apical-oxygen distance becomes slightly longer in the tetragonal to orthorhombic transition at about 200K. The distance is 2.404(5)Å at 150K but becomes 2.409(9)Å at 15K, staying almost constant at low temperatures.

In addition, the shape of oxygen atoms located in the Cu-O layer seems to be influenced by the hole density. Figures 7(a) and 7(b) show ORTEP drawings of $La_{1.92}Sr_{0.08}CuO_{4-y}$ in the as-grown and quenched states, respectively. The thermal ellipsoids of O_1 atoms located in the Cu-O layer are vertically elongated, u_{33}=0.036(2)Å^2, in the as-grown sample but the same atoms are not much elongated, u_{33}=0.016(2)Å^2, in the sample quenched from 1373K. These vertical vibration of O_1 atoms which bridges two copper atoms in the Cu-O layers may indicate some structural instability, which seems common in the superconducting cuprates. Detailed structure data have been reported elsewhere [19].

IV. Summary

4.1 HR-TEM study

The high-J_c CVD-YBC films were examined by HR-TEM to search for flux-pinning-centers. The cross-section images taken with the electron beam parallel to [010] indicate that there are many faulted regions where extra-layers, possibly CuO, about 20~ 50 nm in diameter, are grown parallel to ab-planes and are surrounded by deformed regions. Those faulted regions may act as the flux-pinning-centers. X-ray diffraction studies are underway to determine the average YBC-domain size.

Relatively large precipitates about 30nmφ × 20nm in size are embedded parallel to the ab-planes. In the low magnification TEM image taken with the election beam parallel to [001] direction, we could see round precipitates covered with Moiré fringes. Possibly different but similar precipitates were reported by Watanabe et al. [2] who suggested that these precipitates could act as the flux-pinning-centers. Since the Moiré fringes were seen on the surfaces of precipitates, the crystal structure is similar to the matrix. It is suggested that the interfaces between the precipitates and the matrix could also act as the flux-pinning-centers.

4.2 X-ray and neutron diffraction study on $Nd_2Ce_xCuO_{4-y}$ and $Pr_{2-x}Ce_xCuO_{4-y}$

The crystal structures of the electron-doping-type superconducting cuprates $Nd_2Ce_xCuO_{4-y}$ and $Pr_{2-x}Ce_xCuO_{4-y}$ were elucidated by the X-ray and partly by the neutron single-crystal diffraction studies. Before the present study, the Cu-O layer was considered to be very stable due to the crystal symmetry which does not allow any rotation or displacement of the Cu-O layer. However, the thermal ellipsoids as well as the electron density distribution in the Cu-O layer show interesting Ce^{4+}-substitution-level-dependent and oxygen-content-dependent variations. An elongation of the in-layer oxygens in the direction perpendicular to the Cu-O plane was noticed in these systems. The neutron diffraction data indicate that the elongation is mostly due to the vibration or local displacement of the nuclei of oxygen atoms. Present structural work suggests that above type of movement may be common to both the T'-phase, (Nd_2CuO_{4-y}-type structure), and the T-phase, (K_2NiF_4-type structure). At present, however, we are unable to determine whether the movement is static or dynamic. It is also found that the substitution of Ce^{4+} for Nd^{3+} or Pr^{3+} leads to an elongation in the shape of the bridging oxygens toward the nearest copper atoms.

4.3 X-ray diffraction study on $Tl_2Ba_2CuO_{6-y}$

The $Tl_2Ba_2CuO_{6-y}$ crystal is an interesting cuprate since its superconducting critical temperature can be controlled from OK to 85K easily by controlling the oxygen content. At the highly hole-doped state where no oxygen deficiency is expected and T_c=OK, the CuO_6 octahedron is short but becomes elongated at the appropriately hole-doped state where the oxygen deficiency, y, is 0.15 and T_c=85K. Anisotropic thermal ellipsoids and anisotropic electron density distributions at the oxygen site in the Cu-O layer is confirmed again as in the previous paragraph. The disk-like shape of the oxygen atom, relatively thin in the Cu-O-Cu direction and expanded perpendicular to this direction, is noticed in the high-T_c sample.

4.4 X-ray and neutron diffraction study on $La_{1.92}Sr_{0.08}CuO_{4-y}$

The copper/apical-oxygen distance in this lanthanum system becomes longer in the high-T_c samples as in $Ta_2Ba_2CuO_{4-y}$. A quenched $La_{1.92}Sr_{0.08}CuO_{4-y}$ sample from 1373K, being oxygen poor and less hole-doped, shows the shortest distance, 2.376(7)Å, at room temperature. The copper/apical-oxygen distance increases slightly by the tetragonal to orthorhombic structural transition but it stays almost constant with decreasing temperature in the low temperature range below 150K.

The thermal ellipsoid of oxygen located in the Cu-O layer is anisotropic in the oxygen saturated (as-grown) $La_{1.92}Sr_{0.08}CuO_{4-y}$ sample but becomes isotropic in the quenched sample. Since the electric conductivity becomes higher in the strontium-doped and oxygen-saturated samples, the above anisotropy of the bridging oxygens may contribute positively to the electric conductivity in the Cu-O layers.

Acknowledgment

This work has partly been supported by a Grant-in-Aid for Scientific Research on Priority Area "Mechanism of Superconductivity" from the Ministry of Education, Science and Culture of Japan.

Reference

[1] H. Yamane, H. Kurosawa, T. Hira, K. Watanabe, H. Iwasaki, N. Kobayashi and Y. Muto: Supercond. Sci. Technol. 2 (1989) 115.

[2] K. Watanabe, H. Yamane, H. Kurosawa, T. Hirai, N. Kobayashi, H. Iwasaki, K. Noto and Y. Muto: Appl. Phys. Lett. 54 (1989) 575.

[3] K. Watahnabe, T. Matsushita, N. Kobayashi, H. Kawabe, E. Aoyagi, K. Hiraga, H. Yamane, H. Kurosawa, T. Hirai and Y. Muto: Appl. Phys. Lett. 56 (1990) 1490.

[4] J. S. Satchell, R. G. Humphreys, N. G. Chew, J. A. Edwards and M. J. Kane: Nature 334 (1988) 331.

[5] K. Heine, J. Tenbrink and M. Thoner: Appl. Phys. Lett. 55 (1989) 2441.

[6] K. Sato, T. Hikata, H. Mukai, T. Masuda, M. Ueyama, H. Hitotsuyanagi, T. Mitsui and M. Kawashima: Proc. ISS '89, Tsukuba. Published as 'Advances in Superconductivity II, Springer-Verlag, 1990, p.335.

[7] N. Uno, N. Enomoto, H. Kikuchi, K. Matsumoto, M. Mimura and M. Nakajima: ibid p.341.

[8] C. Murayama, N. Mori, S. Yomo, H. Takagi, S. Uchida and Y. Tokura: Nature 339 (1989) 293.

[9] E. Kaldis, P. Fischer, A. W. Hewat, E. A. Hewat, J. Karpinski and S. Rusiecki: Physica C159 (1989) 668.

[10] R. J. Nelmes, N. B. Wilding, P. D. Hatton, V. Caignaert, B. Raveau, M. I. McMahon and Rj. O. Piltz: Physica C166 (1990) 329.

[11] G. H. Kwei, A. C. Larson, W. L. Hults and J. L. Smith: Physica C169 (1990) 217.

[12] S. Kimura and I. Shindo: J. Cryst. Growth 41 (1977) 192.

[13] K.Hiraga, T.Oku, D.Shindo and M.Hirabayashi: J. Electron Microsc. Technique 12(1989) 228.

[14] T. Kajitani, K. Hiraga, S. Hosoya, T. Fukuda, K. Oh-Ishi, M. Kikuchi, Y. Syono, S. Tomiyoshi, M. Takahashi and Y. Muto: Physica C169 (1990) 227.

[15] Y. Kubo, Y. Shimakawa, T. Manako, T. Satoh, S. Iijima, T. Ichihashi and H. Igarashi: Proc. M^{2}S-HTSC, Stanford (1989), published as Physica C162-164 (1989) 991.

[16] M. Kikuchi, S. Nakajima, Y. Syono, K. Nagase, R. Suzuki, T.

Kajitani, N. Kobayashi and Y. Muto: Physica C166 (1990) 497.

[17] R. J. Birgeneau, C. Y. Cheu, D. R. Gabbe, H. P. Jenssen, M. A. Kastver, C. J. Peters, P. J. Picone, Tineko Thio, T. R. Thurstom, H. J. Teller, J. P. Axe, P. Boni and G. Shirane: Phys. Rev. Letters 59 (1987) 1329.

[18] T. Kajitani, T. Onozuka, Y. Yamaguchi, M. Hirabayashi and Y. Syono: Jpn. J. Appl. Phys. 26 (1987) L1877.

[19] T. Kajitani, K. Hiraga, T. Sakurai, M. Hirabayashi, S. Hosoya, T. Fukuda and K. Oh-Ishi: Physica C (submitted).

Figure Captions

Photo.1 High resolution cross-section image of CVD-YBC film taken with the electron beam parallel to [010]. Unidentified precipitates about 30nm × 20nm in size are shown in upper part of the photograph.

Photo.2 High resolution cross-section image of CVD-YBC film taken with the electron beam parallel to [010]. The complicated nature of the stacking sequence is shown and indicates that homogeneous nucleation in the half-melted YBC layer occurred on the substrate surface.

Photo.3 Low magnification TEM image of the CVD-YBC film taken with the electron beam parallel to [001]. Round precipitates with Moiré fringes on them are shown.

Fig. 1 Nd_2CuO_{4-y}-Ce_2CuO_{4-y} pseudo-binary phase diagram. The resistivity minimum points observed for the sintered pellet samples are plotted with shaded zone.

Fig. 2 Lattice constants vs. temperature relationship of as-sintered $Nd_{2-x}Ce_xCuO_{4-y}$ with x=0, 0.05, 0.10, 0.13, 0.15 and 0.20. Open inverted-triangles and squares are the lattice constants of the reduced samples.

Fig. 3 ORTEP-drawing of Nd_2CuO_{4-y}. Note the oval shape of the O_1 atom.

Fig. 4 Three dimensional representation of the electron density distribution in the Cu-O_1 layers, obtained by the Fourier- and differential-Fourier-syntheses techniques. The electron densities in the Cu-O_1 layers were calculated from X-ray diffraction data taken from

(a) oxygen-saturated Nd_2CuO_{4-y},

(b) quenched Nd_2CuO_{4-y},

(c) oxygen-saturated $Nd_{1.83}Ce_{0.17}CuO_{4-y}$ and

(d) quenched $Nd_{1.83}Ce_{0.17}CuO_{4-y}$.

Fig. 5 Three-dimensional representation of the electron density in the Cu-O_1 layers in the partly reduced i.e. quenched $Pr_{1.85}Ce_{0.15}CuO_{4-y}$.

Fig. 6 Three-dimensional representation of the electron density belonging only to oxygen atoms, obtained by the differential-Fourier-synthesis technique. CuO_6 octahedra in the single crystals at the as-grown state (T_c=OK), annealed states at 810K (T_c=65K) and 870K (T_c=85K) are shown.

Fig. 7 ORTEP drawings of a tetragonal $La_{1.92}Sr_{0.08}CuO_{4-y}$. (a) sample in the as-grown state. (b) sample annealed at 1373K in air and then quenched in liquid N_2.

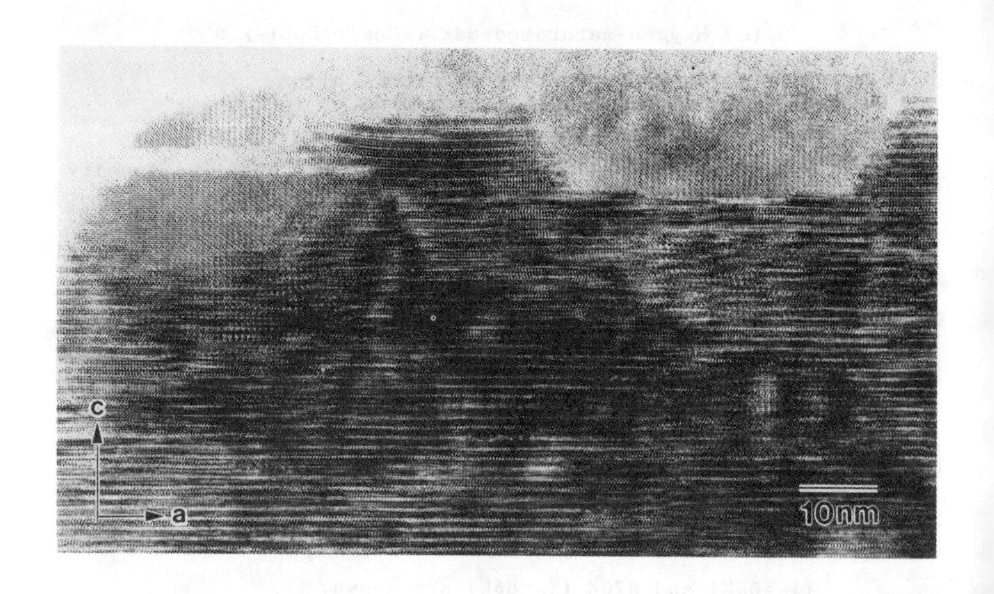

Photo 1

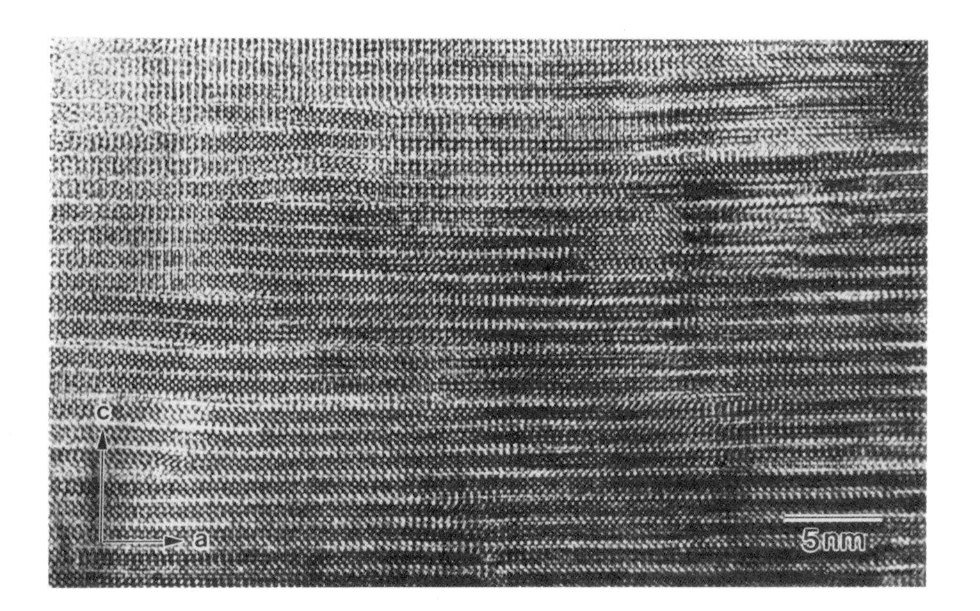

Photo 2

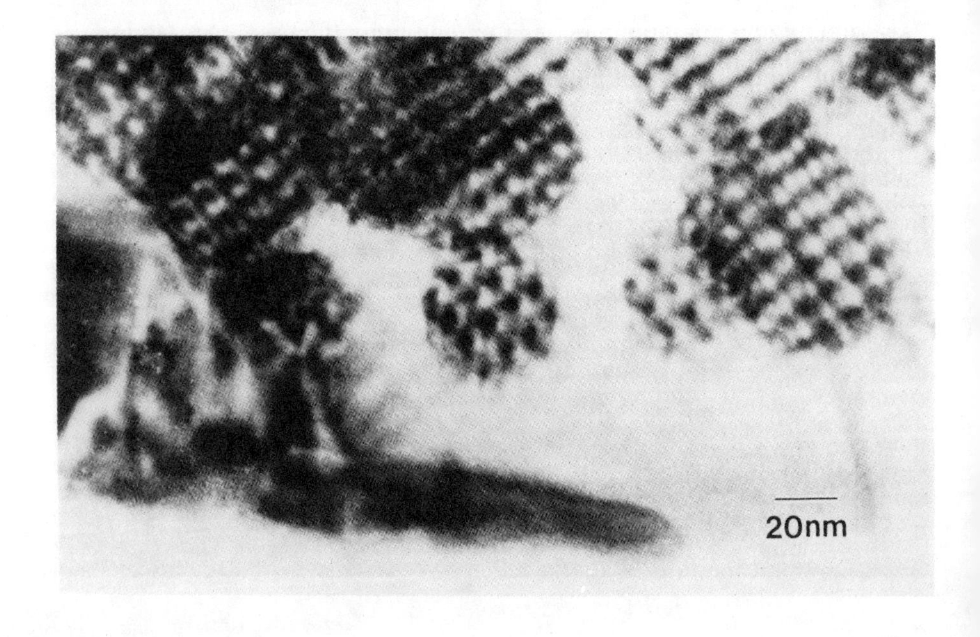

Photo 3

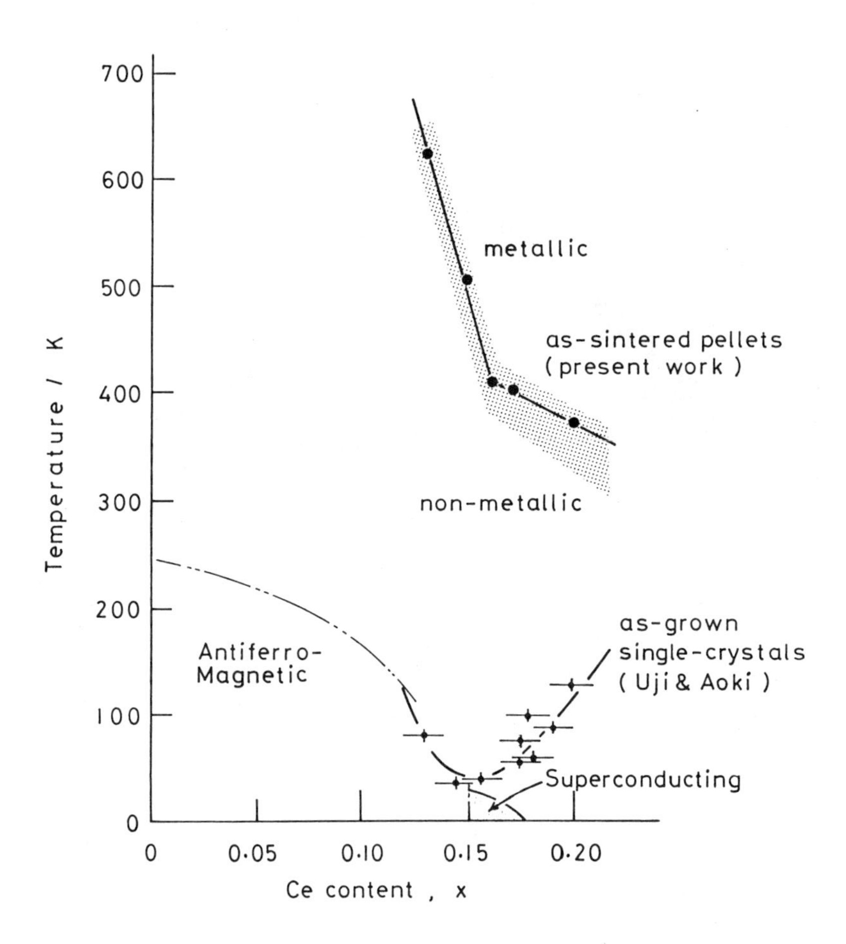
700
600
500
400
300
200
100
0
Temperature / K
metallic
as-sintered pellets
(present work)
non-metallic
Antiferro-
Magnetic
as-grown
single-crystals
(Uji & Aoki)
Superconducting
0
0·05
0·10
0·15
0·20
Ce content , x

Figure 1

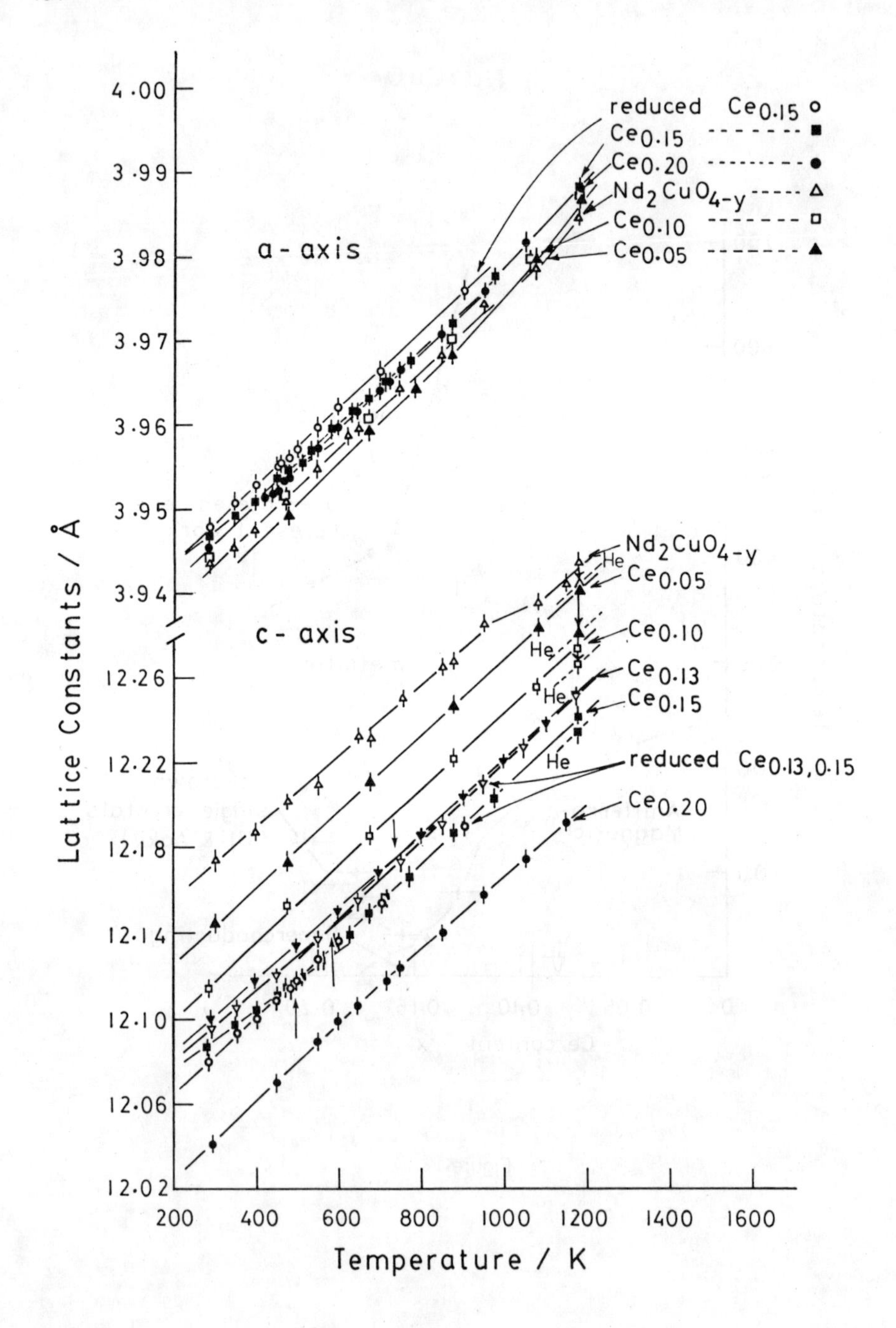
reduced Ce0.15
Ce0.15
Ce0.20
Nd2CuO4-y
Ce0.10
Ce0.05
a- axis
c- axis
Nd2CuO4-y
He
Ce0.05
Ce0.10
He
Ce0.13
He
Ce0.15
He
reduced Ce0.13,0.15
Ce0.20
4.00
3.99
3.98
3.97
3.96
3.95
3.94
12.26
12.22
12.18
12.14
12.10
12.06
12.02
200
400
600
800
1000
1200
1400
1600
Lattice Constants / Å
Temperature / K

Figure 2

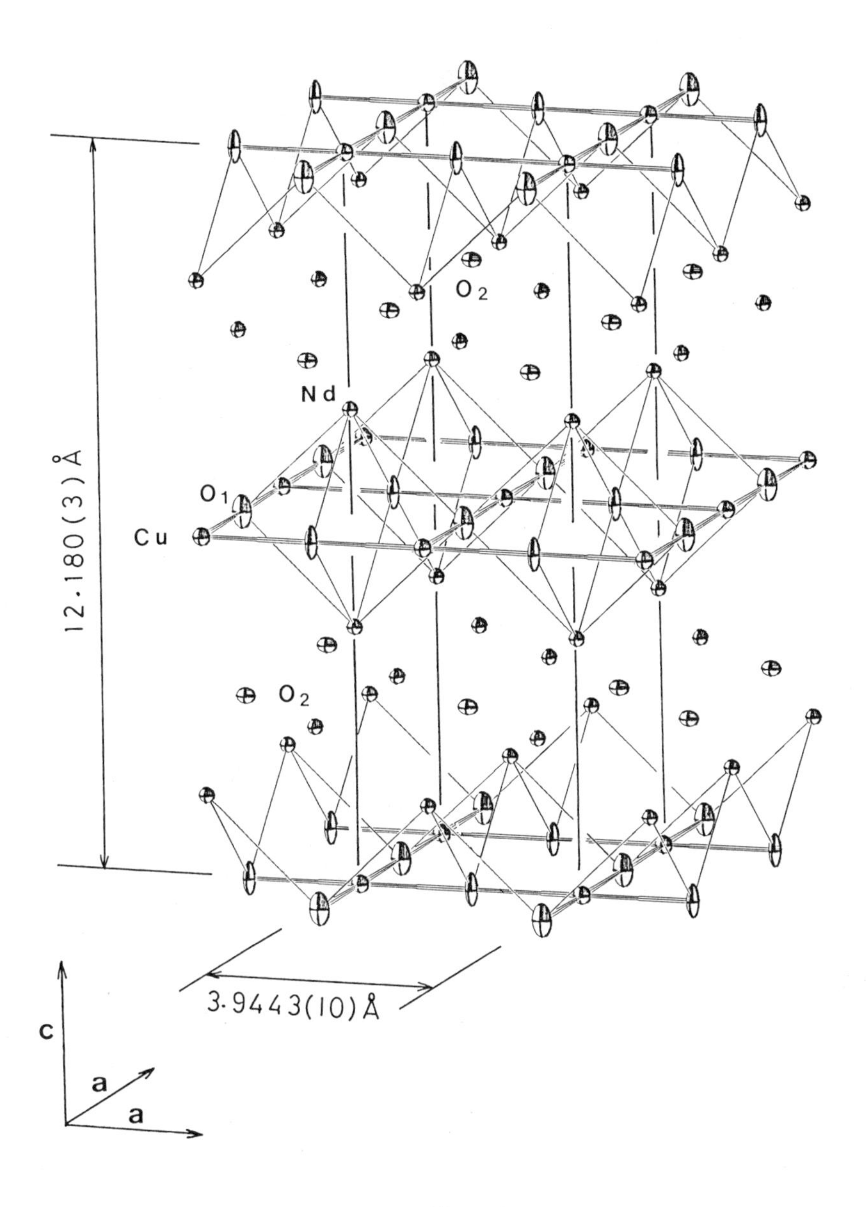

Figure 3

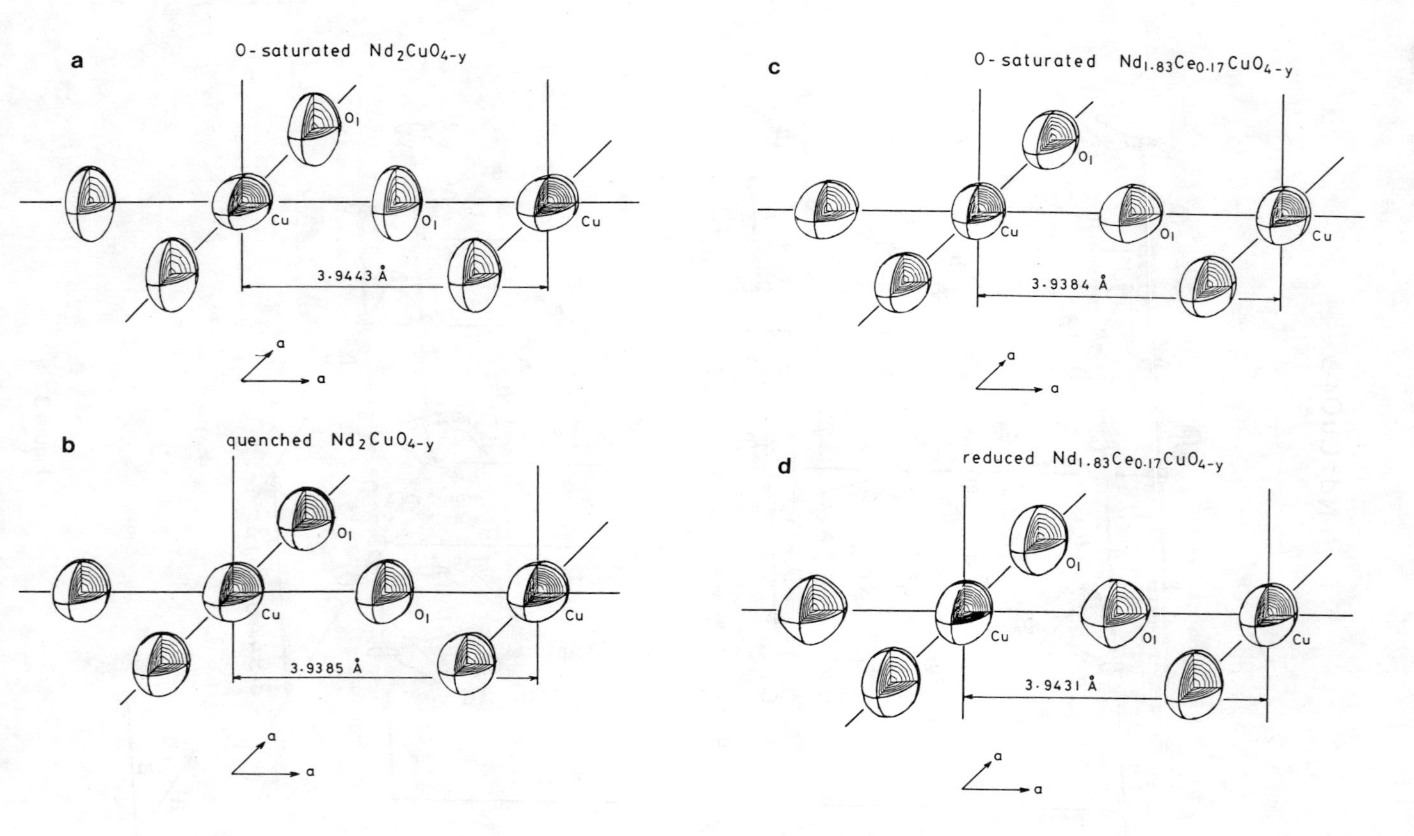
a
O-saturated Nd2CuO4-y
O1
Cu
3·9443 Å
a
b
quenched Nd2CuO4-y
3·9385 Å
c
O-saturated Nd1.83Ce0.17CuO4-y
3·9384 Å
d
reduced Nd1.83Ce0.17CuO4-y
3·9431 Å

Figure 4

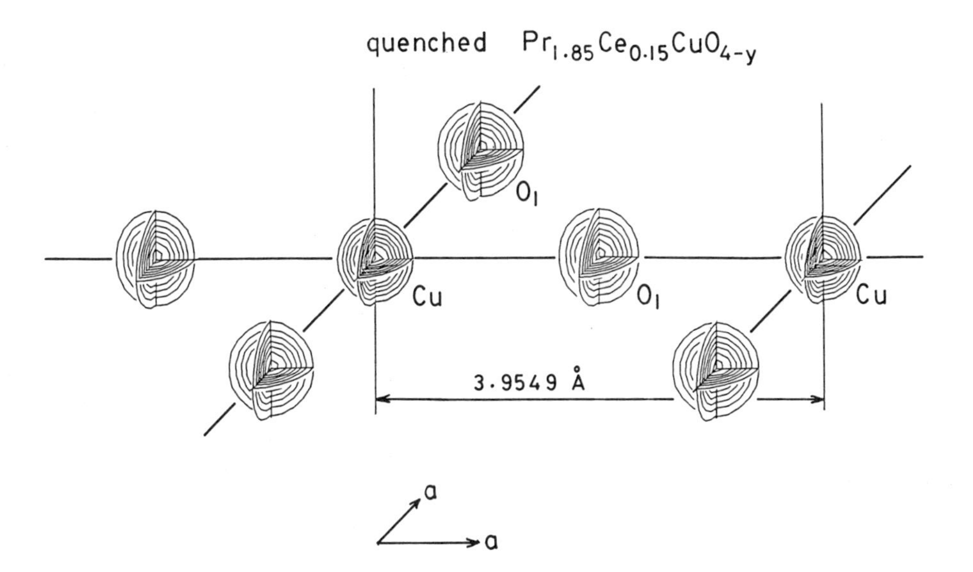

Figure 5

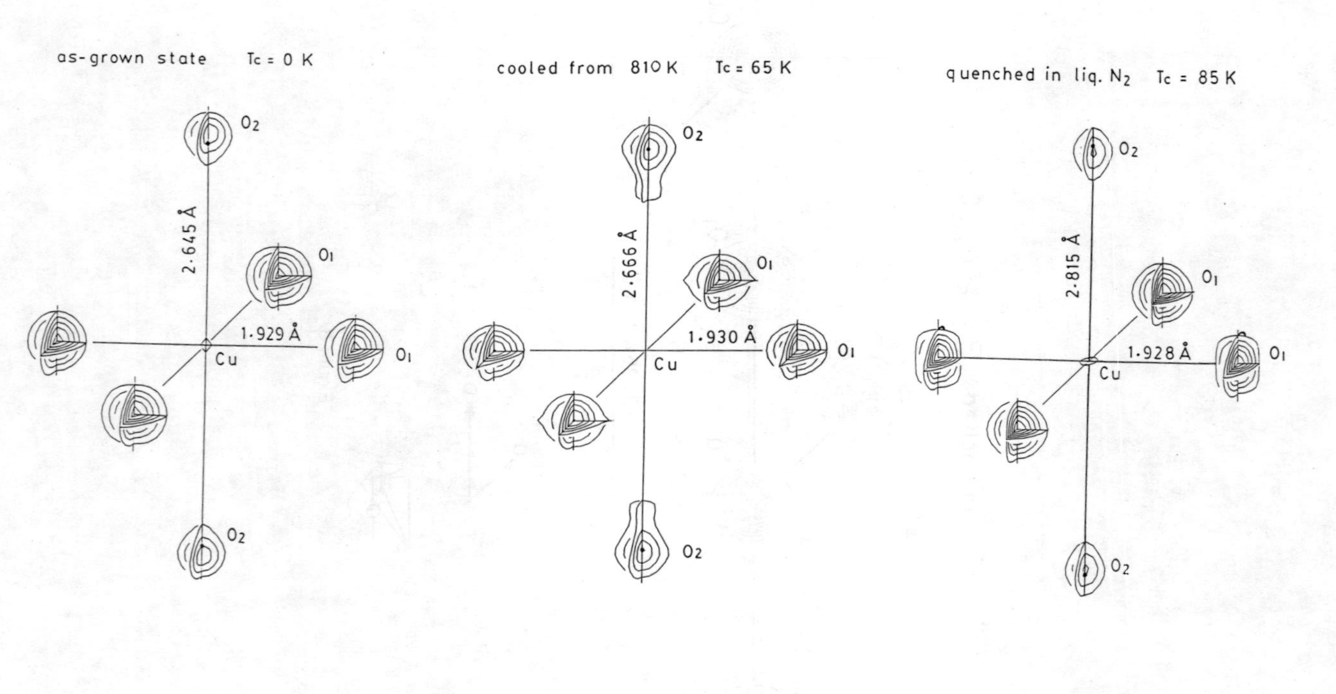

as-grown state
Tc = 0 K
O2
2.645 Å
O1
1.929 Å
Cu
O1
O2
cooled from 810 K
Tc = 65 K
O2
2.666 Å
O1
1.930 Å
Cu
O1
O2
quenched in liq. N2
Tc = 85 K
O2
2.815 Å
O1
1.928 Å
Cu
O1
O2

Figure 6

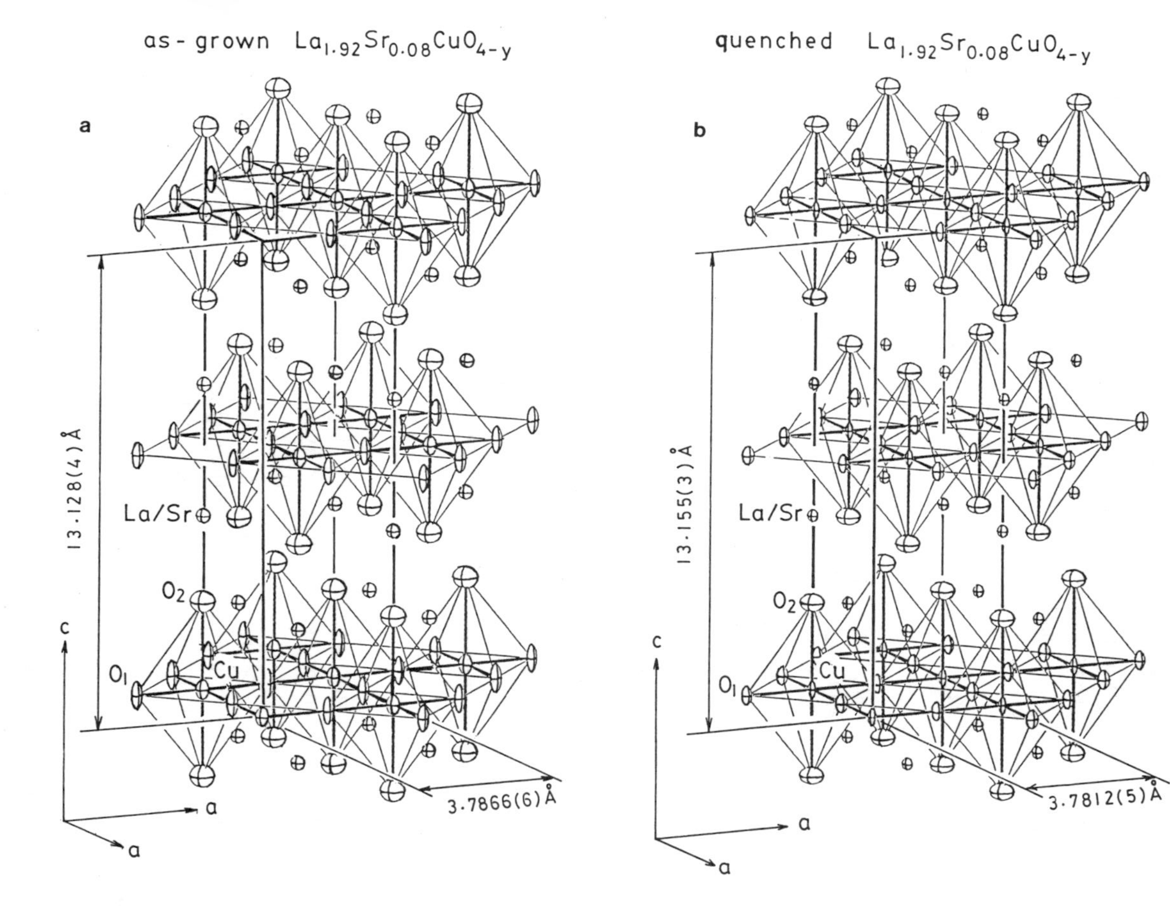
as-grown La1.92Sr0.08CuO4-y
quenched La1.92Sr0.08CuO4-y
a
b
13.128(4) Å
13.155(3) Å
La/Sr
O2
O1
Cu
c
a
a
3.7866(6) Å
3.7812(5) Å

Figure 7

CHARACTERIZATION OF HIGH TEMPERATURE SUPERCONDUCTORS WITH MUON SPIN ROTATION

T. M. Riseman, J. H. Brewer, J. F. Carolan,
W. N. Hardy, R. F. Kiefl, D. Ll. Williams and H. Zhou
Dept. of Physics, Univ. of British Columbia,
Vancouver, B.C., Canada V6T 2A3

L. P. Le, G. M. Luke, B. J. Sternlieb and Y. J. Uemura
Dept. of Physics, Columbia Univ., New York, NY 10027, U.S.A.

B. Hunter and K. N. R. Talyor
A.E.M., Dept. of Physics, Univ. of New South Wales,
P.O. Box 1, Kensington, NSW, 2033, Australia

H. Hart and K. W. Lay
G.E. Corp., Research and Development, Schenectady, N.Y., U.S.A.

Abstract

The positive Muon Spin Rotation (μ^+SR) technique is an extremely versatile tool for the study of magnetic phenomena. In an external field, the muon precession signal is analogous to NMR's free induction decay. Additionally, magnetic ordering phenomena may be studied in zero applied field. This paper will illustrate some recent μ^+SR studies in high-T_c superconductors (HT$_c$SC) which have measured magnetic penetration depths and magnetic properties throughout the electronic phase diagrams.

1 Muons

Muons are leptons with a lifetime of 2.2 μs. The positive muon (μ^+) can be thought of as a light proton or as an isotope of hydrogen. It has a spin of 1/2 and a gyromagnetic ratio (γ_μ) of $2\pi\times0.01355$ MHz/G. The μ^+ and a muon neutrino are the decay products of a positive pion. Since neutrinos can only be left handed, the μ^+'s produced are 100% spin polarized, with their spins antiparallel to their momentum. This means that, unlike MNR, large magnetic fields and low temperatures are not required to align the probe (μ^+) spins in μ^+SR. The μ^+ later decays into two neutrinos and a positron. The positron is emitted in a direction strongly correlated with the muon's spin at the time of decay; this allows observation of the precession of an ensemble of muons in a local magnetic field. The combined effects of the μ^+ lifetime and the timing resolution of the decay positron counters leads to an the experimental time "window" of approximately 10 ns – 100 μs. (For a more detailed explanation of μ^+SR, see Ref. [1].)

"Surface" muons (energy = 4.1 MeV) have a stopping range of about 140 mg/cm^2 in most materials, which corresponds to about 200 μm for $YBa_2Cu_3O_7$. The μ^+ comes to rest in the sample in about 10 ps, which is essentially instantaneously for our purposes. Once at rest in oxide materials like HT$_c$SC, it forms a hydroxyl-like bond with an oxygen ion with a bond length of approximately 1 Å.[2]

2 μ^+SR in Superconducting Samples

After the exponential decay due to the muon lifetime has been removed, the μ^+SR signal is often fit with

$$S(t) = N\cos(\gamma_\mu H t)\exp\left[-\frac{1}{2}(\sigma_t\, t)^2\right], \tag{1}$$

where $\gamma_\mu = 2\pi \times 0.013554$ MHz/G $= 0.085162(\mu\text{s}\ G)^{-1}$, t is time and H is the average magnetic field. In an HT$_c$SC above T_c, this gaussian function fits well with a typical relaxation rate of $\sigma_t = 0.07\ \mu\text{s}^{-1}$. Below T_c, the relaxation rate increases dramatically. The Fourier transform of the time spectrum gives the distribution of muon precession frequencies (ν). This is proportional to the distribution of local fields h—the "lineshape" $A_{\text{exp}}(h)$—experienced by the muon in the sample ($\nu\,[\text{MHz}] = 0.013554\,h\ [\text{G}]$).

What causes this relaxation and its corresponding lineshape? In a type II superconductor, a triangular vortex lattice is formed when the applied field is above H_{c1} and below H_{c2}. In this lattice, the point between three vortices is where the minimum field occurs. The midpoint between adjacent vortices is a saddle point which leads to a cusp in the lineshape. (See Figs. 3 and 6 in Ref. [3].) In the London approximation (appropriate when $\kappa \gg 1/\sqrt{2}$, where $\kappa = \lambda/\xi$, λ is the magnetic penetration depth and ξ is coherence length), the field diverges at the vortex cores, leading to a tail in the lineshape at higher fields. However, in reality, there is a maximum field at the vortex core, which is observable in lower κ type II superconductors such as V and Nb but not in high κ materials such as HT$_c$SCs. These distinctive features (the minimum and cusp fields and the tail at higher fields) are known as the "Redfield" or "Abrikosov" NMR/μ^+SR lineshape—which have been clearly observed in classic superconductors such as V[4] and Nb.[5]

The most easily observed feature is the tail at higher fields. It has been seen in HT$_c$SCs with μ^+SR (Fig. 1b and 1d in this paper and in Refs. [6,7]) but, to the best of our knowledge, not with other techniques such as NMR or EPR.

The lineshape has an overall width which is due to the vortex lattice's inhomogeneous distribution of magnetic fields [$A_{\text{th}}(h)$]. In the high field limit, this width is proportional to the inverse square of the penetration depth. To describe the experimental data empirically, it is reasonable to convolute $A_{\rm th}(h)$ with a gaussian of width σ_h [G]:

$$A_{\rm tot}(h,\lambda) = \int A_{\rm th}(h',\lambda)\exp\left[-\frac{(h'-H)^2}{2\sigma_h^2}\right] dh'. \tag{2}$$

It is logical to suppose that σ_h could be due to the copper nuclear dipole fields,

$$\sigma_t(T > T_c) = \gamma_\mu \sigma_h, \tag{3}$$

where $\sigma_t(T > T_c) \approx 0.8$G is obtained by fitting data taken above T_c with the gaussian relaxation function given in Eq. 1.

Fig. 1b shows the local field distribution in a sample of oriented single crystals of $YBa_2Cu_3O_7$ (produced at the University of New South Wales) at 6.0 K in an applied field of 1.75 kG parallel to the crystalline $\hat{c}$-axis. Fig. 1a shows the theoretical lineshape for a penetration depth (λ_{ab}) of 1900 Å and H_o = 1.78 kG To fit the data, the theoretical lineshape was convoluted with a gaussian of width 15.6 G and a background signal at 1.75 kG was added. The convoluted lineshape is shown as a dashed line in Fig. 1a and as a solid line in Fig. 1b. Fig. 1d shows the local field distribution in an unoriented sintered powder of $YBa_2Cu_3O_7$. Its lineshape is different from than that of the single crystal data because the penetration depth is highly anisotropic—$\lambda_c/\lambda_{ab} \approx 5(1)$,[8] where the subscript refers to the crystal axes along which the supercurrents $\vec{\imath}$ are flowing, with $\vec{\imath} \perp \vec{H}$. Figure 1c shows a model lineshape for a powder-averaged extremely anisotropic superconductor[9] with $\lambda_{ab} = 920$ Å. This lineshape was convoluted with a gaussian of 22 G to fit the data.

Why is the gaussian broadening (16 G and 22 G) greater than that expected from the copper nuclear dipole fields (0.8 G)? Most likely, it is because the flux lattice is not perfectly triangular.[10]

Note that the unoriented sample's lineshape (Fig. 1d) is qualitatively gaussian. In general, it is rarely worth the investment of beam time to take very low background, high statistics runs on unoriented samples in order to try to discern the penetration depth from the lineshape's tail. The more common μ^+SR method for it is to compare the second moment of the theoretical lineshape and the second moment of a gaussian fit (see Eq. 1) to the time spectra $S(t)$. This implies that the relaxation rate σ_t is proportional to the inverse square of the effective penetration depth ($\lambda_{\rm eff}$):[11]

$$\sigma_t = \sqrt{0.00371}\,\gamma_\mu \frac{\Phi_o}{\lambda_{\rm eff}^2} = \gamma_\mu \sqrt{\Delta H^2}. \tag{4}$$

In the case of an isotropic superconductor, $\lambda_{\rm eff}$ is the penetration depth. In an extremely anisotropic uniaxial superconductor ($\lambda_c > 5\,\lambda_{ab}$), $\lambda_{\rm eff} = 1.23\,\lambda_{ab}$.[12] This

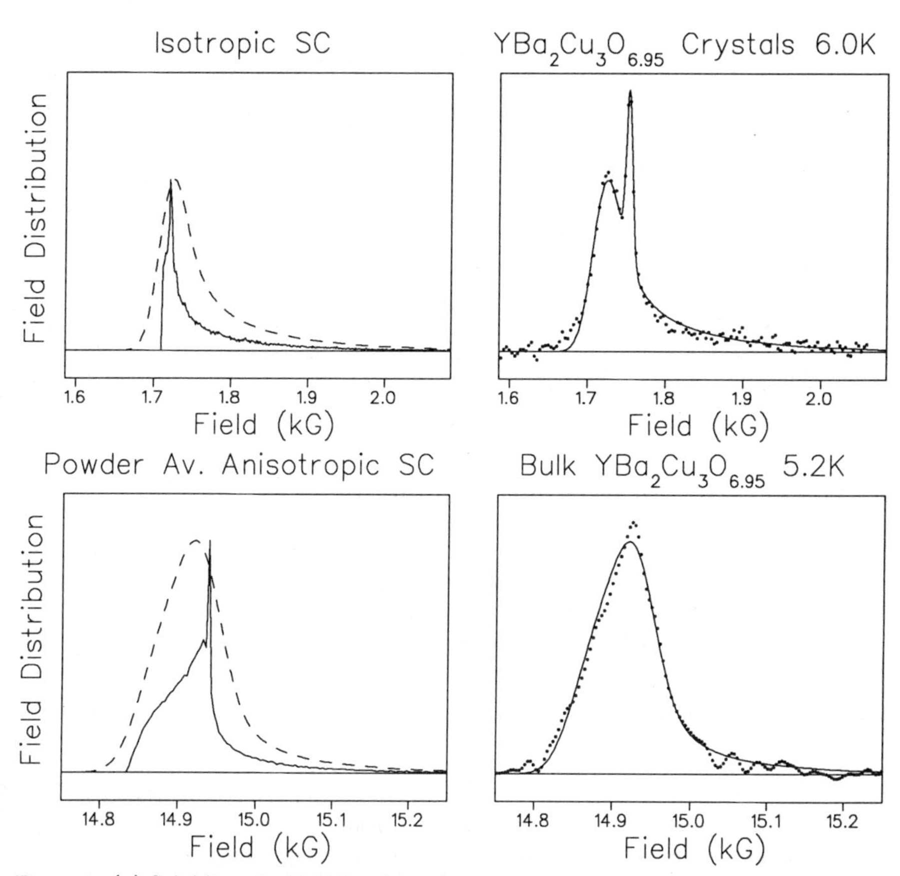

Figure 1: **(a)** Solid line: Redfield lineshape for an isotropic superconductor with $H_o = 1.78$ kG and $\lambda_{ab} = 1900$ Å. Dashed line is the Redfield lineshape convoluted with a gaussian of width 15.6 G. **(b)** Points: $YBa_2Cu_3O_x$ crystals at 6.0 K with $\vec{H} \parallel \hat{c}$. Solid line is the same as the dashed line in (a). The spike at 1.75 kG is a background signal. **(c)** Solid line: Redfield lineshape for a powder averaged extremely anisotropic superconductor with $H_o = 14.94$ kG and $\lambda_{ab} = 920$ Å. Dashed line is the Redfield lineshape convoluted with a gaussian of width 22 G. **(d)** Points: Bulk Sintered $YBa_2Cu_3O_{6.95}$ at 5.2 K. Solid line is the same as the dashed line in (c).

method allows for easy, systematic comparison of the penetration depth in many unoriented sintered powder samples.

Since λ is equal to $\sqrt{m^*c^2/(4\pi n_s e^2)}$, the relaxation rate σ_t is proportional to n_s/m^*, where n_s is the superconducting carrier density and m^* is the effective mass of the carriers. Thus, measurements of the relaxation rate give insight into the carrier density as a function of doping in a given HT_cSC family and also the relative carrier density between families.[13,14,15] In each class of HT_cSC compounds (with either one, two or three CuO_2 planes per unit cell), T_c increases and then saturates with increasing σ_t. What is striking is that all three classes have the same initial linear behavior with T_c directly proportional to σ_t. Why is T_c proportional to n_s/m^*? If the energy scale of the bosons mediating the superconducting pairing is larger than the Fermi energy E_f, one can expect $T_c \propto E_f$.[16] The high T_c cuprate systems are known to have highly two dimensional electronic structure. The Fermi energy of a non-interacting 2-D electron gas is proportional to n/m^*. Then, the linear relation $T_c \propto n_s/m^*$ may be interpreted as suggesting $T_c \propto E_f$. One can, however, expect the linear relation also in some other scenarios: the overall understanding of the observed correlations between T_c and n_s/m^* is open to future development of theories for HT_cSCs.

3 Observation of Magnetic Moments

A muon will precess in any magnetic field—regardless of whether it is applied externally or produced internally by antiferromagnetic (AFM), ferromagnetic (FM) or other static order. Since fields are not required to polarize the muon, μ^+SR is as easy to do in zero applied field (ZF) as in transverse field (TF).

3.1 *Zero Field μ^+SR: Superconducting Samples:*

Static disordered nuclear moments yield a distinctive ZF time spectrum where the signal dips and then recovers to one third of its intial value. In the case of a HT_cSC containing only ^{16}O, the ZF relaxation arises almost exclusively from the nuclear magnetic moments of copper (see Fig. 2). With the addition of ^{17}O, the ZF relaxation increases because the muon sites are closer to oxygen moments than to copper moments[17] (see Fig. 2).

3.2 *Zero Field μ^+SR: AFM Samples:*

In $YBa_2Cu_3O_6$, there are muon precession frequencies corresponding to 330 G and 1400 G local fields at two different muon sites (see Figure 3). These fields are produced by the AFM ordering of the copper electronic moments. In an oriented sintered powder of $YBa_2Cu_3O_6$[18] (produced at G.E.), comparison of the signal

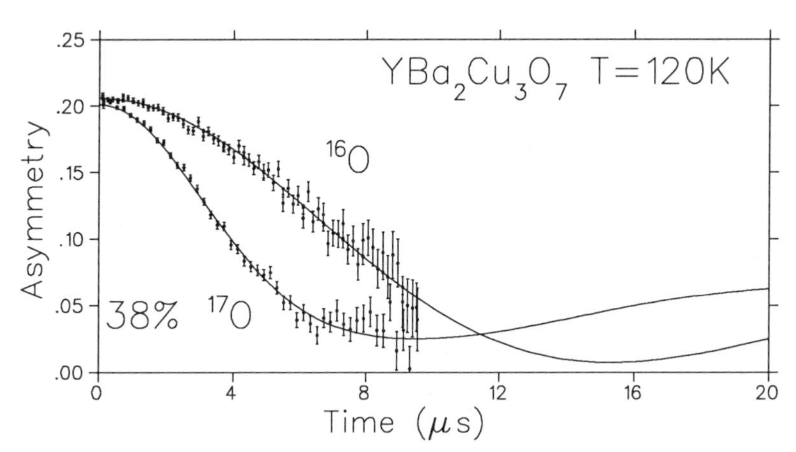

Figure 2: Asymmetry versus time plots of $YBa_2Cu_3O_{6.95}$ sintered powder in ZF with 100% ^{16}O (top) and 38% ^{17}O and 62% ^{16}O (bottom). The relaxation observed is due to Cu and ^{17}O nuclear dipole moments.

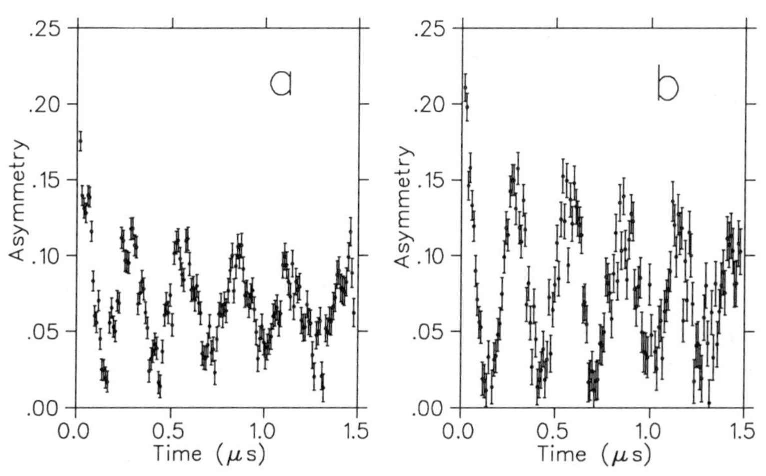

Figure 3: Asymmetry versus time plots of oriented sintered powder of $YBa_2Cu_3O_6$.
(a) Muon spin is perpendicular to the $\hat{c}$ axis. **(b)** Muon spin is parallel to the $\hat{c}$ axis.

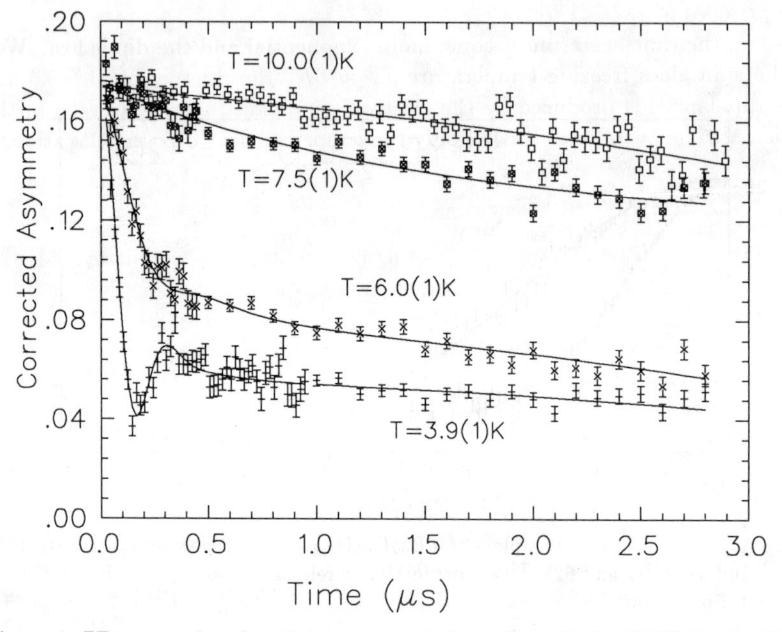

Figure 4: ZF muon spin relaxation spectrum in $La_{1.84}Sr_{0.06}CuO_4$ (NTT # 10). The relaxation at 10 K and 7.5 K can be attributed to fluctuating local fields. The increase in the relaxation rate at $T_f^{\mu} = 6.0$ K is due to the onset of spin freezing. By 3.9 K, fits indicate that the muon depolarization is mostly due to static fields.

amplitudes when the muon spin is parallel to and perpendicular to the hard axis ($\hat{c}$) shows that the 330 G field is about 60° away from the $\hat{c}$ axis while the 1400 G field is roughly parallel. In general, as the doping is increased and AFM is increasingly frustrated, μ^+SR observes increasing relaxation rates associated with the local fields at the muons sites.[19] The muon is truly a point probe, and in this respect, μ^+SR is complementary to neutron scattering.

3.3 *Zero Field μ^+SR: Disordered Magnetism*

In a frustrated AFM such as $La_{1.84}Sr_{0.06}CuO_4$, μ^+SR can be used to observe the transition to static disordered magnetism (spin glass magnetism) at low temperatures (Fig. 4, which is from Ref. [20]). In $La_{1.84}Sr_{0.06}CuO_4$, the electronic moments on the copper atoms are nearly static at 3.9 K. The data resemble those for the nuclear dipolar fields (Fig. 2); the much higher relaxation rate is due to the fact that electronic moments produce much larger magnetic fields at the muons sites than do nuclear moments. As the temperature increases and the electronic moments start to fluctuate, the time spectrum become more exponential and the dip is lost. Well above the spin glass freezing temperature $T_f^\mu \approx 6K$, the fluctuations are so fast that the dipolar fields produced by the electronic moments average to zero in the experimental time "window," leaving only the copper nuclear dipolar relaxation.

3.4 *μ^+SR Characterization of Multi-Phase Samples*

When new species of the 2-D superconducting perovskite genus are discovered, it is difficult to quickly and accurately characterize multi-phase samples with standard measurement techniques. With resistance and Meissner effect measurements, percolative superconductivity can leave the impression that a larger proportion of the sample is superconducting than is in fact the case. With susceptibility measurements, extremely broad weakly AFM transitions make it difficult to determine the transition temperature T_N.

μ^+SR can accurately measure the percentage of a sample (e.g., a bulk unoriented sintered powder) which has AFM or spin glass magnetism. When external field which is smaller than the average internal field is applied, the non-magnetic portion has almost no muon spin relaxation while the magnetic portion relaxes quickly. Because the local fields are randomly oriented, the total fields have greatly varying magnitudes. Comparing the amplitude of the nonmagnetic signal to the total amplitude of the magnetic and non-magnetic signals give the percentage of the sample which is non-magnetic (see Fig. 1 in Ref. [21]). The Néel temperature T_N is usually defined as the temperature at which 50% of the signal is magnetic. This method has been used to explore if samples with compositions near a phase transition (for example, $x = 6.4$ in $YBa_2Cu_3O_x$) are multiphased or exhibit true coexistence of superconductivity and magnetism.[22]

4 Phase Diagrams Using μ^+SR

By combining the properties measured by μ^+SR it is easy to compile phase diagrams of HT$_c$SCs. For instance, Fig. 5 (from Ref. [23]) shows the transition temperatures T_c and T_N on phase diagrams for the hole doped system $La_{2-x}Sr_xCuO_4$ and the electron doped system $Nd_{2-x}Ce_xCuO_4$. Since both systems have approximately the same optimal doping concentration for superconductivity ($x = 0.15$), it is tempting to suppose that they might be "mirror" images of each other. However, the AFM portions of the phase diagrams are strikingly different. In $Nd_{2-x}Ce_xCuO_4$, the AFM phase takes up most of the phase diagram, and there is no sign of a spin glass phase. In $La_{2-x}Sr_xCuO_4$, the AFM phase is rapidly destroyed and replaced first by spin glass magnetism and then by superconductivity.

This can be understood by recognizing that the addition of a hole destroys an electron at an oxygen in the CuO_2 planes in the case of $La_{2-x}Sr_xCuO_4$. This creates

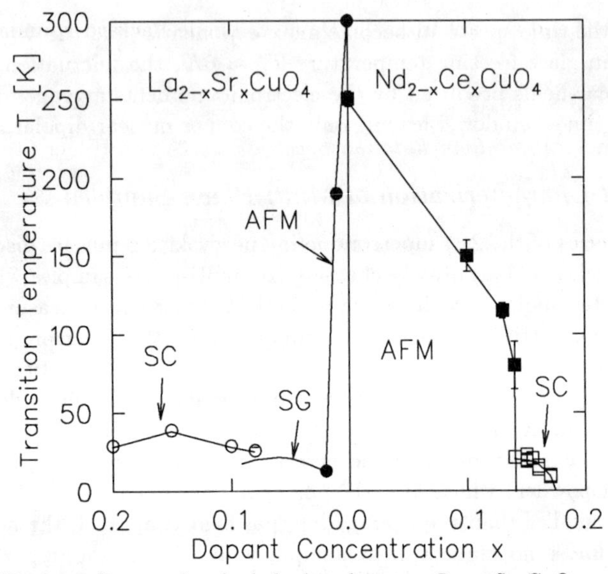

Figure 5: μ^+SR phase diagram for the hole doped system $La_{2-x}Sr_xCuO_4$ and the electron doped system $Nd_{2-x}Ce_xCuO_4$.

an unpaired spin on the oxygen which makes an effective ferromagnetic coupling between the copper atoms on either side of the oxygen, thus frustrating the long range AFM ordering. The spin glass (SG) magnetism observed is frustrated AFM ordering. In the case of $Nd_{2-x}Ce_xCuO_4$, the addition of an electron to a copper in the CuO_2 planes gives the copper a net spin of zero. This is a dilution of the magnetism, which is not as effective as frustration in the destruction of long range AFM order. Consequently, no spin glass phase is formed in this system.

5 Conclusion

This paper has illustrated some of ways in which μ^+SR can be used to characterize HT$_c$SCs and related compounds. The penetration depth can be measured in superconducting samples and the critical temperatures T_N and T_f can be easily measured in samples with AFM and spin glass magnetism. These measurements can be done simply with bulk polycrystalline samples, which makes μ^+SR an excellent tool for characterizing newly created superconducting compounds. With single crystals and oriented powders, additional information, such as the anisotropy of the penetration depth and the crystalline direction of the local fields in AFM samples, can be measured.

This work was supported by NSERC of Canada, (through TRIUMF) by the

Canadian NRC and a grant from the David & Lucile Packard Foundation.

References

[1] A. Schenck. *Muon Spin Rotation Spectroscopy.* Adam Hilger Ltd., 1985.

[2] W. Dawson et al. *Journal of Applied Physics*, **64**, 5809, (1988).

[3] S. L. Thiemann et al. *Physical Review*, **B39**, 11406, (1989).

[4] A. Kung. *Physical Review Letters*, **25**, 1006, (1970).

[5] D. Herlach et al. *to be published in Hyperfine Interactions*, (1990).

[6] T. M. Riseman et al. *Physica*, **C162-164**, 1555–1556, (1989).

[7] T. M. Riseman et al. *to be published in Hyperfine Interactions*, (1990).

[8] B. Pümpin et al. *Physica*, **C162-164**, 151, (1989).

[9] M. Celio et al. *Physica*, **C153-155**, 753, (1988).

[10] E. H. Brandt. *Journal of Low Temperature Physics*, **75**, 355, (1988).

[11] E. H. Brandt. *Physical Review*, **B37**, 2349, (1988).

[12] W. Barford and J. M. F. Gunn. *Physica*, **C153-155**, 691, (1988).

[13] Y. J. Uemura et al. *Physical Review*, **B38**, 909, (1988).

[14] Y. J. Uemura et al. *Physical Review Letters*, **62**, 2317, (1989).

[15] Y. J. Uemura et al. In *Proceedings of the International Conference on Organic Superconductors. Lake Tahoe.* Plenum Press, May 1990.

[16] V. J. Emery and G. Reiter. *Physical Review*, **B38**, 4547, (1988).

[17] J. H. Brewer et al. *to be published in Hyperfine Interactions*, (1990).

[18] J. H. Brewer et al. *Physica*, **C162-164**, 157, (1989).

[19] R. F. Kiefl et al. *Physica*, **C162-164**, 161, (1989).

[20] B. J. Sternlieb et al. *Physical Review*, **B41**, 8866, (1990).

[21] J. H. Brewer et al. *Physical Review Letters*, **60**, 1073, (1988).

[22] J. H. Brewer et al. *Physica*, **C162-164**, 33, (1989).

[23] G. M. Luke et al. *to be published in Physical Review*, **B42**, (1990).

The Effects of Oxygen Doping on the Electronic Properties and Microstructure of $Bi_2Sr_2CaCu_2O_x$ Superconductors Determined by Scanning Tunneling Microscopy

Z. Zhang, Y. L. Wang, X. L. Wu, J. L. Huang, and C. M. Lieber*
Department of Chemistry, Columbia University,
New York, New York 10027 USA

*To whom correspondence should be addressed

Abstract

The effects of oxygen doping on the structural, electronic, and superconducting properties of single crystal $Bi_2Sr_2CaCu_2O_x$ (Bi-2212) superconductors have been characterized by scanning tunneling microscopy (STM), scanning tunneling spectroscopy (STS), and magnetization and resistivity measurements. Removal of oxygen from superconducting single crystals by vacuum annealing systematically reduces the superconducting transition temperature (T_c) until the samples eventually become semiconducting. STS studies of the oxygen deficient materials demonstrates that the density of electronic states are reduced about the Fermi level. STM structural studies also demonstrate that the one-dimensional superstructure and Bi-O layer atomic structure are essentially the same for the superconducting and oxygen deficient nonsuperconducting crystals. These results indicate that vacuum annealing does not remove oxygen from the Bi-O layer as previously suggested, but rather from the Cu-O or Sr-O layers. Additional bias-voltage dependent STM studies show new electronic features near the Fermi level that reflect local changes in the electronic structure due to oxygen loss. It is suggested that oxygen loss from axial or equatorial copper positions causes carrier localization in the Cu-O layer and drives the material to a semiconducting state.

INTRODUCTION

The relationship between the superconducting transition (T_c) and hole concentration in the copper oxide high-temperature superconductors has been widely studied through investigations of oxygen and metal doped materials.[1-5] For example, early studies of $La_{2-x}Sr_xCuO_{4-y}$ materials showed that variations in the Sr and O concentrations could be associated with systematic changes in T_c.[1] The case of oxygen doping is especially important since it serves as a common method to vary the hole concentration and T_c in all of the high-T_c oxide materials. It is clear, however, from recent studies of oxygen deficient $YBa_2Cu_3O_{7-y}$ that subtle adjustments in the oxygen positions can lead to variations in T_c even when the average hole concentration is apparently constant.[6] These studies highlight the importance that variations in local microstructure (and presumably electronic structure) play in determining T_c, although in most instances the microscopic nature of the structural and electronic changes that occur with oxygen doping have not been elucidated. Characterization of the local microstructure including oxygen atomic positions by diffraction techniques is inherently difficult due to crystal disorder and the small scattering cross section of oxygen, although neutron diffraction[6] and X-ray absorption spectroscopy[7] have begun to provide data that address these structural issues. Local variations in the electronic properties due to oxygen doping have yet to be addressed. Herein, we report new results obtained using the real-space techniques scanning tunneling microscopy (STM) and scanning tunneling spectroscopy (STS) that address for the first time the local structural and electronic effects of oxygen doping in $Bi_2Sr_2CaCu_2O_x$ (Bi-2212) single crystals. We find in oxygen deficient nonsuperconducting materials that the one-dimensional superstructure and atomic structure of the Bi-O layer are essentially the same as in superconducting (T_c = 85 K) crystals. These data thus suggest that oxygen loss (doping) does not occur from the Bi-O layer, but rather from the Sr-O/Cu-O layers. STS and bias voltage dependent imaging data provide further insight into the electronic origin of oxygen doping in this system. Lastly, possible microscopic origins for the observed changes in T_c will be discussed.

EXPERIMENTAL

Single crystals of Bi-2212 were grown from melts rich in CuO as described elsewhere.[8,9] Oxygen was removed from the superconducting crystals ($T_c \approx$ 85 K) by annealing in vacuum (P = 10^{-2} torr) at 400 °C. This annealing procedure yields reproducible and reversible decreases in T_c. The single crystal samples were characterized by dc resistivity measurements and magnetic measurements. The magnetic measurements were made using a commercial SQUID based magnetometer (MPMS, Quantum Design).

All of the STM and STS studies were carried commercial and modified instruments (Nanoscope, Digital Instruments, Inc.) in an argon filled glove box equipped with a purification system that reduced the concentrations of water and oxygen to below 1 ppm; mechanically sharpened platinum-iridium tips (80%-20%) were used in all of the experiments. The annealed samples were stored in vacuum and then transferred to glove-box. After mounting the crystals in the glove box they were cleaved and then examined by STM and STS. Magnetic measurements made on several samples subsequent to the imaging and spectroscopy studies demonstrated that T_c and superconducting fraction was unchanged on storage in the glove box over the lifetime of an experiment. Additional experimental details have been described previously.[8,10]

RESULTS and DISCUSSION

The field-cooled magnetization curves for as-grown Bi-2212 and annealed samples are shown in Figure-1.

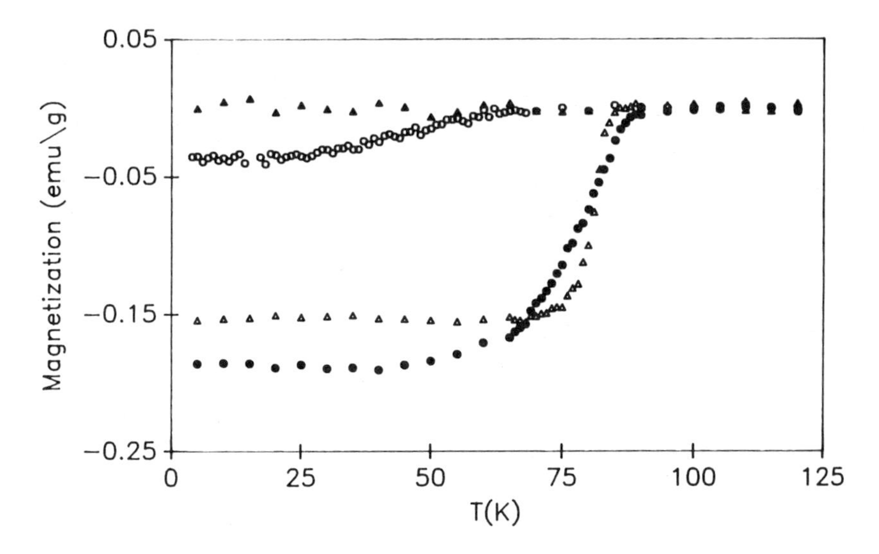

Figure 1. Magnetization vs. temperature data for T_c=85 K (•), T_c≈50 K (∘), and nonsuperconducting (▴) samples recorded using a SQUID magnetometer (MPMS, Quantum Design); the samples were cooled in a field of 5 oersteds from 298 K. The T_c≈50 K sample was prepared by annealing at 400 °C in vacuum for 2.5 h and the nonsuperconducting sample was prepared by annealing in vacuum for 30 h. Reannealing the nonsuperconducting sample at 400 °C in air for 9 h restores the superconducting properties as shown (△).

We find that T_c is reduced to ≈50 K and the fraction of superconducting material significantly decreases after only 2.5 h of vacuum annealing at 400 °C. Additional annealing in vacuum ultimately yields samples with no diamagnetic signal between 300 and 4.2 K (Fig. 1). This observation suggests that these oxygen deficient crystals produced by vacuum annealing for 30 h are nonsuperconducting. Resistivity measurements confirmed that these latter samples are indeed semiconducting with an energy gap of ≈0.15 eV (Fig. 2).

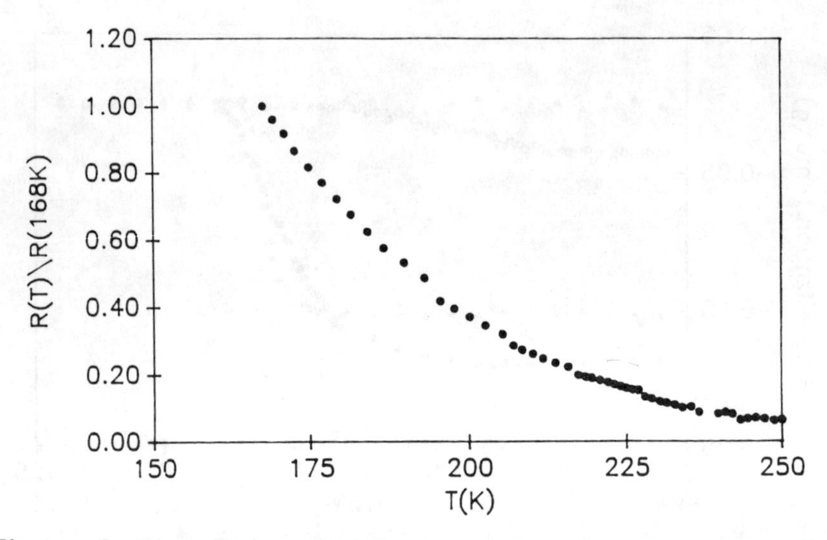

Figure 2. Plot of the normalized resistance versus temperature recorded on a Bi-2212 crystal vacuum annealed for 30 h in vacuum. The material is semiconducting with a energy gap of 0.13 eV.

The observed transformation of the Bi-2212 materials from a metal/superconductor to a semiconductor by vacuum annealing is also completely reversible. Annealing nonsuperconducting crystals in air at 400 °C for ≥8 h yields T_c's and superconducting fractions virtually the same as the starting crystals (Fig. 1). From these magnetization and resistivity measurements we can thus conclude that under mild conditions sufficient oxygen can be removed (reversibly) from Bi-2212 single crystals to drive the material from a metal/superconductor to a semiconductor. These measurements do not, however, elucidate how the local electronic structure changes with doping nor do they show from which layer (i.e., Bi-O, Sr-O, or Cu-O) oxygen is lost.

To characterize changes in electronic structure associated with doping we have made STS measurements. The current (I) versus voltage (V) data in these experiments were acquired by interupting the STM feedback loop, and then ramping the sample-tip bias voltage while digitally storing the resulting changes in tunneling current. The I-V data shown are the average of 20-40 curves obtained at a selected surface site, although these measurements likely represent an average over several atomic positions due to drifts during data acquisition. Representative data obtained on cleaved superconducting and oxygen deficient semiconducting samples of Bi-2212 are plotted as the normalized conductivity (V/I)dI/dV in Figure 3 since it has been well established that (V/I)dI/dV is proportional to the local density of electronic states (DOS) at the sample surface.[11]

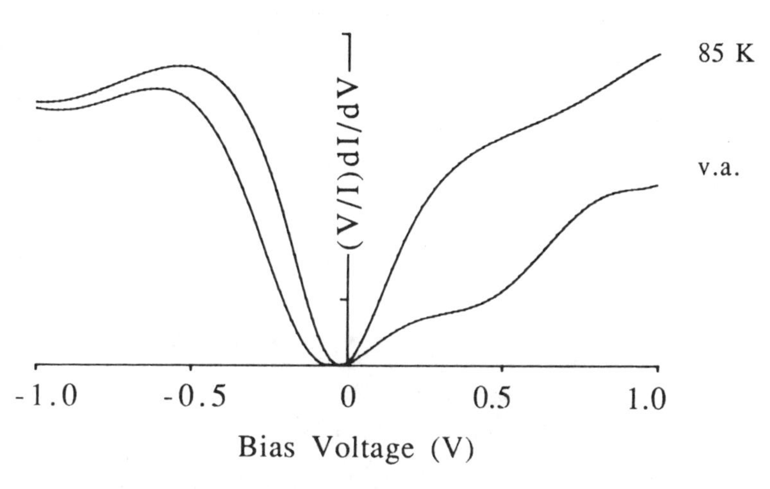

Figure 3. Plots of the normalized conductivity versus bias voltage for a superconducting (T_c=85 K) and a vacuum annealed (v.a.) nonsuperconducting samples. The bias voltage corresponds to the energy relative to the Fermi level (V = 0).

The Bi-2212 crystals cleave preferentially along the weakly interacting Bi-O/Bi-O double layers to yield Bi-O surfaces that have a structure similar to the bulk.[8] The STS and STM experiments thus probe directly the Bi-O layer. Comparison of the curves in

Figure 3 shows that the oxygen deficient material has a significantly lower DOS near the Fermi level than the superconducting material. Since STS is most sensitive to the surface DOS one interpretation of these results is that the Bi-O layer has become more semiconducting in the vacuum annealed samples due to oxygen loss from the Bi-O layer. We have recently shown, however, in Bi-2212 superconducting materials that STS may be sensitive to the metallic Cu-O layer.[10] These latter results suggest an alternative explanation; that is, the reduced DOS in Figure 3 could reflect a modification of the Cu-O layer electronic structure near the Fermi level due to loss of axial and/or equatorial oxygen.

To determine from which layer the oxygen is removed in these vacuum annealed samples we have carried out extensive STM imaging experiments. A series of grey-scale images of the superconducting (T_c = 85 K) and nonsuperconducting samples are shown in Figure 4. These images are typical of the data obtained on a number of samples and thus we believe that they reflect the intrinsic properties of the Bi-2212 crystals.

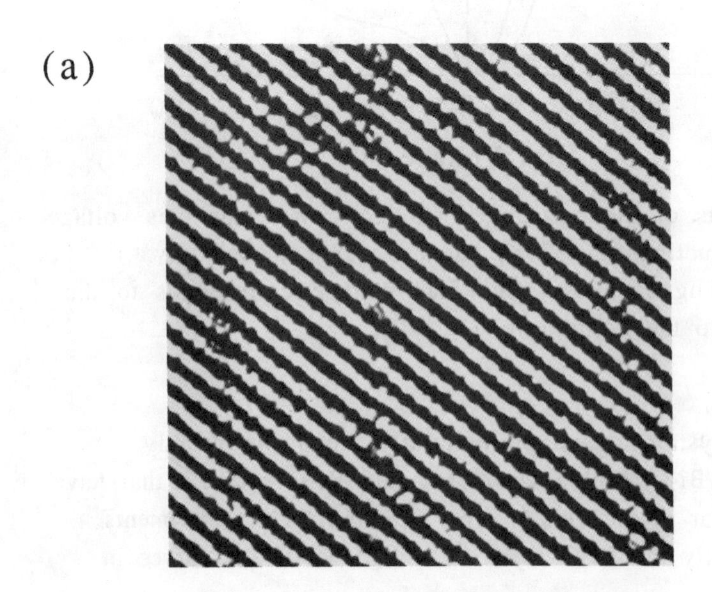

(b)

(c)

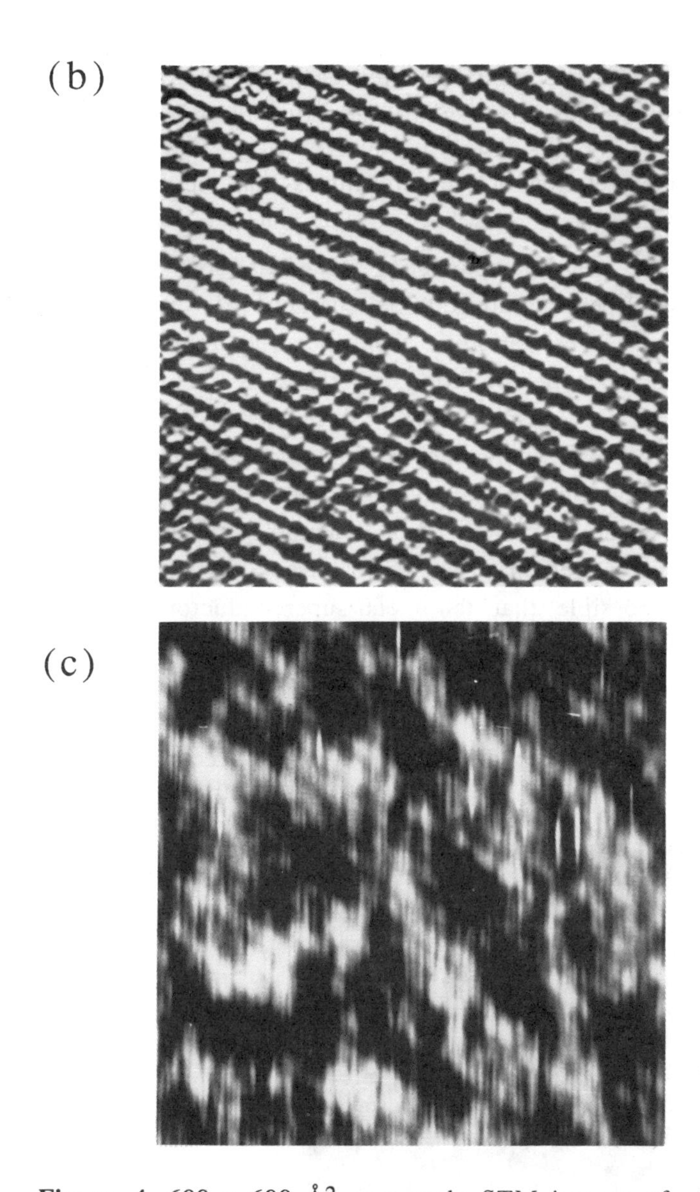

Figure 4. 600 x 600 $Å^2$ grey-scale STM images of (a) T_c=85 K and (b,c) nonsuperconducting samples recorded with bias voltages/tunneling currents of 500/1.2, 980/1.0, and 500 mV/1.0 nA, respectively. The superstructure modulation occurs along the a-axis and is seen in the images as alternating light and dark lines

in (a) and (b). Irregular electronic features detected at low bias voltage in the nonsuperconducting sample are also observed in (c).

The images of the Bi-O layer of the 85 K samples and the high bias voltage images of the nonsuperconducting materials both exhibit the one-dimensional superstructure that is characteristic of the Bi-2212 materials.[8,12-14] The period of the superstructure in both superconducting and nonsuperconduting materials is virtually the same: 25 ± 1 and 26 ± 1 Å, respectively. Since there is considerable evidence[8,13,14] indicating that extra oxygen in the Bi-O layer causes the one-dimensional superstructure the absence of change in the superstucture period of the vacuum annealed samples suggests that oxygen may not be lost from the Bi-O layer. Alternatively, it is possible that the metal/superconductor to semiconductor transformation in the Bi-2212 materials is due to a sufficiently small loss of oxygen such that no structural changes are observed.

To resolve further these possibilities we have recorded atomic resolution images of the Bi-O layer (Fig. 5).

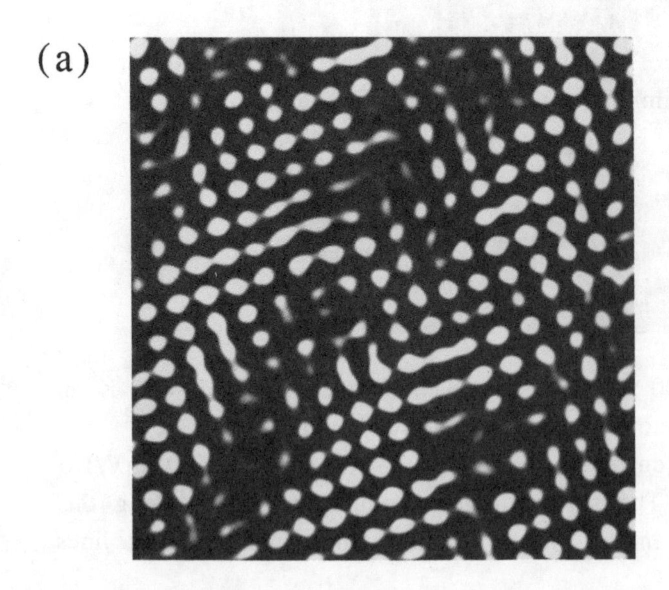

(b)

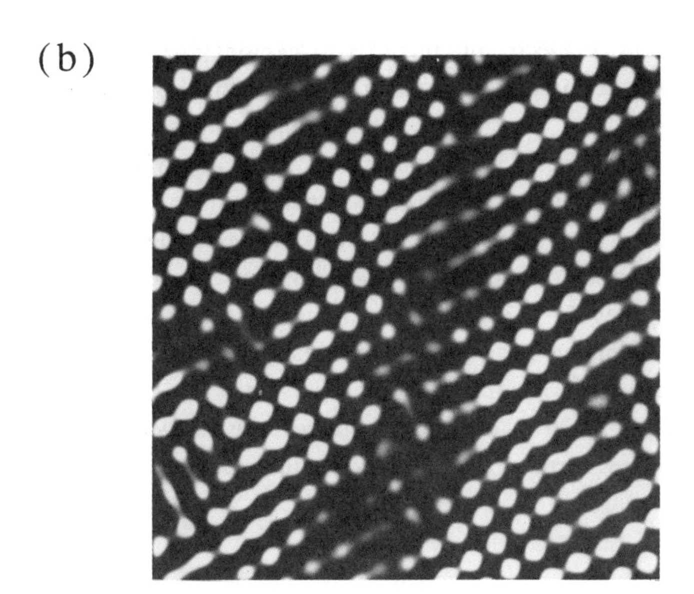

Figure 5. 60 x 60 Å^2 grey scale images of (a) superconducting and (b) nonsuperconducting crystals recorded with bias voltages/tunneling currents of 300/0.9 and 362 mV/2 nA, respectively. The atomic lattice in both images is tetragonal with an average spacing of 3.8 Å between the oxygen sites.

The STM images of the superconducting and nonsuperconducting material exhibit similar tetragonal atomic structure as reported previously.[8,15,16] Notably, we have been unable to detect obvious missing atomic sites in the vacuum annealed samples. It is possible, however, that if the oxygen concentration changed by less than ≈ 1% that the missing oxygen atomic sites would not have been detected in our experiments. Nevertheless, these data do suggest that the metal/superconductor to semiconductor transformation is not caused by significant oxygen loss from the Bi-O layer but rather oxygen loss from either the Sr-O and/or Cu-O layers.

Since the atomic resolution images indicate that oxygen is removed from the Sr-O or Cu-O layers we cannot characterize the changes in oxygen microstructure by imaging the Bi-O layer. Our bias-voltage dependent imaging of this layer demonstrate, however, that we are sensitive to the spatial variations in the electronic states (Fig. 4c). Images of the nonsuperconducting samples exhibit new nonperiodic features between approximately -300 and +600 mV in addition to the one-dimensional superstructure modulation. As discussed in detail elsewhere[17] these new features can be attributed to variations in the electronic states near the Fermi level. Assuming that oxygen is lost from the axial (or equatorial) copper positions, then these large modulations in the DOS may reflect the localization of carriers in a "semiconducting" Cu-O layer. Additional studies are needed, however, to investigate further the detailed nature of these interesting results.

CONCLUSIONS

In summary, we have used STM and STS to characterize the microscopic effects of oxygen doping in Bi-2212 single crystals. We have shown that superconducting and vacuum annealed nonsuperconducting samples exhibit the same superstructure and atomic structure in the Bi-O layer. These data suggest that vacuum annealing does not remove oxygen from Bi-O layers in this system. STS and bias dependent STM measurements show, however, that these techniques are very sensitive to variations in the DOS that result from oxygen doping. Hence, STM can provide heretofore unavailable data addressing the spatial variations in electronic structure caused by oxygen doping. Analysis of these latter results suggests that the metal/superconductor to semiconductor transformation in the vacuum annealed samples is due to localization of the carriers in the Cu-O planes following loss of axial or equatorial oxygen.

REFERENCES

1. Tarascon, J. M., Greene, L. H., McKinnon, W. R., Hull, G. W. and Geballe, T. H., Science 235, 1373 (1987).

2. Tokura, Y., Torrance, J. B., Huang, T. C. and Nazzal, A. I., Phys. Rev. B 38, 7156 (1988).
3. Shafer, M. W., Penney, T., Olson, B. L., Greene, R. L. and Koch, R. H., Phys. Rev. B 39, 2914 (1989).
4. Suzuki, M., Phys. Rev. B 39, 2312 (1989).
5. Wang, Z. Z., et al., Phys. Rev. B 36, 7222 (1987).
6. Jorgensen, J. D. et al., Physica C 167, 571 (1990).
7. Conradson, S. D., Raistrick, L. D. and Bishop, A. R., Science 248, 1394 (1990).
8. Wu, X. L., Zhang, Z., Wang, Y. L. and Lieber, C. M., Science 248, 1211 (1990).
9. Wang, Y. L., Wu, X. L., Chen, C. C. and Lieber, C. M., Proc. Natl. Acad. Sci. 87, 7058 (1990).
10. Zhang, Z., Wang, Y. L., Wu, X. L., Huang, J.-L. and Lieber, C. M., Phys. Rev. B 42, 1082 (1990).
11. Feenstra, R. M., Stroscio, J. A. and Fein, A. P., Surf. Sci. 181, 295 (1987).
12. Bordet, P. et al., Studies High Temp. Supercond. 2, 171 (1989).
13. Zandbergen, H. W., Groen, W. A., Mijlhoff, F. C., van Tendeloo, G. and Amelinckx, S., Physica C 156, 325 (1988).
14. Le Page, Y., McKinnon, W. R., Tarascon, J. M. and Barboux, P., Phys. Rev. B 40, 6810 (1989).
15. Kirk, M. D. et al., Science 242, 1673 (1988).
16. Shih, C. K., Feenstra, R. M., Kirtley, J. R. and Chandrashekhar, G. V., Phys. Rev. B 40, 2682 (1989).
17. Wu, X. L., Wang, Y. L., Zhang, Z. and Lieber, C. M., submitted for publication.
18. This work was supported by the David and Lucile Packard and National Science Foundations.

CHARACTERIZATION OF THE HYSTERETIC MAGNETORESISTANCE BEHAVIOUR OF YBCO AND BPSCCO AND THE MEASUREMENT OF FLUX PINNING ENERGIES

D.N. Matthews, G.J. Russell, K.N.R. Taylor, A. Donohoo, S.X. Dou* and K.H. Liu*

Advanced Electronic Materials Group
School of Physics
The University of New South Wales
P.O. Box 1, Kensington NSW 2033 Australia

*Dept. of Materials Science and Engineering
The University of New South Wales
P.O. Box 1, Kensington NSW 2033 Australia

ABSTRACT

The observation of hysteresis-like behaviour of the magnetoresistance has been reported by a number of workers. Our recent investigations have been directed to fully characterise this phenomenon for the first time in order to develop a more quantitative model to describe the observations. Observation of the time dependence of the magnetoresistance, which occurs as a result of flux motion in the system, has been used to determine flux pinning energies in both yttrium and bismuth based materials. The behaviour of the two compounds appears to be quite different.

INTRODUCTION

One of the major problems facing the development of the high critical temperature superconductors is the ease with which magnetic flux can enter and subsequently move through the crystalline or granular materials. This occurs at two levels, first at low magnetic field strengths the interaction of the magnetic field with the poorly coupled grain boundary system allows Josephson vortices to permeate sintered materials resulting in the well known non-linear behaviour [1] which is readily accounted for by critical field methods[2]. At slightly higher fields (typically ≅ 100 Oe) the magnetic field enters the crystalline grains as Abrikosov vortices which may be trapped within the lattice. Because of the high anisotropy of these cuprate materials, and the extremely short coherence lengths the nature and behaviour of these vortices is unlike those of the classical low temperature superconductors, as has been discussed by Clem[3].

Since the early work on these materials, which showed large time dependent effects of the magnetization, and a high sensitivity of the resistive transition to the presence of external magnetic fields, numerous workers have investigated the energetics of the vortex system in an effort to gain some insight into the mechanics of the flux motion and of the nature of the weak pinning sites in these solids.

Measurements of the time dependent magnetization[4)] and of the magnetic field dependent resistive transition[5)] have provided values for the flux pinning energies and it is suggested that the observed behaviour is best described by a range of pinning energies which extends to approximately 1eV.

In granular materials the trapped flux leads to hysteretic magnetoresistance behaviour in the mixed state in which the voltage-magnetic field strength characteristics are non-reversible. This behaviour has been reported by a number of workers[6-9)] and may be interpreted in terms of either energy dissipation within the sample or field dependent critical currents, the latter probably being a secondary effect rather than the primary mechanism. The hysteresis like V-H loops (see Fig.1) which occur in this phenomenon are characterized by a number of features, most of which are strongly dependent upon both the transport current used in the measurements and the maximum magnetic field strength used in the field cycle. Despite the increasing number observations of this phenomenon, no attempt seems to have been made to define the overall behaviour. Intrinsic to this behaviour is a time dependent change in the hysteresis loop, which is related to trapped flux and hence may be used to determine the flux pinning energies in these materials. In this paper we have attempted to establish the factors controlling this behaviour and have shown that the time dependence may be used to determine pinning energies.

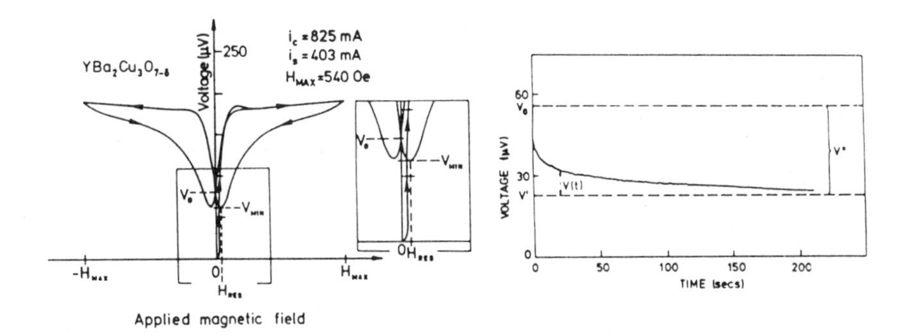

Figure 1: (a) A typical hysteretic V-H characteristic for YBCO obtained at constant sample current for the applied field swept to a value H_{max}. The insert shows the region of the characteristic about $H_{appl} = 0$.

(b) The time decay curve of the residual voltage V_o after a field cycle. Note the definitions of V_o, V' and V''.

EXPERIMENTAL

The superconducting $YBa_2Cu_3O_{7-\delta}$ and $Bi_{1.8}Pb_{0.4}Sr_2Ca_{2.2}Cu_3O_x$ bulk samples used in this study were prepared by our conventional sintering routes[7,10]. Each sample of YBCO ($T_c = 91K$) and BPSCCO ($T_c = 107K$) had silver paste contacts attached as for the conventional 4-point probe technique. The samples were cooled directly in liquid nitrogen and the applied d.c. magnetic field, provided by a Helmholtz coil system, was oriented parallel to the direction of the current flowing through the sample during the magnetoresistance measurements. Measurements with the field perpendicular to the current are essentially the same provided that demagnetizing effects are allowed for.

Figure 1 shows a typical YBCO 'butterfly' characteristic obtained when the sample current, i_s, and H_{max} exceed $i_c(H)$ for the sample. Associated with this characteristic is the initial, virgin, up curve and the symmetrical right and left branches for the complete field cycle, an almost constant output voltage after ~100 Oe when the magnetic field enters the grains, a minimum sample voltage, V_{min}, as the external field strength is decreased from its maximum value and a residual voltage, V_o, at zero applied field strength. This residual voltage depends upon H_{max} and i_s respectively, and this variation, given for E_o (to remove dimensional effects) is shown in Figure 2. Further, the voltage, V_o (i_s, H_{max}) is readily observed to be time dependent, decaying rapidly initially to a new value and subsequently very slowly. To observe these time decays of V_o, the output from the potential contacts on the sample was taken directly to an x-t recorder and observations started when the applied field strength became zero. A typical time dependent variation of V_o, for a YBCO sample, is shown in Figure 1, where the variables V_o, V' and V" are defined and in which the observed is interpreted V_o in terms of two components; V" associated with the gradual decay and V' (= V_o- V"). Because of the difficulty in obtaining a precise value of V' due to the continuous decay over the time window studied, we have used the arbitrary definition V' ≡ V_o at = 200 s. Note the initial, fast decay below 0.5 second (the limiting accuracy of t=0 for these measurements) which is followed by the gradual decay region extending to a third, quasi-constant voltage value. Similar effects have been observed in time dependent magnetization studies[4].

RESULTS AND DISCUSSION

To remove dimensional effects on the data, voltages are represented in terms of the electric field strengths E and currents in terms of current densities. Figure 2 gives, for a typical sample, the sample current density, j_s, dependence of E_o and E" for a number of H_{max} values as well as the variation of E with H_{max} at fixed j_s values. It is clear that both E_o and E" vanish at a common current density j_o (fixed H_{max} value) below which the sample remains fully superconducting.

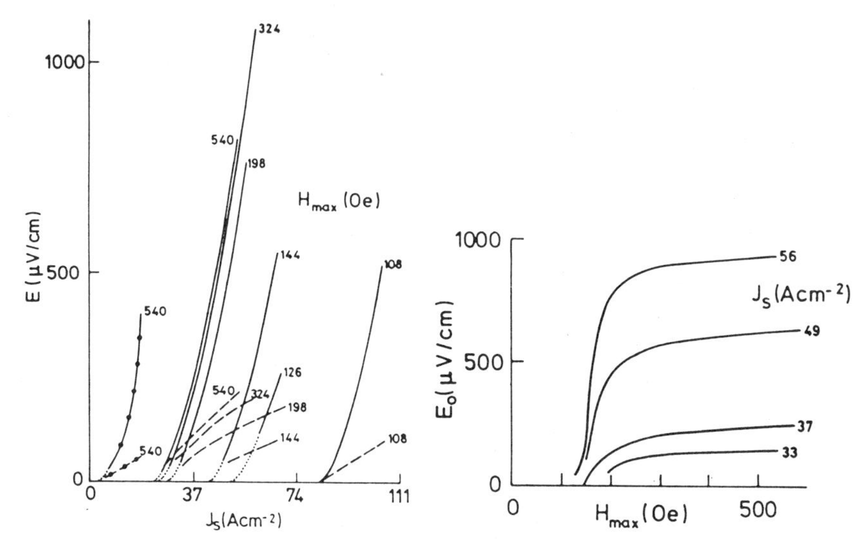

Figure 2: (a) Variation of E_o(——) and E" (- - -) with sample current density for different H_{max} values and 85% dense material. For 65% dense material the variations are shown by (-•-•-) and (-•-•-). The value j_o is defined as the point of intersection of the extrapolated (dotted) curves and the j_s axis.

(b) Variation of E_o with H_{max} for different sample current densities.

The variation of the different electric field strengths with both current density and applied magnetic strength have been established at liquid nitrogen temperatures for a number of samples, and the results are presented in Figure 3 as functions of a reduced excess current $(j-j_o)/j_c$, where j_c is the critical current density in zero field. As may be seen, these changes can all be represented by power law variations with the relationships shown on each diagram. With the exception of E" these relations appear to be independent of the magnetic field strength, however the established power laws are found to depend on the sample quality.

On the assumption that the time dependence of E_o is related to the flux creep associated with time dependent magnetization effects[11)], and since there appears to be a major difference between the decays of E" and E', then $E(t) \equiv E''(t)$ is plotted as a log t

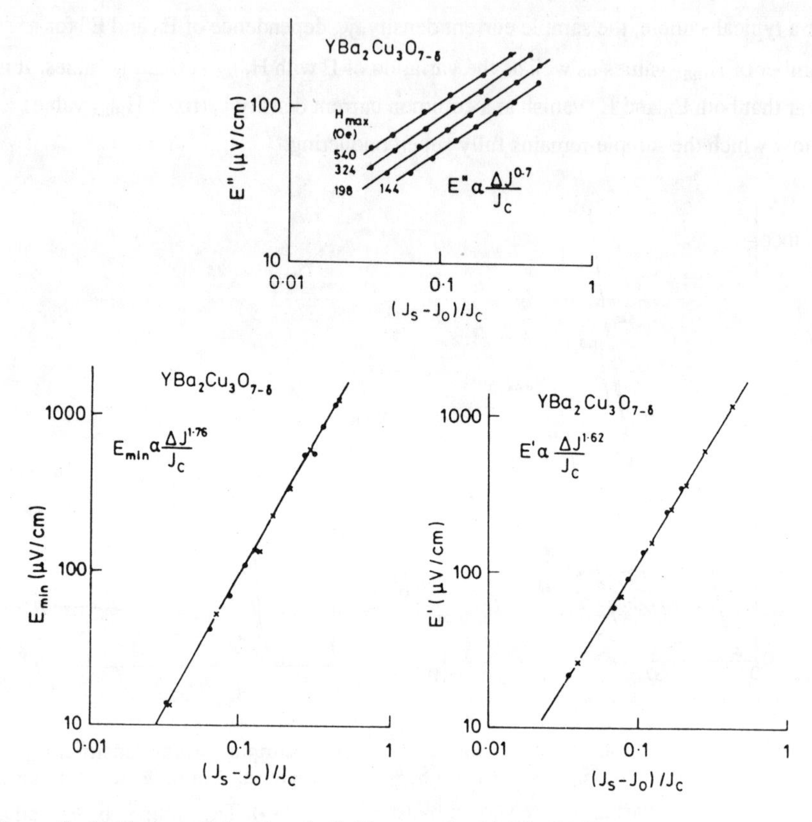

Figure 3: Variation of the electric field strengths (a) E"; (b) E_{min}; (c) E' as functions of the reduced excess current $(j_s-j_o)/j_c$ for the constant H_{max} values shown in (a). The curves for both E_{min} and E' are independent of the magnetic field strength. Each electric field variation can be represented by the power low relationship shown on the diagram.

variation in Figure 4. As may be seen the E(t) variation follows a logarithmic decay curve over the entire time range studied and we believe that this further supports the view that this contribution may be associated with a discrete mechanism.

Figure 5 shows typical hysteretic V-H characteristics for BPSCCO samples that have significantly different j_c values. For low j_c values V_{min} occurs at H~O while for j_c's ~ 100 a cm^{-2} V_{min} occurs at H≠0. In both cases the time dependent residual voltage V_o is found to be similar to the YBCO decay curves and E(t) follows a logarithmic decay for the time range studied.

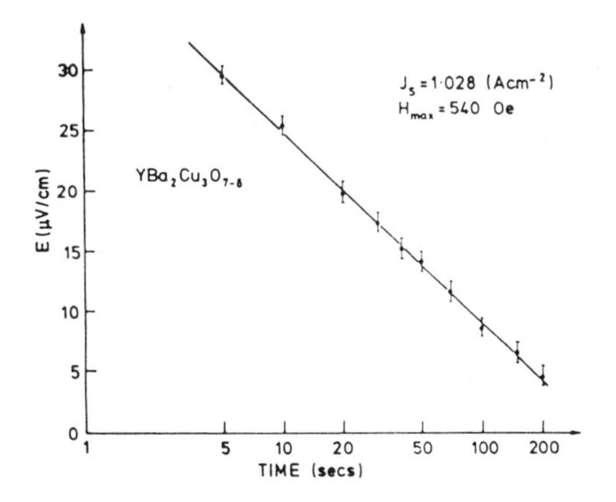

Figure 4: The electric field strength E(t) ≡ E"(t) plotted as a log t variation.

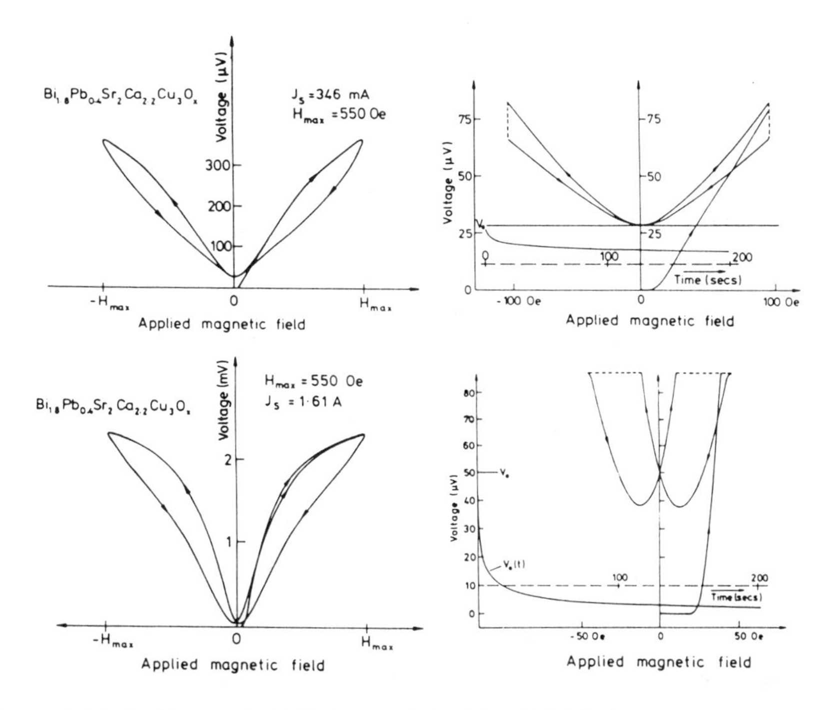

Figure 5: Typical hysteretic V-H characteristics for BPSCCO samples that have significantly different j_c values.

(a) For low j_c values V_{min} occurs at $H_{appl} \sim 0$ as shown by the expanded characteristic, which also shows the form of the V_o decay over a time of 200 s.

(b) For higher j_c values (≥ 100 A cm^{-2}) V_{min} occurs at $H_{appl} \neq 0$.

Following the time dependent magnetization studies of Hagen and Griessen[11], it is assumed that the logarithmic decay of E(t) may be represented by an equation of the form

$$E(t) = E''[1 - A \ln(t/t_o)]$$

where the constant A is given by kT/U, for the time variation of trapped flux. Here U is the activation energy associated with flux trapping and $1/t_o$is an attempt frequency. Following the assumptions of Griessen et al[4] that t_o must lie in the range $10^{-6} < t_o < 10^{-12}$ s and without significantly affecting the derived values of U we will adopt a value of $t_o = 10^{-9}$ s.

Figure 6 shows the excellent fit of E(t) to a log (t/t_o) variation for both YBCO and BPSCCO samples. The values of U, determined from such graphs, as a function of the sample current density j_s relative to the critical current density j_c, are given in Table 1.

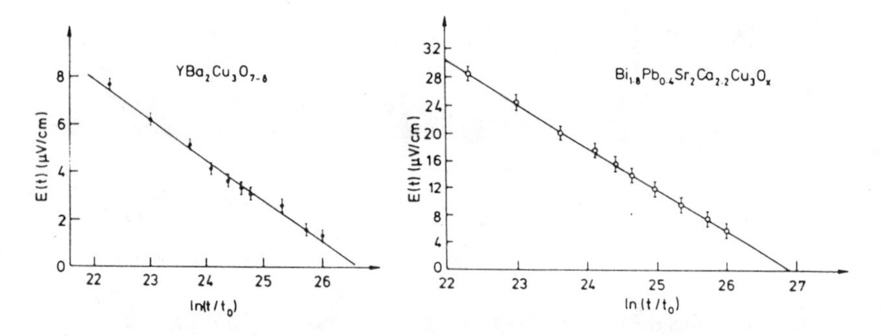

Figure 6: Typical logarithmic time decay curves for YBCO and BPSCCO samples. Note: $t_o = 10^{-9}$ s has been used to define the x-axis scale.

The pinning energies determined by this method have a value of ~ 50 meV for both materials. The magnitude of the pinning energy appears to show some dependence on sample current for large j_c values.

The range of pinning energies found in this work are comparable to the range covered by the observations of other workers. In particular, for thin films of YBCO with H⊥c-axis and H||c-axis the pinning energies range from 40 to 120 meV for magnetic fields of 40 kOe[12], while single YBCO crystals with H||c-axis have a pinning energy of ~80 meV at a field of 50 kOe[13]. The article by Deutscher[14] also places derived pinning energies in the range 60 to 100 meV for YBCO and 10 to 40 meV for bismuth cuprates.

TABLE 1: Pinning Energies

$YBa_2Cu_3O_{7-\delta}$

90% dense J_c = 219.5 A/cm²

Measuring Current j_s/j_c	U (meV)
0.525	84.69
0.474	88.07
0.436	89.1
0.409	89.58
0.383	133.42
0.364	162.7

$YBa_2Cu_3O_{7-\delta}$

85% dense J_c = 47.1 A/cm²

Measuring Current j_s/j_c	U (meV)
0.37	73.5
0.41	64.3
0.33	65.7
0.25	72.86

$YBa_2Cu_3O_{7-\delta}$

65% dense J_c = 16.59 A/cm²

Measuring Current j_s/j_c	U (meV)
0.199	54.65
0.242	61.82
0.465	64.38
0.417	65.19
0.376	53.9

$Bi_{1.8}Pb_{0.4}Sr_2Ca_{2.2}Cu_3O_x$

J_c = 12.7 A/cm²

Measuring Current j_s/j_c	U (meV)
0.91	49.42
0.75	49.0
0.547	74.0
0.438	48.64
0.629	39.29
1.1	55.55

Since the voltage V(t) observed in this work is associated with the intergranular magnetic field component of flux trapped within the grains, the energies derived in this work relate to the intragranular flux pinning energies of YBCO and BPSCCO. Thus, it is expected that the values of U will show little variation from sample to sample, as was found by Matthews et al[15)] for YBCO material, since in all cases we are observing properties of the grains themselves rather than the weak-link intergranular system.

CONCLUSION

The observation of time dependent voltages associated with trapped magnetic flux in ceramic samples of YBCO and BPSCCO has allowed estimates of the flux pinning energies for the grains of these materials to be obtained. The values derived for U are consistent with those determined by other workers for these materials. While the observations are carried out in the presence of a transport current through the sample, the derived values of U show little, if any change, with current density, even though under appropriate conditions should show a variation of U with current due to the effect of the Lorentz force. We believe that this method of determining pinning energies offers a significant advantage over the more conventional techniques in that it employs the simpler experimental method of magnetoresistance measurement.

REFERENCES

1. Nieuwenhuys, G.J., Friedman, T.A., Price, J.P., Gehring, P.M., Salamon, M.B. and Ginsberg, D.M., "Low Field Hysteresis and Loss in Sintered Samples of $YBa_2Cu_3O_{7-\delta}$", Solid State Common. 67, 1253 (1988).
2. Chaddah, P., Ravi Kumar, G., Grover, A.K., Radhakrishnamurty, C. and Subba Rao, G.V., "Critical State Model and the Magnetic Behaviour of High T_c Superconductors", Cryogenics 29, 907, (1989).
3. Clem, J.R., "Theory of AC Losses in Type-II Superconductors with a Field-Dependent Surface Barrier", J. Appl. Phys. 50, 3518, (1979).
4. Griessen, R., Lensink, J.G., Schröder, T.A.M. and Dam, B., "Flux Creep and Critical Currents in Epitaxial High-T_c Films", presented at Critical Currents in High T_c Superconductors Conference, Karlsruhe 24-25 October, (1989).
5. Yeshurun, Y., Malozemoff, A.P., Wolfus, Y., Yacoby, E.R., Felner, I. and Tsuei, C.C., "Flux Creep and Related Phenomena in High Temperature Superconductors", Physica C 162-164, 1148, (1989).
6. Nojima, H., Tsuchimoto, S. and Kataoka, S., "Galvanomagnetic Effect of an Y-Ba-Cu-O Ceramic Superconductor and its Application to Magnetic Sensors", Jap. J. Appl. Phys. 27, 746, (1988).
7. Russell, G.J., Matthews, D.N., Taylor, K.N.R. and Perczuk, B., "Intergranular Flux Trapping Effects in Yttrium Barium Cuprate Superconductors", Mod. Phys. Lett. B3, 437, (1989).
8. Evetts, J.E. and Glowacki, B.A., "Relation of Critical Current Irreversibility to Trapped Flux and Microstructure in Polycrystalline $YBa_2Cu_3O_7$", Cryogenics 28, 641, (1988).
9. Kwasnitza, K. and Widmer, Ch., "New Programmable Switching and Storage Effects in Superconducting Polycrystalline $YBa_2Cu_3O_7$ due to Hysteretic Critical Intergrain Transport Current", Cryogenics 29, 1035, (1989).
10. Dou, S.X., Liu, H.K, Bourdillon, A.J., Kviz, M., Tan, N.X. and Sorrell, C.C., "Stability of Superconducting Phases in Bi-Sr-Ca-Cu-O and the Role of Pb Doping", Phys. Rev. B40, 5266, (1989).
11. Hagen, C.W. and Griessen, R., "Thermally Activate Magnetic Relaxation in High T_c Superconductors", to be published in "Studies of High Temperature Superconductors", Ed. A.V. Narlikar, Nova Science Publishers.
12. Smith, G.B., Bell, J.M., Filipczuk, S.W. and Andrikidis, C.,"Temperature, Field and Grain Size Dependence of Flux Pinning in High T_c Superconductors",

preprint, submitted to Physica C (1990).
13. Enpuku, K., Yoshida, K., Takeo, M. and Yamafuji, D., "Evaluation of Temperature and Magnetic-Field Dependence of Pinning Potential of High T_c Superconductors from Flux-Creep Resistivity Experiments", Jap. J. Appl. Phys. 28, L2171, (1989).
14. Deutscher, G., "The Potential for Large Critical Currents in the High T_c Oxides", reprint, (1990).
15. Matthews, D.N., Russell, G.J. and Taylor, K.N.R., "Flux Trapping Energies in YBCO in the Presence of a Transport Current", submitted to Physica C, (1990).

5. FABRICATIONS

HIGH CURRENT CAPACITY THICK YBCO FILMS OPTIMIZED PROCESSING AND APPLICATION RELEVANT CHARACTERISTICS

A. Bailey, G.J. Russell, K.N.R. Taylor, D.N. Matthews, G. Alvarez and J. Ceremuga*

Advanced Electronic Materials ,
School of Physics, University of New South Wales,
P.O. Box 1, Kensington, NSW, 2033, Australia.

* Department of Electrical Engineering,
James Cook University of North Queensland,
Townsville, Q'land, 4811, Australia.

ABSTRACT

By optimizing the processing variables, it has been possible to obtain critical current densities of 3000 Acm^{-2} in 7-10 micron thick films of YBCO on YSZ. The fabrication route is discussed, and the effects of changing some of the variables considered.

Measurements of the flux pinning energies have been made using time dependent magnetoresistance observations. Microbridge weak links have been shown to exhibit strong chaotic effects which may have important consequences for the high frequency applications envisaged for these thick films. Finally, the 10GHz shielding performance shows a 19dB improvement over copper at 77K.

INTRODUCTION

Despite the present successful development of commercial, passive microwave devices using thin film high temperature superconductors[1)] the technology is essentially complex and at present is unable to fabricate single circuit boards larger than a few square centimetres. Traditionally, thick film processing has been used for the production of such systems as this method of fabrication is essentially simple, low cost and offers greater versatility in developing large areas of circuit.

Unfortunately, attempts to manufacture thick films of high temperature superconductors during the past three years have generally been unsuccessful[2-4)]. At the high processing temperatures necessary, there are excessive interactions between the superconducting cuprate and substrate materials leading to many problems including loss of adhesion, impurity phases and poisoning. The most attractive substrates from the microwave viewpoint (alumina, sapphire etc) are generally affected most. Where reasonable transition-temperatures (>90K) have been achieved by the use of multiple layers, the final thickness is generally too great for normal application and the layer is both low density and friable. Finally, even when otherwise acceptable films have been fabricated on alternative substrates, the granular nature of the sintered film leads to low critical currents with a strong dependence on external magnetic fields.

As with bulk ceramic materials, there is a clear need for minimizing the number of intergranular contacts, improving the grain connectivity and establishing a high degree of grain alignment (texturing). In bulk materials, it was first demonstrated by Jin et al[5]) that many of these features could be achieved by 'melt' processing at temperatures above the peritectic temperature of these incongruently melting solids. While this would appear to be an attractive possibility for thick film materials, it must be remembered that the increased processing temperatures will lead to enhanced interfacial reactions and consequently to more significant structural degradation and poisoning effects. Attempts to use this method on alumina[6]) confirm these problems.

These effects could be inhibited, if the interfacial reaction products involved a compound which was chemically and thermally compatible with both the substrate and the superconducting material and which formed a stable barrier layer during the processing.

In a recent investigation of melt processing YBCO on a number of substrate materials we have shown that the interfacial reaction between the liquid phase and yttrium stabilized zirconia (YSZ) leads to the formation of a mechanically stable, bonded barrier layer based on $BaZrO_3$ and at the same time to highly textured YBCO thick films of controllable thicknesses in the range 7-40 microns[7-8]). The processing conditions for these films have now been optimized and critical current densities of the order of 3000 A cm^{-2} have been obtained.

The details of this optimized route are given in the following, along with the dependence of the superconducting parameters on the processing conditions. Measurements of the nature of the intergranular coupling, flux pinning and microwave attenuation have also been made and these are briefly reported.

EXPERIMENTAL

Thick film formation is a complex route to obtaining surface coatings of thickness < 5 microns to > 200 microns. While it consists essentially of spreading a suitably prepared ink over the surface, and subsequently thermally processing the layers, the final products depend critically upon a number of parameters. These include the form of the solid material (to be coated) in the inks, the nature of the carrier liquids and the presence of additives. It also includes the processing temperatures and times once the surface has been coated, the layer thickness and the substrate material.

The production of YBCO thick films is no exception to this complexity and in the following we describe the optimization of a number of these parameters for films formed on YSZ substrates. Specifically, we have studied the effects of the reactivity of the precursor powder on the final film performance, the effects of processing temperature and the enhancement of superconducting parameters which can be obtained by silver addition.

i. Precursor powder reactivity was controlled by varying the sintering times for the YBCO used to prepare the final inks. A sintering temperature of 900°C was employed and firing times varied from 6 hours to 48 hours. The sintered pellets were then ground to a particle size suitable for the inks, mixed ultrasonically in triethanolamine and used to coat YSZ substrates.

ii. The film firing temperatures were varied over the range 940°C - 1050°C in air, so as to include the peritectic. The inks used were prepared from powders made using the optimized conditions determined from i) above,

iii. Silver-addition was examined over a concentration range 0-15%Ag, again using optimized powder at the optimized processing temperature of ii) above. Once the optimum silver concentration was determined, the firing temperatures were again varied to determine if the optimum had changed.

For the optimization of the precursor, ad hoc processing conditions were used which involved heating to 1015°C at 5° C/hour, firing at 1015°C for 5 mins. and cooling to room temperature at 2°C/hour. These conditions had previously been shown to give a reasonable film quality.

3.1. Precursor Powder

A double processing cycle (reaction and sintering) was used with equal times for both cycles at 900°C. The individual times varied from 6-48 hours (i.e. a total processing time range of 12-96 hours). Fig. 1 shows the properties of the final films prepared in this way. As may be seen (Fig. 1a), the resistive transition improved markedly to the 2×24 firing time, above which the normal state resistivity increases slowly. This is accompanied by an improvement in the texturing and the corresponding critical current density (j_c) variation is shown in Fig. 1b along with the normal state resistivity (100K) and the critical temperature.

The well defined peak in j_c close to 2×24 hours firing time we believe to be due to optimizing both the reactivity of the powder and its conversion to the YBCO compound.

This improved reactivity enables the formation of the $BaZrO_3$ barrier layer and the decay at shorter times is almost certainly due to inadequate reaction of the starting compounds.

The high value of critical current density obtained after this initial optimization step is already close to an order of magnitude improvement in performance over previous work.

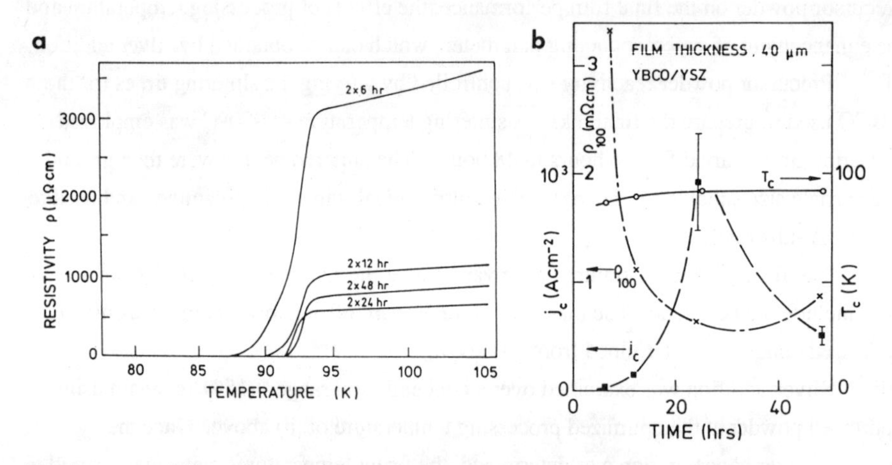

Fig. 1a. The variation of the resistive transition in thick films fabricated from YBCO powder processed for various times.
b. Enhanced critical current densities resulting from variation of the powder processing times. The normal state resistivities and critical temperatures are also shown.

3.2. Film Firing Temperature

Using ink prepared from the optimized precursor powder, an investigation of the effect of firing temperature on the film properties again showed a well developed maximum in the critical current, in this case close to the ad hoc temperature of 1015°C used in optimizing the precursor materials. This variation, shown in Fig. 2a, has three clearly defined regions. Below the peritectic (T = 1000°C in air) the film is granular and the associated performance is poor, being dominated by intergranular effects. Once melt processing is involved (T > 1000°C) the critical current density rapidly increases to a maximum value of 1000 Acm^{-2} at 1015°C. Below this temperature, microstructural analysis shows that grain growth and crystallization effects are dominating the improvement while above 1015° the decrease in the performance is associated with increased (211) phase formation and gradual degradation of the YBCO phase, presumably as a result of increased interfacial reactions. Fig. 2b shows the corresponding variation of the texturing figure of merit (= I_{006}/I_{103}) and the normal state resistivity.

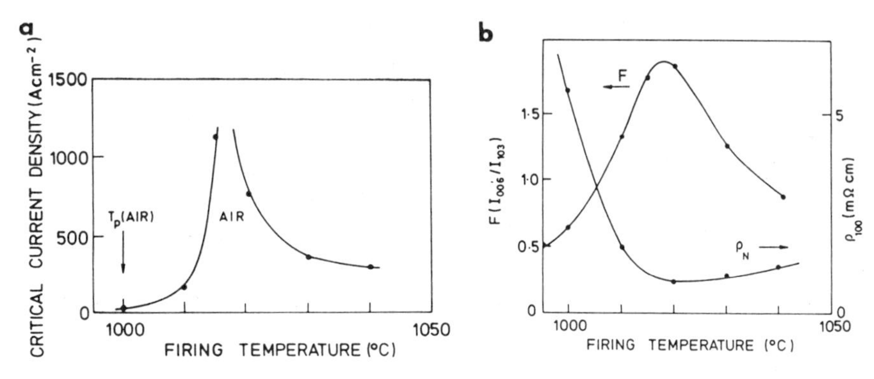

Fig.2a. The optimization of the critical current density with film firing temperature.

b. The variation of the texturing figure of merit and the normal state resistivity corresponding to a).

3.3 Silver Additives

In an attempt to enhance grain growth and improve the intergranular connectivity, increasing amounts of silver powder were added to the ink before film fabrication. As may be seen from Fig. 3, this had a dramatic influence on both the degree of texturing and the critical current density, the latter reaching values close to 3000 Acm^{-2} at 77K for 10% Ag addition. Despite these high values, this current capacity is still strongly dependent on external magnetic fields and as Fig. 4 shows, falls to approximately 200 Acm^{-2} above 0.3T.

Inspection of the effects of varying the firing temperature for this optimum silver concentration showed that 1015°C remained the best condition.

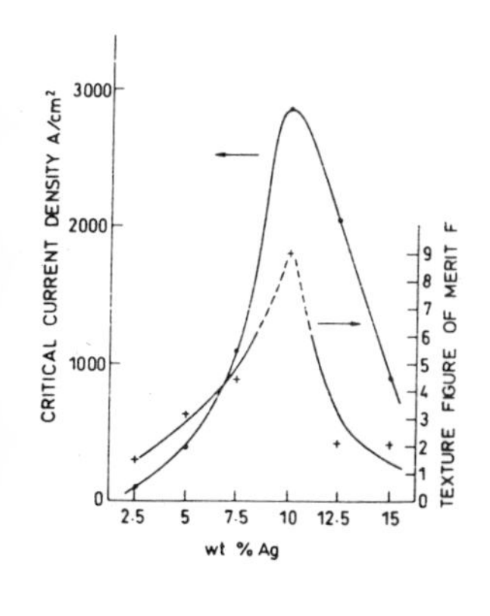

Fig.3. The variation of a) texturing figure of merit and b) critical current density with silver concentration.

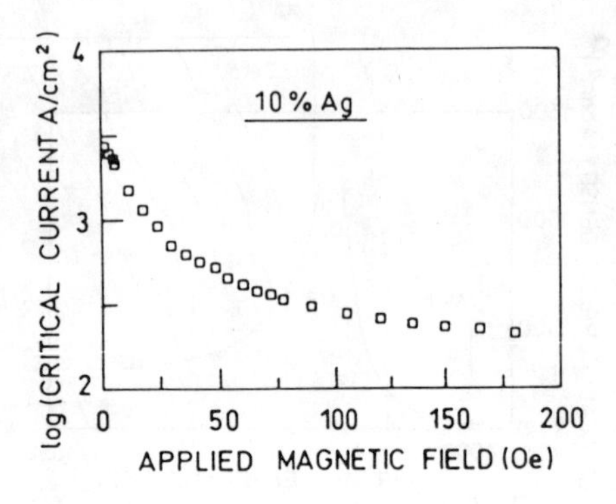

Fig.4. The magnetic field dependence of the critical current density at 77K for the 123 + 10wt %Ag sample.

3.4. Film Properties

3.4.1. Granular junctions Variable thickness microbridges were fabricated in the films in an attempt to isolate small groups of grains so as to allow an investigation of the granular junctions. Current-voltage characteristics at 77K generally revealed well defined Josephson-like critical currents, and in the presence of microwave radiation the anticipated Shapiro steps as shown in Fig. 5.

To our surprise, in some bridges, these characteristics developed high levels of voltage noise at increasing microwave power levels. This took two forms as shown in Fig. 6, and we believe that this is a manifestation of chaotic noise. Such chaotic behaviour is known to be possible for weak link bridges, and arises from one of two sources. For single junctions biassed with both dc and rf currents, the junction equation contains chaotic solutions for a limited range of junction properties[9)] which in general seem to be applicable to these bridges[10)]. Alternatively, it is also possible to obtain chaotic behaviour in systems of coupled junctions[11)] which would certainly apply to these materials, and we are presently continuing our study to model these junctions.

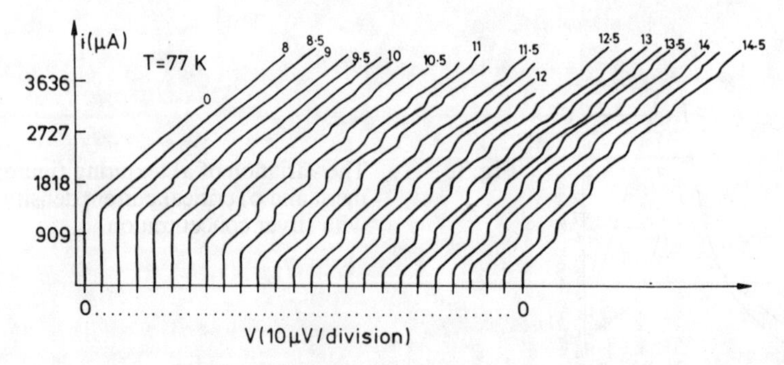

Fig.5. Well defined Shapiro steps in a microbridge fabricated from YBCO films.

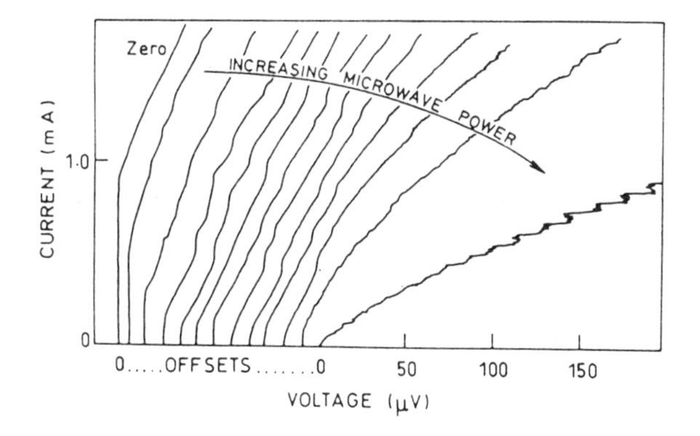

Fig.6. Chaotic voltage noise features of two kinds which appear to be characteristic of microbridges cut in melt processed thick films.

3.4.2. Flux pinning Measurements of the flux pinning energies by the observation of time dependent magnetoresistance voltages[12)] show that the energies in the YBCO + 10%Ag films are essentially the same as those of pure YBCO[13)]. This suggests that the improvements introduced by the presence of silver arise dominantly from improvements in the microstructure of the films (grain size, connectivity, texturing etc) rather than from changes in the flux pinning mechanisms.

3.4.3. Microwave screening Measurements of the propagation of 10GHZ microwaves through the optimized films have been compared with the transmission of copper films at 77K. The results, given in Table 1 show a marked improvement for the superconducting films, despite the big difference in thickness. Also given in this table are the observational results for other films.

TABLE 1: Screening Performance of YBCO Thick Films at 10GHz and 77K.

Material	YBCO/YSZ	YBCO(10%Ag)/YSZ	YBCO/Al_2O_3
A/A_{Cu}(300K)	1.0	0.16	0.14
A/A_{Cu}(77K)	8.9	0.35	0.04
S rel to Cu_{77} (dB)	19.0	7 (9 at 8 GHz)	-10.0

Note: Thickness of Films 10 μm; Copper 500 μm; A Attenuation of transmitted signal

CONCLUSIONS

It has been shown, that by careful and systematic optimization of the processing parameters it is possible to improve the current carrying capacity of YBCO thick films by at least on order of magnitude.

The properties of these films show good screening performance at 77K, with almost 20dB improvement over a thicker copper film.

REFERENCES

1. Hammond, R.B, Second World Congress on Superconductivity, Houston (1990) Proceedings to be published.
2. Takahashi, K., Shimura, S., Tsutsumi, M., Seidoh, M. and Kakegawa, Physica C, 385 153-155 (1988).
3. Barboux, P., Tarascon, J.M., Bagley, B.G., Greene, L.H., Hull, G.W., Meagher, B.W., and Eom, C.B, Mat.Res.Soc.Symp.Proc. 99, 49 (1988)
4. Miller, J.H., Holder, S.L., Hunn, J.D. and Holder, G.N, Appl.Phys.Letts. 54 2256 (1989).
5. Jin, S., Tiefel, T.H., Sherwood, R.C., Davis, M.E, van Dover, R.B., Kammlott, G.W., Fastnacht, R.A. and Keith, H.D, Appl.Phys.Lett. 52 2075 (1988).
6. Bailey, A., Russell, G.J. and Taylor, To be published.
7. Shields, T.C., Wellhofer, F., Abell, J.S., Taylor, K.N.R. and Holland, D, Physica C, 162-164, 1265(1989).
8. Bailey, A., Town, S.L., Alvarez, G., Taylor, K.N.R. and Russell, G.J, Physica C, 161, 347 (1989).
9. Kautz, R.L. and Monaco, R, J.Appl.Phys. 57, 875 (1985).
10. Alvarez, G., Russell, G.J. and Taylor, K.N.R, Physica C to be published (1990.
11. Nerenberg, M.A.H., Blackburn, J.A. and Jillie, D.W, Phys.Rev.B21, 118 (1980).
12. Matthews, D.N., Russell, G.J. and Taylor, K.N.R. Physica C, Accepted (1990).
13. Matthews, D.N., Russell, G.J., Taylor, K.N.R., Donohoo, A., Dou, S.X. and Liu, H.K, Third International Symposium on Superconductivity, Japan (1990).

SOME CURRENT ISSUES IN THE DEVELOPMENT OF PROCESSES FOR OXIDE SUPERCONDUCTOR SYNTHESIS

M.S. Chandrasekharaiah, R.G. Bautista*, and J.L. Margrave
Houston Advanced Research Center
4802 Research Forest Drive
Woodlands, Texas 77381

Visiting Scientist, from the Department of Chemical & Metallurgical Engineering, Mackay School of Mines, University of Nevada, Reno.

ABSTRACT

The large scale engineering applications of the recently discovered high T_c oxide superconductors hinge largely on the development of a production process for these materials with quality assured properties. In spite of a large scale frantic scientific effort during the past two years, there are several critical questions yet unanswered. The presently available literature shows that the J_c values attainable in any of the production processes currently in vogue are many orders lower than the required values in the range of 10^5 to 10^6 A/cm^2. Everyone working in this field is aware of the importance of oxygen content in determining the superconductive properties. Yet there is no consensus regarding the actual role of oxygen or its chemical state in these oxide superconductors. Further, in establishing a checklist of quality control parameters, there are gaps which will hinder further development of good quality assurance procedures both in the bulk production of powdered oxide superconductors and also during storage. In this presentation, some of these issues will be discussed.

1. INTRODUCTION

The progress in the development of high T_c ceramic superconductors during the past two years is admirable. The most studied and probably the most promising material for any large scale bulk applications is the $YBa_2Cu_3O_{7-x}$ (123-superconductor). So far, no such application has been demonstrated. Probably the absence of a production process for bulk superconductive material with quality assured properties may be the limiting step in this direction.

A number of properties of the 123-superconductor have been well characterized[1)] and yet there is no generally accepted checklist of quality control parameters that can be used in the development of a production process. There exist many gaps in such a checklist. One or two critical issues in the development of a process for bulk production will be highlighted in this presentation.

1.1 Properties of 123-Oxide Superconductor

Before discussing these issues, it will be fruitful to recapitulate the relevant properties of 123-oxide. The compound $YBa_2Cu_3O_{7-x}$ $(0<x<0.5)$ prepared by any one of several methods[1-5)] exhibits a value of 90K for the T_c (onset) and a T_c (zero resistance) above 77K if the sample has an orthorhombic crystal structure, is single phase and has the cationic ratio of Y: Ba: Cu = 1:2:3. The superconductive sample loses oxygen in two steps, beginning at 673K[6-10)]. Most of the oxygen loss in the first step is recovered almost completely on cooling. Both the T_c and the J_c are sensitive functions of the oxygen content. The presence of pure oxygen during the sintering process is essential in producing samples with acceptable properties. The superconductive properties of the samples degrade on standing at or near ambient temperature.

These observations are now accepted and hence are necessary quality assurance parameters, but they are not sufficient. Among others, the oxygen chemical state and the control of J_c are two issues for which there is yet no consensus. Solutions to these two issues will go a long way in the development of processes for synthesis of assured quality materials.

1.2 Preparative Methods

Laboratory methods of preparing phase pure 123-oxide superconductor are well documented[1-5]. Production processing of bulk samples are not that well reported. Recently, Bautista[9] has reviewed the production processes under current consideration. Large scale production schemes demand that one solves the problems of inhomogeneous mixing of the components, control of stoichiometry and the particle distribution, completion of reaction and contamination of the secondary phases.

2. THE CHEMICAL STATE OF OXYGEN

It is generally accepted now that the superconductive properties of the ceramic oxide material are very sensitive functions of its oxygen content[1-10]. Yet, our understanding of the exact role of oxygen in determining the J_c and T_c of the samples is incomplete[1]. Even the exact chemical state of oxygen is not established unambiguously.

2.1 Oxygen Stoichiometry and Superconductive Properties

The $YBa_2Cu_3O_x$-sample as prepared before suitable annealing in pure oxygen is a nonsuperconductor[1]. The orthorhombic phase is stable only with an oxygen stoichiometry above a value of x=6.5. The superconductive sample exchanges a fraction of its oxygen with molecular gaseous oxygen reversibly above 400°C[7,8]. Results of some thermogravimetric analysis[8] are indicative of two orthorhombic 123-oxides, differing from each other only slightly in their oxygen contents. It is probable, therefore, that all of the oxygen present in superconductive 123-oxide samples may not be in the same chemical state.

In spite of the data accrued so far, there are several unanswered questions regarding the role of oxygen[11]. Is all of the oxygen in a superconductive sample present in the same chemical (thermodynamic) state? The universally assumed oxidation state of oxygen as O^{2-} is in conflict with the results of thermogravimetric studies. Is there an oxygen compositional width (i.e., compositional homogeneity) in the samples? If so, at what oxygen stoichiometry do the samples exhibit the best superconductive properties? If there are two orthorhombic phases, which of them is a better material? What is the relation between T_c and J_c values and the corresponding oxygen chemical potential? These are some of the unanswered questions.

2.2 Oxidation States of Copper

In solving this oxygen problem, the theoretical models so far have been unhelpful. The chemical state of oxygen is somehow tied up with a proper description of the oxidation state of copper[11] in the ceramic oxide superconductors. Most of the theoretical models[1] have assumed that a fraction of copper ions are in the Cu^{3+} state. All experimental attempts to detect the presence of Cu^{3+} in the oxide superconductors have been unsuccessful[11]. In fact under the usual processing conditions of one atmosphere of oxygen, thermodynamic analysis shows that the Cu^{3+} state for any copper-oxygen system is highly improbable[11]. Yet, the theoretical models do not consider the possibility of some oxygen in oxidation states other than O^{2-}. The recent discovery of oxide superconductors without copper[12] casts doubt on the necessity of Cu^{3+} presence. The chemistry of alkaline earth metal oxides strongly suggests the existence of either superoxide (O_2^-) or peroxide ($O_2^=$) ions or O^- ions or even ozonide ($O3^-$) ions.

The oxygen chemical potential is a unique function of its chemical state in an oxide sample. An accurate measurement of oxygen chemical potentials in these oxide superconductors and establishing a correlation between the oxygen chemical potential and the T_c (and J_c) of the samples should help to elucidate the role of oxygen.

3. CRITICAL CURRENT DENSITY (J_c)

The critical current density values (J_c) for bulk oxide superconductors (123-oxide included) are several orders of magnitude lower than the requirements of most projected applications[1]. There is a frantic effort throughout the world to improve J_c values of the bulk samples. A complete understanding of this problem is far from satisfactory, but one or two critical parameters can be identified.

3.1 Oriented Grain Sintering and the J_c Values

Single crystal samples and epitaxial films of 123-oxide superconductors with J_c values of 10^6 A/cm^2 have been synthesized. But bulk samples with J_c higher than 10^3 A/cm^2 have not been prepared yet. Thus there is clear indication that low J_c values of bulk samples are traceable to the processes across grain boundaries[13]. Following are a few problems that deserve further consideration:

(i) severe anisotropy in conductivity (high in ab plane, low along c direction) causing weak connections across randomly oriented grains,

(ii) the presence of impurity layers such as carbonates at the grain boundaries and

(iii) the presence of porosity/microcracks.

Realizing the importance of the alignment of grain boundaries in enhancing the J_c values, various methods of synthesizing the samples with definitely oriented grains have been reported[14,15]. Among them, melt-textured growth using directional solidification and mechanical deformation through pressing, forging and rolling[15] have resulted in considerable enhancement of J_c values.

However, the expectation that alignment of grains would result in "exponential" improvement in J_c as a result of better coupling across grain boundaries has not materialized[14]. It appears that some other metallurgical or chemical characteristics of ceramic grain boundaries are also contributing significantly to the J_c values of bulk samples.

3.2 Secondary Impurity Phases at the Grain Boundaries

The phase composition (or phase purity) of the samples is invariably ascertained using x-ray powder diffraction analysis. It is possible to miss the presence of another phase (or phases) if the relative amounts of the phases are less than about 5%. But, the presence of such secondary phases at the grain boundary surfaces may severely contribute to the weak links at the grain boundaries. Experimental observations are accumulating to support the view that such secondary phases are present at the grain surfaces in samples prepared in the usual ways[16,17]. The partial equilibrium diagram at 900°C shows[18] that the 123-oxide co-exists with CuO, $BaCuO_2$, and Y_2BaCuO_5. Even unreacted $BaCO_3$ may also be present. If one or more of them are present as grain surface layers, then their presence may seriously affect J_c values.

A number of observations using different electron microscopic techniques have confirmed the presence of $BaCuO_2$, $BaCO_3$, CuO and Y_2BaCuO_5 phases as minor, impurity phases[15-20]. Usually they are segregated at the grain boundary surfaces. A recent study by Gao et al.[20] has reiterated the issue of process parameters and the presence of secondary phases at the grain boundaries. They have observed that several samples sintered in oxygen containing a small percent of CO_2 (<1%) exhibited J_c=0 even though the bulk of the samples was essentially pure 123-superconductor (according to magnetization measurements). When these samples were carefully examined by TEM, HREM, and SEM, the results showed the presence of secondary phases ($BaCO_3$, $BaCuO_2$, carbon and the tetragonal phase). This study focuses the importance of controlling the gaseous atmosphere, particularly in the vicinity of grain-grain interphase and also the necessity of rapidly and completely removing the CO_2 formed during calcination and sintering. Carrying out the calcination in reduced pressure or in dynamic vacuum and/or in two steps may eliminate the problem of CO_2.

3.3 Other Methods to Improve J_c

To circumvent this intergrain weak link limiting the J_c values of bulk superconductors, a number of alternate approaches have been reported. A small quantity of silver addition has resulted in enhancing the critical current density by providing paths to shunt current across the grains. Alternately, one can use various mechanical means to generate atomic scale defects in the sample to enhance flux pinning and thus to improve J_c values. This approach has been beneficial. Solving this problem should precede the large scale applications of high T_c superconductors.

4. CONCLUSIONS

Several experimental data strongly indicate that not all oxygen present in the orthorhombic, superconductive

$YBu_2Cu_3O_{7-x}$-phase is in the same chemical state (viz. O^{2-}) as it is assumed at present. A careful measurement of oxygen chemical potential should establish unambiguously the chemical state of oxygen, and probably also the chemical state of copper. The oxygen chemical potential and the microstructure of the 123-oxide have to be included, along with chemical composition, crystal structure, oxygen annealing temperature and time as quality assurance parameters in developing a synthesis process for bulk oxide superconductors.

Preventing the formation and segregation of secondary phases (carbon induced tetragonal phase, $BaCO_3$, $BaCuO_2$, etc) at the grain boundaries appears to be the first step in improving the critical current densities of bulk samples. The importance of removing completely and rapidly the CO_2 formed during the pyrolysis step in preventing the formation of grain boundary segregation of carbon cannot be ignored in designing the process flow sheet. All the production processes under current consideration involve $BaCO_3$ as one of the intermediate compounds. It is one of the more stable metal carbonates and is responsible for carbon segregation at the grain surfaces. A production process without the intermediate $BaCO_3$ as an alternate should be examined to avoid this problem.

REFERENCES

1. Poole, Jr.,C.P., Datta, T., and Farach, H.A., "Copper Oxide Superconductors", John Wiley, NY, (1988).
2. Katayama, S. and Sekine, M., J. Mater. Res., 5, 683 (1990).
3. Zheng, H., and Mackenzie, J.D., Mater. Letters, 7, 182 (1988).
4. Merkle, B.D., Kniseley, R.H., Schmidt, F.A., and Iverson, I.E., Mater. Sci. Eng., A124 (1989).
5. Horowitz, H.S., McLain, S.J., Steight, A.W., Druliner, J.D., Gai, P.L., Vankavelaar, M.J., Wagner, J.L., Biggs, B.D., and Poon, S.J., Science, 243, 66 (1989).
6. Strobel, P., Capponi, J.J., Marezio, M., and Monod, P., Solid State Commun., 64, 513 (1987).
7. Shelby, J.E., Bhargava, A., Simmins, J.J., Corah, N.L., McCluskey, P.H., Sheckler, C., and Snyder, R.L., Mater. Letters, 5, 420 (1987).
8. Swaminathan, K., Janaki, J., Rao, G.V.N., Sreedharan, O.M., and Radhakrishnan, T.S., Mater. Letters, 6, 261 (1988).
9. Bautista, R.G., J. Metals, 42, 23 (1990).
10. Tarascon, J.M., McKinnon, W.R., Greene, L.H., Hull, G.W., and Vogel, E.M., Phys. Rev., B36, 226 (1987).
11. Chandrasekharaiah, M.S., High Temp Sci, 24, 185 (1989).
12. Sharp, J.H., Br. Ceram. Trans. J., 89, 1, 1990.
13. Chaudhari, P., Mannhart, J., Dimos, D., Tsuei, C.C., Chi, J., Opryski, M.M., and Scheuermann, M., Phys. Rev. Letters, 60, 1658; 61, 219 (1988).
14. Tkaezyk, J.E., and Lay, K.W., J. Mater. Res., 5, 1368 (1987).
15. Jin, S., Tiefel, T.H., Sherwood, R.C., Davis, M.E., Van Dover, R.B., Kammlott, G.W., Fastnacht, R.A., and Keith, H.D., Appl. Phys. Letters, 52, 2074 (1988).
16. Nakahara, S., Fisanick, G.J., Yan, M.F., van Dorver, R.B., Boone, I., and Moore, R., J. Crystal Growth, 85, 639 (1987).
17. Verhoeven, J.D., Bevalo, A.S., McCallum, R.W., Gibson, F.D. and Noack, M.A., Appl. Phys. Letters, 52, 745 (1988).
18. deLeeuw, D.M., Mutsaers, C.A.H., Langereis, C., Smoorenburg, H.C.A., and Rommers, P.J., Physica C, 152, 39 (1988).
19. Ginslay, D.S., Venturini, E.L., Kwak, J.F., Baughman, R.J., and Morosin, B., J. Mater. Res., 4, 496 (1989).
20. Gao, Y., Li, Y., Merkle, K.L., Mundy, J.N., Zhang, C., Balachandrar, U., and Peoppel, R.B., Mater. Letters, 9, 347 (1990).

CONTROL OF $YBa_2Cu_3O_{7-x}$ DEGRADATION DURING HEAT TREATMENT

S. E. Dorris, R. B. Poeppel, J. J. Picciolo, U. Balachandran,
M. T. Lanagan, C. Z. Zhang, K. Merkle, Y. Gao, and J. T. Dusek
Argonne National Laboratory, 9700 S. Cass Ave., Argonne, IL 60439

H. E. Jordan, R. F. Schiferl, and J. D. Edick
Reliance Electric Co., 24800 Tungsten Road, Cleveland, OH 44117

ABSTRACT

The importance of closely controlling the furnace atmosphere during processing of $YBa_2Cu_3O_{7-x}$ (YBCO) was examined. The stability of YBCO during sintering was studied as a function of CO_2 partial pressure in CO_2/O_2 gas mixtures. The zero-field critical current density, J_c, decreased with increasing CO_2 partial pressure, and ultimately reached zero, even though Meissner effect measurements showed that the bulk of the samples with zero transport J_c remained superconducting. Examination of the microstructure and composition of the samples by TEM, AEM, and SIMS showed the presence of $BaCuO_2$ and $BaCu_2O_2$ at a minority of the grain boundaries. Near other grain boundaries, where second phases were not readily evident, the structure was found to be tetragonal for several tens of nm up to the grain boundaries, whereas the grain interiors were found to be orthorhombic. At high partial pressures of CO_2, YBCO completely decomposed to $BaCO_3$, Y_2BaCuO_5 (211), and CuO. In the firing of plastically processed superconductors, such as extruded YBCO coils, it was found that the use of reduced total pressure prevents the decomposition of YBCO.

INTRODUCTION

It is known that atmospheric contaminants such as CO_2 and H_2O can strongly impact the transition temperature, critical current density, and the width of superconducting transition of YBCO superconductors.[1-6] Jahan et al.[7] indicated the formation of insulating phases when YBCO is reacted with water vapor, while several other researchers[1,5,6,8,9] have reported on the reaction of YBCO with CO_2. Gallagher et al.[1] reported that YBCO does not decompose at 1000°C in a 1% CO_2/O_2 mixture, while it decomposes completely in a 10% CO_2/O_2 mixture, forming $BaCO_3$, $Y_2Cu_2O_5$, and CuO. Fjellvag et al.,[8] concluded that the reaction between YBCO and CO_2 occurs in two steps. Below 730°C the reaction products are $BaCO_3$, Y_2O_3, and CuO, while above this temperature the products are $BaCO_3$, $Y_2Cu_2O_5$, and CuO. Because of limitations of the X-ray diffraction technique, however, neither Gallagher et al.[1] or Fjellvag et al.[8] were able to determine the distribution of the reaction products, which may be very important with regard to the low value of the critical current density found in ceramic superconductors.

For many practical applications of the high-T_c superconductors, it will be necessary to make long, continuous lengths of superconductor in a variety of shapes. Plastic extrusion is one versatile technique by which this could be accomplished, however, it leads to the evolution of relatively large amounts of CO_2 and H_2O during heat treatment. It is therefore important to know the extent to which YBCO decomposes in the presence of CO_2 and H_2O and to know if the decomposition of YBCO can be prevented by careful control of the furnace atmosphere.

In this paper, we report on the degradation of properties (critical temperature, T_c, and critical current density, J_c) of YBCO superconductors sintered in CO_2-containing atmospheres. The microstructures and compositions of the samples were investigated by transmission electron microscopy (TEM), analytical electron microscopy (AEM), and secondary ion-mass spectroscopy (SIMS). The relations between the properties and the partial pressure of CO_2 will be discussed in terms of the microstructural changes. In addition, we demonstrate that the firing of YBCO coils at reduced total pressure prevents decomposition of the YBCO.

EXPERIMENTAL

Powders of YBCO and 211 are made by solid-state reaction of the constituent oxides. The appropriate proportions of Y_2O_3, $BaCO_3$, and CuO are mixed and milled in methanol for ~12 h. After drying, the mixtures are calcined in flowing oxygen at a reduced total pressure of ~2 mm Hg. This method leads to decomposition of $BaCO_3$ at lower temperatures, avoids the formation of impurity phases, and, in the case of YBCO, produces phase-pure material after a single calcination at 800°C[10] In the case of 211, the powder is annealed in oxygen for 24 h at 950°C after calcination at 800°C. Silver is added to the YBCO powder before extrusion to give a YBCO/Ag composite superconductor (85 vol.% YBCO/15 vol.% silver), because the addition of silver has been shown to improve the mechanical properties of bulk YBCO.[11] Details of coil fabrication have been given previously[12] so they will not be described here.

In order to examine the reaction between YBCO and CO_2 in detail, pellets were pressed and sintered in the temperature range 900-1000°C for about 5 h in flowing (≈1 atm) O_2/CO_2 gas mixtures. The concentration of CO_2 in the mixtures ranged from 0 to 5%. The samples were cooled slowly to room temperature with a 12 h hold at 450°C to allow for re-oxygenation. J_c was measured by standard four-probe resistivity measurement in liquid nitrogen. A criterion of 1 μV/cm was used for measurement of J_c in pellets; for coils, a criterion of 1 μV across the entire coil was used. T_c values were obtained by resistivity and magnetization techniques. A low field rf SQUID magnetometer was used for the magnetization measurements. Transmission electron microscopy (TEM) specimen discs (3 mm diameter) were cut from the sintered bulk samples, polished, and dimpled from both sides until a thin area at the center obtained. The final TEM specimens were argon-ion thinned at liquid nitrogen temperature.

RESULTS AND DISCUSSION

Table 1 shows that, as the CO_2 partial pressure in the sintering atmosphere increased, J_c (zero field/77 K) of sintered pellets decreased and finally became zero. Resistivity measurements showed that the materials with J_c = 0 were semiconductive. On the basis of such measurements, the stability region for superconducting YBCO was derived with respect to the partial pressure of CO_2 at the four sintering temperatures (See Fig. 1).

Table 1. Relative densities and J_c (zero field/77 K) for samples fired over a range in temperature in various CO_2/O_2 mixtures

Sintering Temp. (°C)	CO_2 (%)	Density (%)	J_c (A/cm^2)
1000	0	93	320
	0.005	92	157
	0.05	92	106
	0.5	92	34
	5.0	93	0
970	0	87	315
	0.005	86	138
	0.05	90	94
	0.5	86	0
	5.0	87	0
940	0	67	128
	0.005	67	21
	0.05	68	0
	0.5	66	0
	5.0	–	–
910	0	63	60
	0.005	62	0
	0.05	63	0
	0.5	62	0
	5.0	–	–

Magnetization measurements indicated that, even in the case of semiconductive samples, the major portion of the sample was still superconducting, and that the temperature for onset of superconductivity was still ≈90 K. This can be seen in Fig. 2, where resistivity and magnetization data are given for two samples fired at 940°C: one being a superconducting sample processed in 100% O_2 and the other a semiconductive sample processed in 0.5% CO_2/O_2. These results suggest that CO_2 reacts with YBCO to form a thin layer of nonsuperconducting second phase at grain boundaries, but does not degrade the superconducting grain interior. When enough of

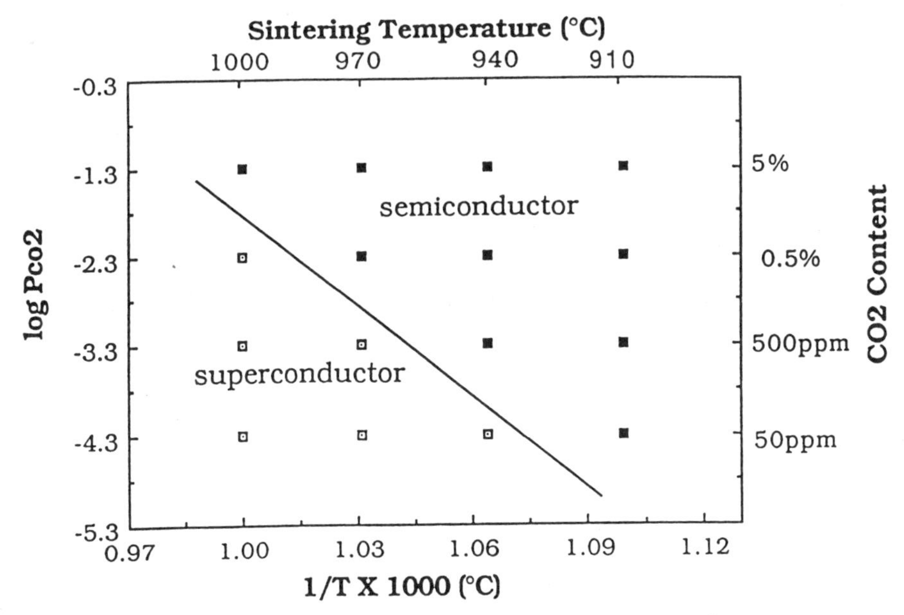

Fig. 1. The stability of YBCO with respect to CO_2 partial pressure over a range of temperature.

the grain boundaries become coated with the second phases, as in the semiconductive samples, the passage of superconducting current is effectively blocked, but because of the superconducting grain interiors, a sharp change in magnetization is still evident.

These arguments are supported by TEM observations,[5] which show the presence of secondary phases at some grain boundaries. An example of one such grain boundary is shown in Fig. 3 for the sample sintered at 970°C in 0.5% CO_2/O_2 gas mixture. X-ray energy-dispersive spectroscopy (XEDS) shows the grain boundary material to consist of $BaCuO_2$ and $Y_2Cu_2O_5$. The width of this grain boundary material is much larger than the coherence length in YBCO, so that it can completely obstruct the superconducting current and cause a reduction in the overall critical current density. However, this type of grain boundary accounts for only about 10% of the observed grain boundaries; the majority of the grain boundaries appear quite sharp with no obvious evidence of a second phase. Because of the multitude of possible percolation paths, the value of

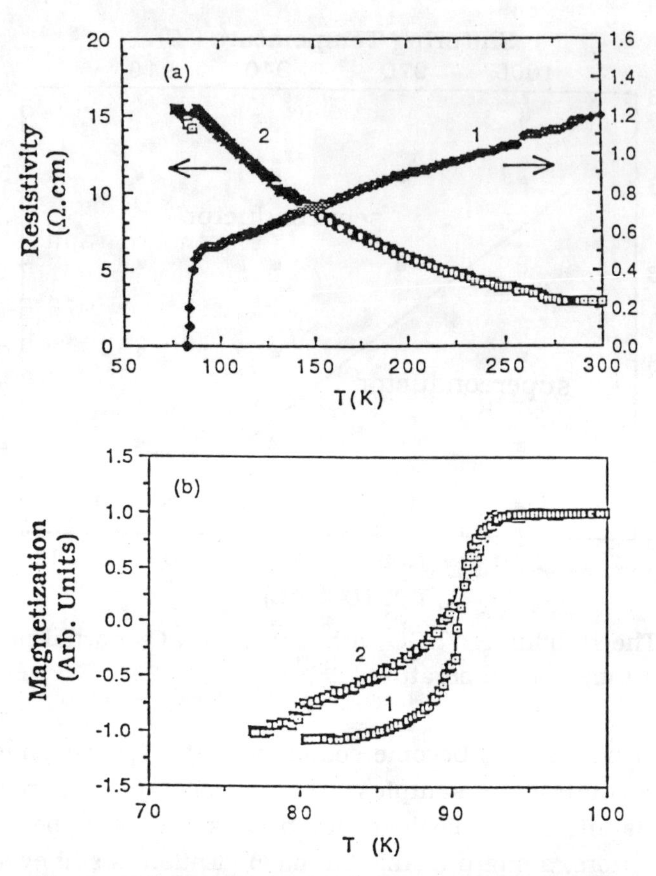

Fig. 2. (a) Resistivity versus temperature, and (b) magnetization versus temperature for two samples fired at 940°C: Sample 1 was fired in 100% O_2 and was superconducting, Sample 2 was fired in 0.5% CO_2/O_2 and was semiconductive.

J_c would not become zero if only 10% of the grain boundaries are coated with a second phase. In order for the superconducting current to be completely blocked, as in the case of the semiconductive samples, a majority of the grain boundaries have to be nonsuperconducting. High-resolution electron microscopy (HREM) images of the grain boundaries suggested that, in fact, the majority of grain boundaries probably are nonsuperconducting.

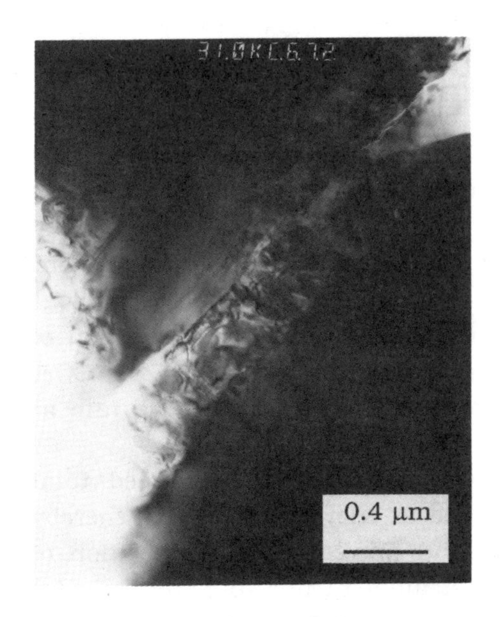

Fig. 3.
TEM micrograph of a grain boundary in a YBCO sample sintered at 970°C in 0.5% CO_2/O_2 atmosphere. The thick second phase layer consisted of $BaCuO_2$ and $Y_2Cu_2O_5$.

Careful study of the HREM images showed that the structure near the sharp grain boundaries is not orthorhombic, but some other phase, possibly tetragonal YBCO. By careful measurement of the inter-planar spacing, it was found that the spacing is about 1.19 nm in the regions near grain boundaries, while the spacing is approximately 1.17 nm in the regions far from the grain boundaries. Neutron diffraction data[13] show that tetragonal YBCO has c = 1.19 nm. Another indication of tetragonal material is the termination of twinning near the grain boundaries, which can be taken as the demarcation line between orthorhombic and tetragonal structures, since the tetragonal structure has no twins. A possible explanation for the phase transformation from orthorhombic to tetragonal structure is the incorporation of carbon into the lattice, due to the presence of CO_2 in the sintering atmosphere.[5,6] Segregation of carbon at grain boundaries or in regions near grain boundaries has been confirmed in a previous study[5] using secondary ion mass spectroscopy (SIMS). Carbon can diffuse into the lattice, and expel oxygen from the orthorhombic structure, thus forming a nonsuperconducting tetragonal structure that can block the superconducting current.

In view of the above, the effect of residual carbon on the processing and properties of extruded coils must be considered. In the green state, superconducting coils contain ≈10 wt.% organics, which must be completely removed in a fashion that does not damage the superconductor. Incomplete removal of the organics can cause decomposition of the superconductor or leave carbon-rich material at the grain boundaries, either of which can degrade superconducting properties. Organics can be easily removed by thermal decomposition in the temperature range of 240–350°C, but if the decomposition proceeds too rapidly, the coils can bloat severely and in some cases even explode. Also, decomposition of the organics produces significant concentrations of H_2O and CO_2, which we have seen can react with YBCO; therefore, the rate at which organics are removed must be carefully controlled.

When superconducting coils are fired at a reduced total pressure, CO_2 and H_2O are removed as they are produced, thereby minimizing their concentrations and preventing decomposition of YBCO. But when coils are fired at ambient pressure, the harmful gaseous products accumulate and lead to the decomposition of YBCO. To demonstrate this, mixtures made from YBCO powder and the same organics used in extrusion were fired in flowing oxygen at either ambient or reduced pressure (≈2 mm Hg). Two different powders were used: powder produced by solid-state reaction at reduced pressure and powder produced by a liquid mix technique. Samples were taken from the mixtures at 240, 300, and 350°C, and their X-ray patterns were obtained. Figures 4 and 5 are schematic illustrations of the major peaks in these patterns. Figure 4 shows that, when the YBCO/organic mixtures were fired at ambient pressure, both samples of YBCO decomposed, the liquid-mix powder at 240°C and the solid-state powder at 300°C. Figure 5 shows, however, that gross decomposition of YBCO did not occur when the mixtures were fired at reduced total pressure, even though the relative intensities of peaks varied as a result of changing oxygen content.

Shown in Table 2 are the firing conditions and properties of five coils made by extrusion and fired at reduced total pressure. Although the critical current densities are well below those necessary for many large-scale applications, it should be noted that the length of continuous superconductor in these coils ranges up to ≈12 m. Moreover, the measurements were made in magnetic fields

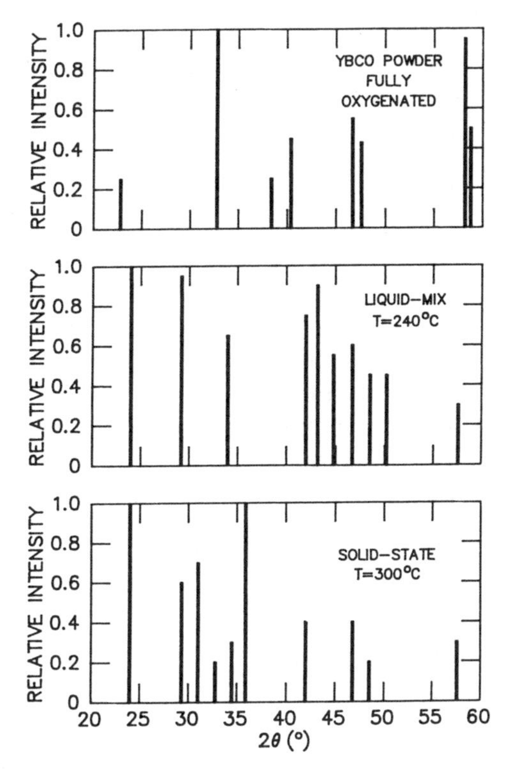

Fig. 4.
Schematic X-ray patterns of fully oxygenated YBCO powder (as reference) and two other YBCO powders that were fired at <u>ambient pressure</u> in contact with the organics used in extrusion of coils. The patterns show that both powders fired with the organics decomposed during firing.

up to 73 Gauss, and fringing effects at the coil ends probably increase the field on the end turns even further. Considering that just a few years ago it was not possible to consistently obtain such performance on even short lengths of superconductor in zero field, these results represent significant improvement in bulk superconductor fabrication. Although large differences in size and geometry make it difficult to compare the J_c results of pellets and coils, a comparison of J_c in Tables 1 and 2 suggests that CO_2 had minimal impact on the superconducting properties of coils that were fired at reduced total pressure. Keeping in mind that J_c of bulk materials drops dramatically with magnetic field, and that the coils were measured in fields up to at least 73 Gauss, it is suggested that the coil results agree most closely with the results for pellets fired in 0% CO_2/100% O_2. This suggests further that large, multilayer, superconducting coils can be successfully fabricated by firing at reduced total pressure, and that these coils have

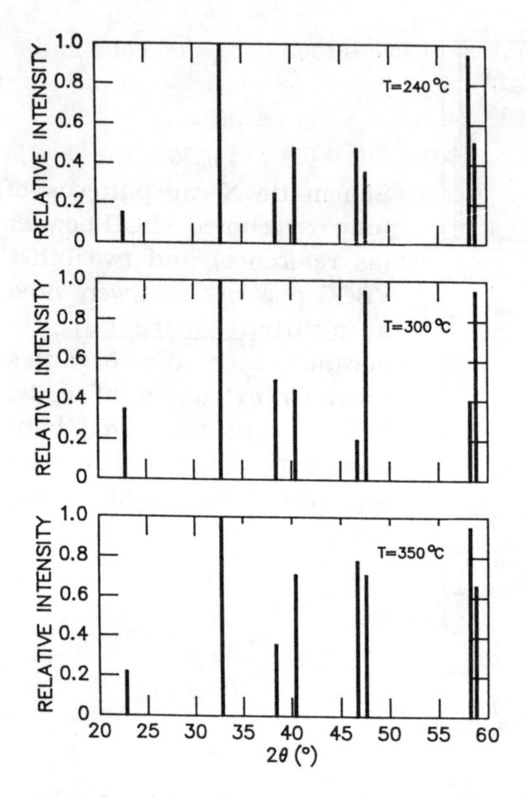

Fig. 5.
Schematic X–ray patterns of YBCO powder as a function of temperature. The powder was fired at reduced total pressure in contact with the organics used in extrusion of coils. The patterns show no signs of decomposition.

superconducting properties that are representative of bulk materials.

CONCLUSIONS

YBCO reacts strongly with CO_2 at high temperatures, leaving superconducting grain interiors encased in nonsuperconducting grain boundary phases. The secondary phases obstruct super-conducting currents and cause a decrease in J_c. At high partial pressures of CO_2, YBCO completely decomposes to $BaCO_3$, Y_2BaCuO_5, and CuO. At lower partial pressures of CO_2, carbon segregates at the grain boundaries and causes the transformation from the orthorhombic phase to the nonsuperconducting tetragonal phase. In the firing of large, multilayer coils, decomposition of YBCO can be avoided and the concentration of CO_2 can be minimized by firing at

Table 2. Description and properties of five superconductor coils and their firing conditions

Coil Description	Firing Conditions	J_c (A/cm^2)	B[a] (Gauss)	B(Fe Core) (Gauss)
1 Uncoated 24 Turns	100% O_2 910°C	120	19	---
2 Coated 211 25 Turns	100% O_2 910°C	130	20	---
3 Uncoated 21 Turns	2 mm Hg 875°C	225	36	---
4 Coated 211 2 Layers 42 Turns	10 mm Hg 875°C	150	42	160 @ 77 K
5 Coated 211 5 Layers 75 Turns	10 mm Hg 875°C	150	73	330 @ 77 K 420 @ 73 K

[a]Magnetic field.

reduced total pressure. As a result, superconducting coils can be fabricated that produce magnetic fields up to 73 Gauss with an air core and 330 Gauss with an iron core.

ACKNOWLEDGMENTS

Work was supported by the U. S. Department of Energy, Conservation and Renewable Energy, as part of a program to develop electric power technology (SED, RBP, JJP, UB, MTL, and JTD), and Office of Basic Energy Sciences–Materials Science (KM), under Contract W–31–109–Eng–38; and the National Science Foundation, Office of Science and Technology Center (CZZ, YG), under Contract

DMR–8809854. The iron core coil test fixture was developed as part of an Electric Power Research Institute (EPRI) contract.

REFERENCES

1. P. K. Gallagher, G. S. Grader, and H. M. O'Bryan, Mat. Res. Bull., 23, 1491 (1988).
2. E. K. Chang, E. F. Ezell, and M. J. Kirschner, Supercond. Sci. Technol., 8, 391 (1990).
3. T. B. Lindemer, C. R. Hubbard, and J. Brynestad, Physica C., 167, 312 (1990).
4. T. M. Shaw, D. Dimos, P. E. Batson, A. G. Schrott, D. R. Clarke, and P. R. Duncombe, J. Mater. Res., 5, 1176 (1990).
5. Y. Gao, K. L. Merkle, C. Zhang, U. Balachandran, and R. B. Poeppel, J. Mater. Res., 5, 1363 (1990).
6. Y. Gao, Y. Li, K. L. Merkle, J. N. Mundy, C. Zhang, U. Balachandran, and R. B. Poeppel, Mater. Lett., 9, 347 (1990).
7. M. S. Jahan, D. W. Cooke, H. Sheinberg, J. L. Smith, and D. P. Lianos, J. Mater. Res., 4, 759 (1989).
8. H. Fjellvag, P. Karen, A. Kjekshus, P. Kofstad, and T. Norby, Acta Chem. Scand., A42, 178 (1988).
9. E. A. Cooper, A. K. Gangopadhyay, T. O. Mason, and U.-Balachandran, submitted to J. Mater. Res., September 1990.
10. U. Balachandran, R. B. Poeppel, J. E. Emerson, S. A. Johnson, M. T. Lanagan, C. C. Youngdahl, Donglu Shi, K. C. Goretta, and N. G. Eror, Mater. Lett., 8, 454 (1989).
11. J. P. Singh, H. J. Leu, R. B. Poeppel, E. Van Voorhees, G. T. Goudey, K. Winsley, and D. Shi, J. Appl. Phys., 66, 3154 (1989).
12. S. E. Dorris, J. T. Dusek, J. J. Picciolo, R. Russell, J. P. Singh, and R. B. Poeppel, Proceedings of the International Conference on Electrical Machines (ICEM), Cambridge, MA, August, 1990, in press.
13. J. D. Jorgensen, M. A. Beno, D. G. Hinks, L. Soderholm, K. J. Volin, R. L. Hitterman, J. D. Grace, I. K. Schuller, C. U. Segre, K. Zhang, and M. S. Kleefisch, Phys. Rev. B36, 3608 (1987).

PROCESSING GRAIN-ORIENTED BULK $YBa_2Cu_3O_x$ BY PARTIAL-MELT GROWTH

Donglu Shi, Justin Akujieze, and K. C. Goretta

Argonne National Laboratory, Argonne, Illinois 60439

ABSTRACT

We have developed a technique to produce a grain-oriented microstructure in bulk $YBa_2Cu_3O_x$. We have found that the textured material possesses high homogeneity over its entire length of 50 mm. Owing to well-controlled cooling rates and partial-melting temperatures, porosity and Y_2BaCuO_x phase in the textured product were greatly reduced compared to previously produced products. Magnetization measurements indicated a sharp superconducting transition near 90 K. Mechanisms of texturing and phase formation during partial melting are discussed.

INTRODUCTION

Practical application of high-T_c superconductors requires their ability to carry large critical current densities of at least 10^4 A/cm^2 in high magnetic fields greater than 1 T. The superconductors must also exhibit high flexibility and chemical stability. Current high-T_c superconductors are, however, brittle and chemically reactive. More important, application of high-T_c superconductors made by conventional sintering of powders is limited by the well-known weak-link effects; critical current density (J_c) values are low.

To enhance the mechanical properties of high-T_c superconductors, researchers have used silver doping and so-called powder-in-tube techniques.[1-4] To increase J_c, researchers have developed various methods of producing grain-oriented bulk superconductors.[5-14] Although significant improvements have been achieved, many problems remain. For instance, although melt-texturing methods can increase J_c in a magnetic field 500 times, only extremely small samples (~10 mm) have been made to date. The small sizes make this type of processing impractical for large-scale wire fabrication. Another problem is occurrence of microcracks in melt-textured samples. The cracks can severely deteriorate electrical-transport and mechanical properties.

In this paper, we present details of a technique for producing grain-oriented bulk $YBa_2Cu_3O_x$ (123) samples. We discuss the mechanisms of partial melting and texturing in Y–Ba–Cu–O system. We also discuss phase decomposition, peritectic reaction, and second-phase formation during the partial-melting process.

EXPERIMENTAL DETAILS

Sample Preparation and Partial-Melting Apparatus

Dense bars 50 mm x 6 mm x 2 mm were made by a conventional method of powder sintering. Pressed bars from well-calcined $YBa_2Cu_3O_x$ powder were sintered at 950°C in air for 8 h and then furnace-cooled to room temperature. The bars were coated by sputtering with a thin layer of silver. Each sintered bar was placed vertically in a ZrO_2 crucible as shown in Figure 1. The crucible was inserted directly into a vertical furnace at a preset temperature of 1150°C. The heating schedule is shown in Figure 2. Each bar sample was heated at 1150°C for 0.2 h and then cooled to 1050°C at 20°C/min. At 1050°C, a temperature gradient of 2°C/mm was established along the length of the bar by adjusting the sample's position at the edge of the furnace hot zone. A type K thermocouple was used to profile the temperature gradient. The temperature was decreased from 1050°C to 850°C at a rate of 1°C/h. From 850°C, the sample was cooled to room temperature at 1°C/min and then oxygenated at 450°C for five days.

Phase Diagram and Transformations

A phase diagram for a Y–Ba–Cu–O system is shown in Figure 3. According to this figure, the phase reaction sequence during partial melting

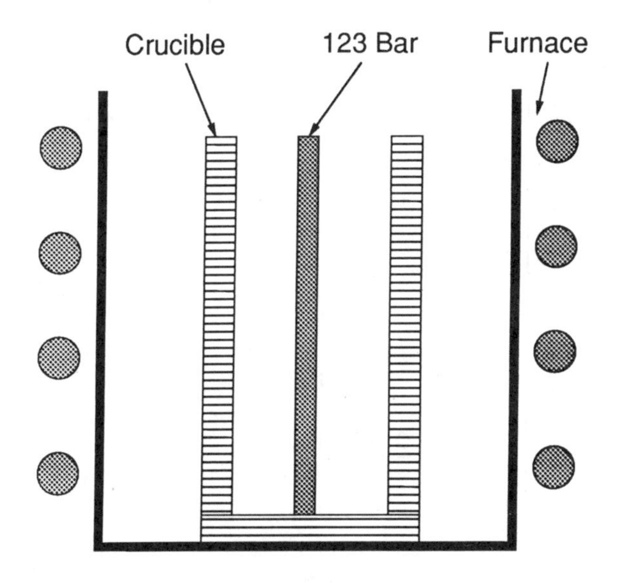

Figure 1. Schematic diagram of melt-texturing process.

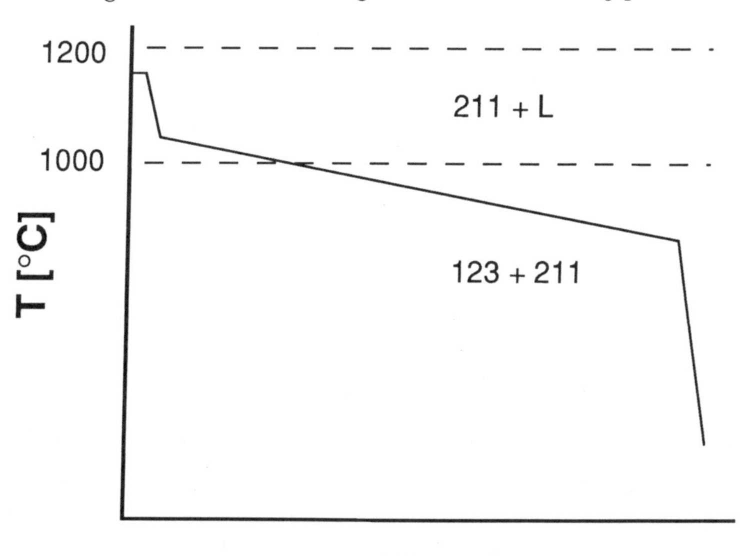

Figure 2. Heating schedule for melt-texturing process.

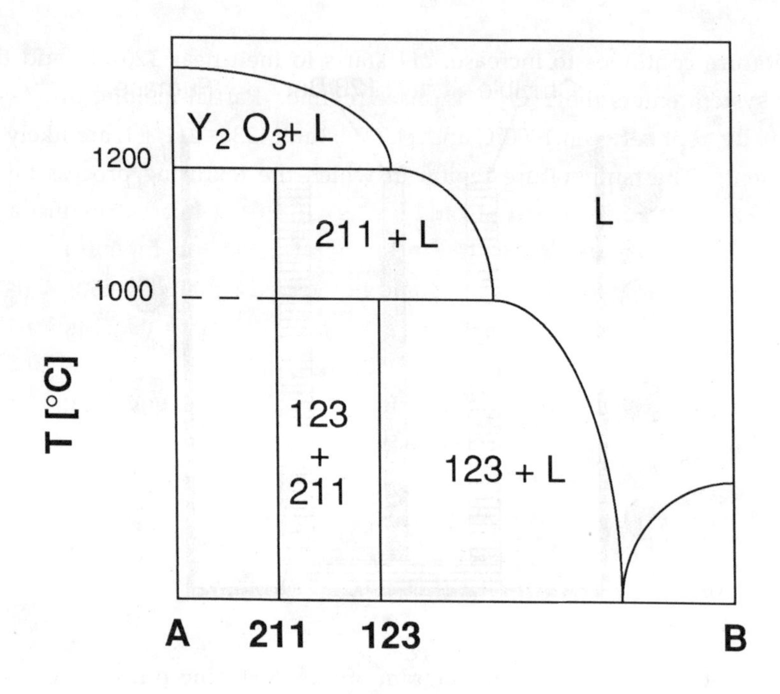

Figure 3. Two dimensional phase diagram for Y–Ba–Cu–O system in air, in which A = $5Y_2O_3$ + 2BaO and B = 3(211) + 2(CuO).[5]

can be described as follows. As the temperature is raised to more than 1000°C, 123 starts to melt incongruently. The composition of the liquid phase (L) may vary greatly, depending on the temperature and the extent of 123 melting. The amount of the liquid phase is proportional to the temperature above the partial-melt point. It should be noted that the precise temperature at which 123 begins to melting is not clear. Generally, the point of significant melting lies between 1000°C and 1020°C. Unless the material is perfectly phase pure, however, melting can occur substantially below 1000°C.[15,16] This uncertainty is due to several eutectics that are associated with the initial stoichiometry and second phase concentrations.[16] Since the system is in a two-phase regime, Y_2BaCuO_x (211) also precipitates from the liquid. Thus, the liquid must be rich in barium and copper. As equilibrium is approached, three phases may coexist: 123 + 211 + L (in equilibrium, only 211 + L will be present). If the temperature continues to increase, 211 starts to melt near 1200°C, and the whole system enters the Y_2O_3 + L phase regime. Partial-melting processes are usually kept between 1000°C and 1150°C; thus 123 + 211 + L are likely to be present. The temperature region in which the texturing process takes place is crucial, since it will strongly affect the final microstructure and second-phase concentrations. If the temperature is too high (say, near 1200°C), a great amount of 211 phase and liquid phase may develop; thus, a long time may be required for the formation of 123 upon cooling. If the temperature is too low (say, near 1000°C), only a small amount of liquid phase may form, thus, microstructural texturing becomes difficult to accomplish during cooling. We have experimentally determined that the optimized partial-melting temperature for texturing is between 1050°C to 1150°C, depending upon the cooling rates and processing speeds.

MICROSTRUCTURAL DEVELOPMENT

Texturing Mechanism

In the traditional directional melting processes, the motion of the solid/liquid interface is established by moving either the hot zone of the furnace or the sample through the hot zone.[11,12] The speed of the solid/liquid interface motion is less than the maximum growth rate of a single crystal, and the growth tends to take place along the direction of the motion (i.e., normal to the solid/liquid interface). In the partial-melt growth process of 123, a peritectic reaction occurs during cooling, and new 123 phase nucleates and grows at the interfaces between 211 and the liquid phase. If the 123 is only partially melted, the new 123 may grow coherently on the remaining 123, along the direction of the temperature gradient. The texturing process is established by the slow motion of the solid/liquid interface, which is similar to the traditional method of directional growth of single crystals. Here, the motion of the solid/liquid interface is established by slowly moving the temperature gradient through the sample bar. This mechanism is illustrated in Figure 4. The sample is placed at the edge of the hot zone so that a temperature gradient can be established. As the temperature of the furnace is slowly lowered, the temperature gradient front is displaced vertically as indicated by the arrows. As crystallization

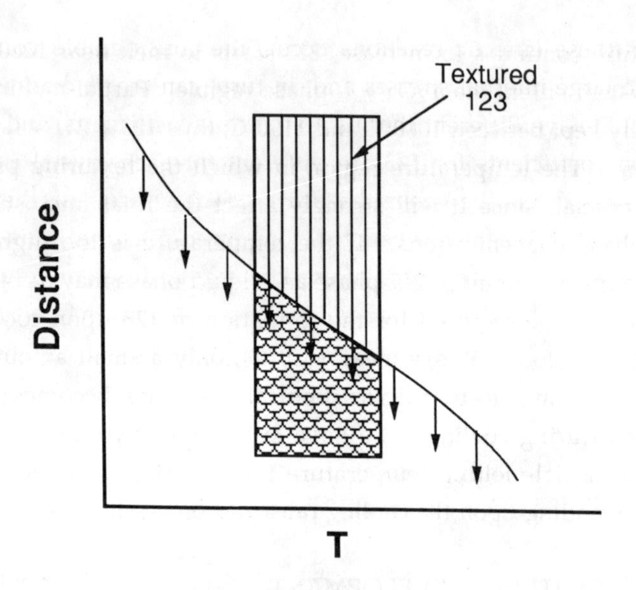

Figure 4. Temperature gradient along bar sample during melt texturing.

occurs, 123 tends to grow along the direction of the moving temperature-gradient front and coherently with the parent 123 phase. As a result, the 123 grains are overall highly textured. Each grain can extend to a quite long distance of a few centimeters.

Problems of Partial Melting: Cracks and Second Phases

Cracks are generated by two main sources: (1) the multiphase nature of the system and (2) 123 crystal anisotropy. We have already discussed the phase diagram and phase transformations during partial melting, in which some eutectic reactions were mentioned. These eutectics for an air atmosphere are indicated below:[16]

$$YBa_2Cu_3Ox + BaCuO_2 + CuO \rightarrow \text{Liquid at } 900°C \ ,$$

$$YBa_2Cu_3Ox + CuO \rightarrow Y_2BaCuO_5 + \text{Liquid at } 940°C \ ,$$

$$YBa_2Cu_3Ox + BaCuO_2 \rightarrow Y_2BaCuO_5 + \text{Liquid at } 1000°C \ .$$

As these eutectic reactions occur, the considerable volume change can cause large internal stresses and, in turn, can create cracks. Moreover, 211 and 123 have different thermal expansion coefficients, and the thermal expansion coefficients for 123 are different for each crystal axis.[17] Large thermal stresses can be generated during cooling, causing the material to crack severely. We have observed some cracks at the grain boundaries along the a–b plane. This type of crack is due to different volume changes along the a–b plane and the c axis. Cracking problems may be alleviated by doping the materials with a ductile metal such as silver. The silver can accommodate the large internal stresses generated by the above mechanisms.[1]

CHARACTERIZATION OF MICROSTRUCTURE AND PROPERTIES

Figure 5 shows a 123 bar processed by partial-melt growth. The bar exhibits a uniformly textured microstructure over its entire length. The more detailed microstructure can be seen in Figure 6, which shows that the superconducting 123 grains are well aligned along the a–b plane (Figure 6b). For comparison, Figure 6a shows the microstructure of a zone-melted sample that had been previously produced.[14] A large amount of 211 phase (irregularly shaped particles) is seen in the zone-melted sample. However, as shown in Figure 6b, no 211 phase exists in most regions of the sample processed by partial-melt growth. The reason for this second-phase reduction is not yet clear. We believe that there are three possible causes: (1) effects of processing temperature and phase decomposition kinetics; (2) differences in control of cooling rate; and, (3) the small addition of silver in the partial-melt sample. Current work on partial melting of samples without silver will determine the validity of possibility (3).

Liquid phase starts to form significantly above about 1000°C. Above 1000°C, the higher the temperature or the longer the time at temperature, the more 211 will be formed. Therefore, heating the sample to a temperature much above the peritectic point will likely result in a large quantity of 211. However, phase decomposition requires diffusion, which means that the kinetics of the melting process also plays an important role in terms of second-phase concentration. The partial-melt growth is done in such a way that an appropriate amount of liquid phase is formed in a

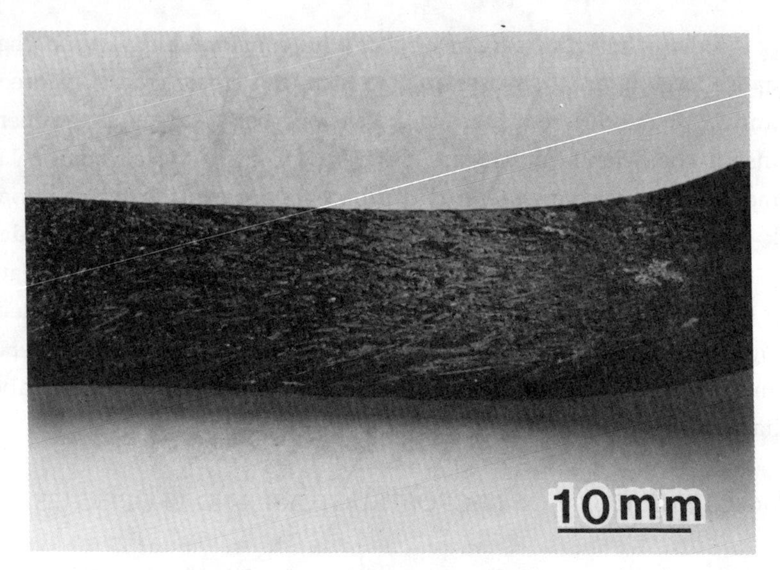

Figure 5. Optical micrograph showing a continuous and uniform microstructure of a textured 123 sample.

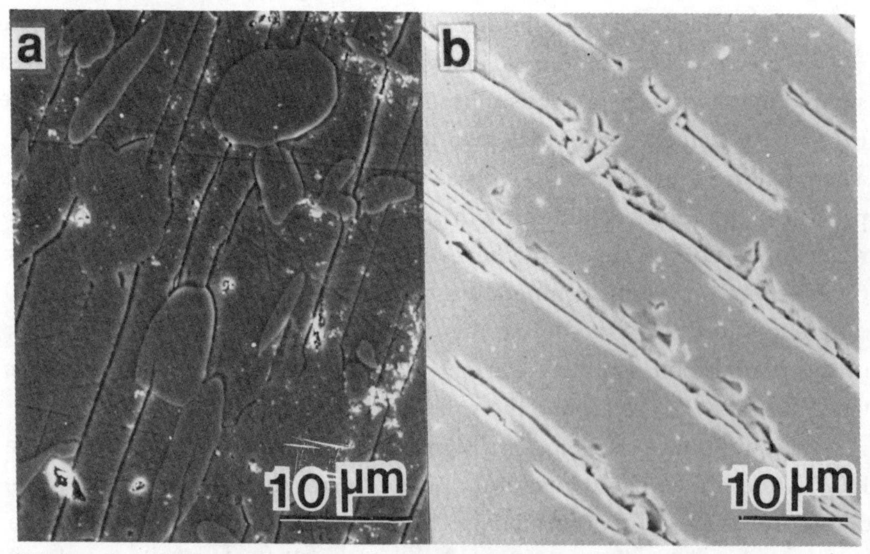

Figure 6. Scanning electron microscopy photograph showing well-textured microstructures of the samples processed by (a) zone melting and (b) partial-melt growth; note that second phases are rarely observed in the sample processed by partial-melt growth.

short time frame. This approach may cause 123 to decompose without much 211 precipitating from the liquid. 123 decomposes incongruently and the composition of the liquid may vary to a large extent.

Since silver has a considerably lower melting point (≈960°C) than 123, it may accelerate the partial melting of 123. Clearly, to confirm this hypothesis, we need to carry out detailed experiments, including differential thermal analysis.

The superconducting properties of the textured samples were examined by magnetization and transport current measurements. Figure 7, a plot of zero-field-cooled magnetization vs. temperature (M vs. T), shows that the sample processed by partial-melt growth exhibits a sharp transition at 90 K, indicating that it is highly homogeneous. In Figure 8, zero-field-cooled magnetization vs. applied field (M vs. H) data are presented. The sample exhibits large magnetic hysteresis at 77 K up to 3 T. By applying the critical-state model on the magnetic hysteresis results from references 18 and 19, we calculated that the textured sample had a J_c of about 10^5 A/cm^2 at 77 K and 2 T. This J_c estimation is based on the assumption that the induced currents are within the superconducting grains. The grain size was estimated from scanning electron microscopy to be on the order of 10 μm. The transport J_c was previously reported:[20] at 77 K and 1.8 T, J_c values in the textured sample reached 4.4 x 10^4 A/cm^2.

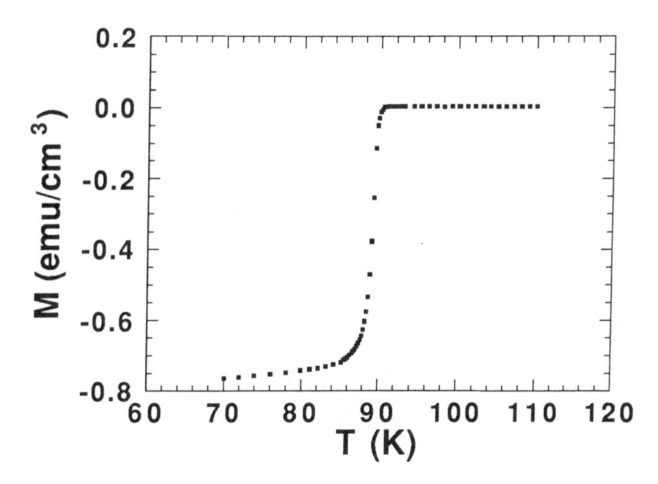

Figure 7. Zero-field-cooled magnetization vs. temperature at 5 G for a 123 sample processed by partial-melt growth.

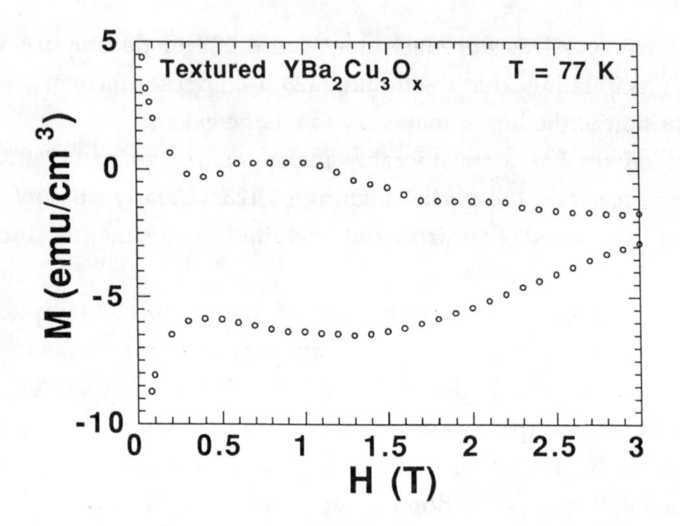

Figure 8. Zero-field-cooled magnetization vs. applied field at 77 K for a 123 sample processed by partial-melt growth.

ACKNOWLEDGMENTS

This work was supported by the U. S. Department of Energy, Basic Energy Sciences–Materials Science, under Contract W-31-109-ENG-38. J.A. is also with Department of Physics, Illinois Institute of Technology, Chicago, Illinois 60616.

REFERENCES

1. J. P. Singh, H. J. Leu, R. B. Poeppel, E. Van Voorhees, G. T. Goudey, K. Winsley, and D. Shi, J. Appl. Phys. **66**, 3154 (1989).
2. D. Shi, M. Xu, J. G. Chen, A. Umezawa, S. G. Lanan, D. Miller, and K. C. Goretta, Mater. Lett. **9**, 1 (1989).
3. D. Shi and K. C. Goretta, Mater. Lett. **7**, 428 (1989).
4. S. X. Dou, H. K. Liu, M. H. Apperley, K. H. Song, and C. C. Song, Supercond. Sci. Technol. **3**, 138 (1990).
5. S. Jin, T. H. Tiefel, R. C. Sherwood, M. E. Davis, R. B. van Dover, G. W. Kammlott, R. A. Fastnacht, and H. D. Keith, Appl. Phys. Lett. **52**, 2074 (1988).

6. K. Salama, V. Selvamanickam, L. Gao, and K. Sun, Appl. Phys. Lett. **54**, 2352 (1989).
7. M. Murakami, M. Morita, K. Doi, and K. Miyamoto, Jpn. J. Appl. Phys. **28**, 1189 (1989).
8. M. Murakami, M. Morita, and N. Koyama, Jpn. J. Appl. Phys. **28**, 1125 (1989).
9. M. Murakami, M. Morita, K. Doi, K. Miyamoto, and H. Hamada, Jpn. J. Appl. Phys. **28**, 399 (1989).
10. U. Balachandran, R. B. Poeppel, J. E. Emerson, S. A. Johnson, M. T. Lanagan, C. A. Youngdahl, D. Shi, K. C. Goretta, and N. G. Eror, Mater. Lett. **8**, 454 (1989).
11. P. J. McGinn, M. Black, and A. Valenzuela, Physica C **156**, 57 (1988).
12. P. J. McGinn, W. H. Chen, N. Zhu, U. Balachandran, M. T. Lanagan, Physica C **165**, 480 (1990).
13. R. L. Meng, C. Kinalidis, Y. Y. Sun, L. Gao, Y. K. Tao, P. H. Hor, and C. W. Chu, preprint (1990).
14. D. Shi, H. Krishnan, J. M. Hong, D. Miller, P. J. McGinn, W. H. Chen, M. Xu, J. G. Chen, M. M. Fang, U. Welp, M. T. Lanagan, K. C. Goretta, J. T. Dusek, J. J. Picciolo, and U. Balachandran, J. Appl. Phys. **68**, 228 (1990).
15. N. Chen, D. Shi, and K. C. Goretta, J. Appl. Phys. **66**, 2485 (1989).
16. T. Aselage and K. Keefer, J. Mater. Res. **3**, 1279 (1988).
17. S. E. Dorris, M. T. Lanagan, D. M. Moffat, H. J. Leu, C. A. Youngdahl, U. Balachandran, A. Cazzato, D. E. Bloomberg, and K. C. Goretta, Jpn. J. Appl. Phys. **28**, L1415 (1989).
18. C. P. Bean, Rev. Mod. Phys. **36**, 31 (1964).
19. D. Shi, M. S. Boley, U. Welp, J. G. Chen, and Y. Liao, Phys. Rev. B **40**, 5255 (1989).
20. D. Shi, M. M. Fang, J. Akujieze, M. Xu, J. G. Chen, and C. Segre, Appl. Phys. Lett., in press (1990).

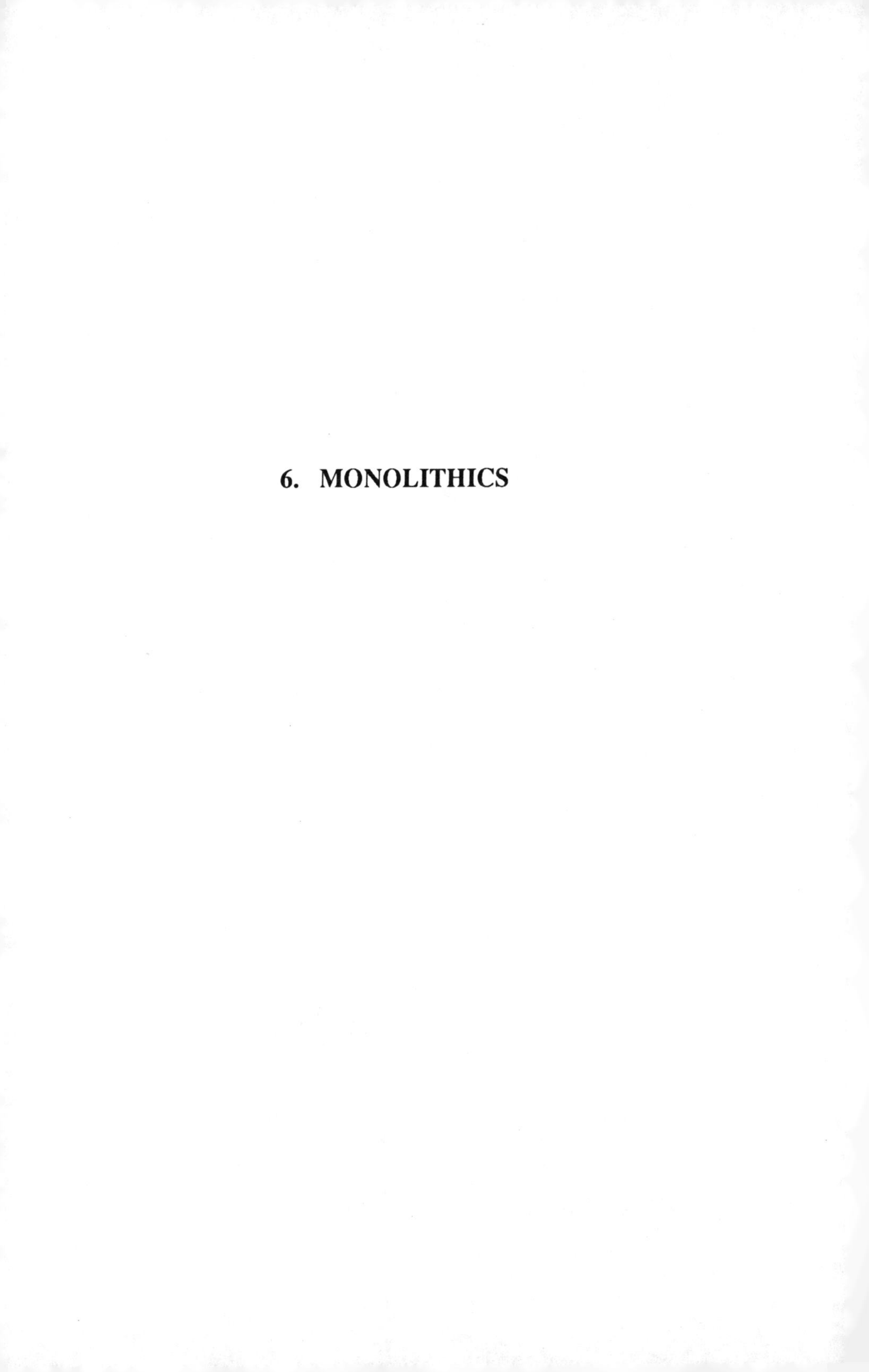

6. MONOLITHICS

A NEW SERIES OF MIXED-METAL CUPRATES IN THE T′ STRUCTURE: $Nd_{2-x-y}A_xCe_yCuO_{4-\delta}$ (A^{II} = Mg and Ca)

S. M. Wang, J. D. Carpenter, M. V. Deaton, and S.-J. Hwu*
Department of Chemistry, P. O. Box 1892, Rice University, Houston, TX 77251, U.S.A.

J. T. Vaughey, and K. R. Poeppelmeier
Department of Chemistry, Northwestern University, Evanston, IL 60208, U.S.A.

S. N. Song, and J. B. Ketterson
Department of Physics and Astronomy, Northwestern University, Evanston, IL 60208, U.S.A.

ABSTRACT

We report the synthesis of a new series of layered copper-oxide compounds $Nd_{2-x-y}A_xCe_yCuO_{4-\delta}$ (A^{II} = Mg and Ca), which are isostructural with the T′-phase of superconducting, n-type $Nd_{2-x}Ce_xCuO_{4-\delta}$. An extended solid solubility range, $x + y \leq 0.60$, is observed. The as-prepared air-quenched samples show semiconducting behavior. These newly prepared compounds raise further questions about electron-doped superconductivity in cuprate oxides.

1. Introduction

Since the discovery[1] of n-type high transition-temperature (high T_c) superconductivity in cerium doped $Ln_{2-x}Ce_xCuO_4$ (Ln = Pr, Nd, and Sm) compounds, the effort to understand the fundamental parameters that describe electron-type superconductivity[2] has been two-fold. Synthetically, the substitution of fluorine for oxygen[3] in Nd_2CuO_4 has been employed as an alternative route to achieving a reduction of copper, electron conductivity and superconductivity ($T_c \approx 27$ K). Chemical substitutions by tetravalent cations[4,5] demonstrate that the superconductivity appears to be associated with electrons in the CuO_2 planes donated by tetravalent (or intermediate valent, +3 and +4) Ce and Th, when substituted for trivalent Nd, *e.g.*, $Eu_{2-x}Ce_xCuO_{4-\delta}$ and $(Nd, Pr)_{2-x}Th_xCuO_{4-\delta}$. Experimental evidence from both photoemission spectroscopy[6] and X-ray Absorption Spectroscopy (XAS)[7] have reportedly confirmed that the system is indeed doped with electrons. Theoretically, the local density functional energy band calculations[8] of T′-$Nd_{2-x}Ce_xCuO_{4-\delta}$ reveal that the electronic structure of this electron-doped superconductor near the Fermi level (E_F) is similar to the hole doped superconductors. The unique feature of the band structure appears to be the additional O(2) p-band that has resulted from the interaction of p_z - p_z orbitals from vertical O(2) - O(1) - O(2) - O(1) chains (see structure description below) lying just below E_F.

The Ce-doped superconductors adopt the Nd_2CuO_4 (T′-phase, as shown in Figure 1) structure, which is composed of a metal framework similar to that of the K_2NiF_4 structure with Cu-O(1) layers. However, the O(2) atoms are in a totally different position than the apical oxygen of Cu in La_2CuO_4 (T-phase). They share the same in-plane symmetry as the O(1) and also connect the O(1) atoms along the z-direction perpendicular to the Cu - O(1) plane. As shown in Figure 1, the A-site cation (Nd in this case) has a coordination number of eight which is the same as that in $YBa_2Cu_3O_7$ but less than the 9 as seen in La_2CuO_4 and $La_{2-x}Sr_xCuO_4$ (T-phase). It is interesting to note that the structure of the Sr codoped n-type $Nd_{2-x-y}Sr_xCe_yCuO_{4-\delta}$ superconductors ($T_c \approx 28$ K)[9~11] no longer retains the T´-phase. The T*-phase has the same metal framework but a different oxygen arrangement. Its structure can be described as a combination of T-phase (*e.g.*, CuO_6 octahedral in $La_{2-x}Sr_xCuO_{4-\delta}$) and T´-phase (*e.g.*, CuO_4 square planar in $Nd_{2-x}Ce_xCuO_{4-\delta}$) to give rise to a square pyramidal coordination at the Cu-sites; *e.g.*, CuO_5. Furthermore, the ordering of the cations Nd(Ce) and Nd(Sr) is evident in T´ and T-type structures. However, the factors that govern the phase formation of T, T* or T´ are not totally understood.

In this paper we describe the synthesis, crystal structure and conductivity of divalent cation doped T´-$Nd_{2-x-y}A_xCe_yCuO_z$ (A^{II} = Mg and Ca). We demonstrate that the relative cation sizes of Mg^{2+} and Ca^{2+} (including Ce^{4+}) vs. Nd^{3+} may be responsible for retaining the T´-structure. Superconductivity appears to be introduced by Ce^{4+}-doping in the compound of $Nd_{2-x}Ce_xCuO_{4-\delta}$ ($\delta \approx 0.04$)[1] but does not exist in the current systems due to an appreciable amount of oxygen deficiency.

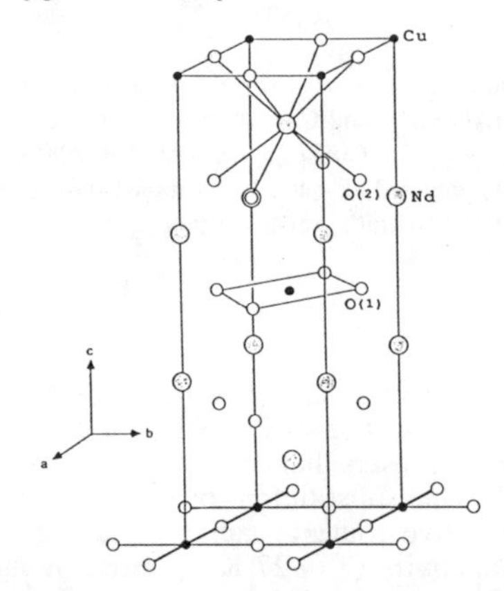

Figure 1. A schematic drawing of the structure of T´-Nd_2CuO_4.

2. Experimental

Polycrystalline samples of $Nd_{2-x-y}A_xCe_yCuO_z$ (A^{II} = Mg and Ca) were prepared by the solid state reaction of Aldrich cupric oxide (99.999%), magnesium oxide (99.99%), calcium carbonate (99.995%), cerium oxide (99.9%), and neodymium oxide (99.9%). Powders were ground with a mortar and pestle and calcined in air at temperatures ranging from 950 °C to 1000 °C three times (12 hours each time) with intermittent grindings. The dark black samples were pressed into pellets before continuing the reaction at higher temperatures, *e.g.*, 1000 °C ~ 1100 °C. The products were reground and repelletized twice. Thermogravimetric studies (with a DuPont 9900 Thermal Analysis System) were performed in a hydrogen atmosphere to determine the oxygen composition.

Disc-shaped specimens, 12 mm in diameter and 1 mm thick, were isostatically pressed at 12 Kbar at room temperature for resistivity measurements. The pellets were sintered at

1000 °C ~ 1100 °C and quenched in air. Discs were checked for metallic behavior using a two-probe voltmeter from room temperature down to liquid nitrogen temperature. Selected discs were cut into rectangular specimens with cross sections of 3 x 7 mm^2 and four leads were attached with silver paint for 4-point resistivity measurements.

X-ray diffraction (XRD) measurements were carried out on pelletized polycrystalline samples. XRD powder patterns were recorded on a Philips PW1840 diffractometer with Cu Kα radiation and an Ni filter. National Bureau of Standards (NBS) silicon was used for an internal standard. The XRD powder patterns (20° ≤ 2θ ≤ 60°) were indexed and refined by the least squares program LATT[12] with constraint to the tetragonal crystal system. The refined lattice parameters of observed XRD powder patterns are tabulated in Table I.

Table I. Cell parameters and TGA results of T′ - $Nd_{2-x-y} A_x Ce_y CuO_z$ (A^{II} = Mg, Ca).

Compositions	a(Å)	c(Å)	c/a	z[a]
System A (y = x - 0.2)				
$Nd_{1.80}Mg_{0.20}CuO_z$	3.9446(7)	12.181(2)	3.09	3.80
$Nd_{1.70}Mg_{0.25}Ce_{0.05}CuO_z$	3.947(2)	12.137(6)	3.08	3.72
$Nd_{1.60}Mg_{0.30}Ce_{0.10}CuO_z$	3.951(2)	12.119(5)	3.07	3.70
$Nd_{1.50}Mg_{0.35}Ce_{0.15}CuO_z$	3.951(2)	12.082(6)	3.06	3.67
"$Nd_{1.40}Mg_{0.40}Ce_{0.20}CuO_z$"	3.952(2)	12.072(6)	3.06	NA[b]
"$Nd_{1.30}Mg_{0.45}Ce_{0.25}CuO_z$"	3.9497(9)	12.057(4)	3.05	NA[b]
System B (y = x - 0.2)				
$Nd_{1.80}Ca_{0.20}CuO_z$	3.944(1)	12.162(4)	3.08	4.06
$Nd_{1.70}Ca_{0.25}Ce_{0.05}CuO_z$	3.9445(6)	12.145(2)	3.08	4.00
$Nd_{1.60}Ca_{0.30}Ce_{0.10}CuO_z$	3.946(2)	12.104(8)	3.07	3.84
$Nd_{1.50}Ca_{0.35}Ce_{0.15}CuO_z$	3.944(2)	12.082(6)	3.06	3.84
$Nd_{1.45}Ca_{0.37}Ce_{0.18}CuO_z$	3.946(1)	12.073(5)	3.06	3.78
$Nd_{1.40}Ca_{0.40}Ce_{0.20}CuO_z$	3.946(1)	12.052(4)	3.05	3.74
"$Nd_{1.30}Ca_{0.45}Ce_{0.25}CuO_z$"	3.948(1)	12.062(4)	3.06	NA[b]
System C (x = 0.20, y = 0.05 ~ 0.25)				
$Nd_{1.75}Ca_{0.20}Ce_{0.05}CuO_z$	3.9468(8)	12.148(3)	3.08	3.87
$Nd_{1.70}Ca_{0.20}Ce_{0.10}CuO_z$	3.946(1)	12.118(4)	3.07	3.88
$Nd_{1.65}Ca_{0.20}Ce_{0.15}CuO_z$	3.9429(9)	12.077(3)	3.06	3.91
$Nd_{1.60}Ca_{0.20}Ce_{0.20}CuO_z$	3.9485(8)	12.070(5)	3.06	3.91
"$Nd_{1.55}Ca_{0.20}Ce_{0.25}CuO_z$"	3.9439(9)	12.064(3)	3.06	NA[b]
Nd_2CuO_4	3.9453(6)	12.174(2)	3.09	3.95
$Nd_{1.85}Ce_{0.15}CuO_{3.93}$[c]	3.946	12.081	3.06	——
$Nd_{1.85}Ce_{0.15}CuO_z$	3.954(1)	12.092(5)	3.06	3.95

[a] The oxygen stoichiometries (± 0.02) are calculated based upon the TGA results.
[b] Not available. The reaction product is not a single phase (see text).
[c] Reference 17.

3. Results and Discussion

The current research was motivated by the following questions: 1) What is required for a structure to retain a T´-phase and 2) What happens to the superconductivity while the degree of orbital overlaps between oxygens p_z - p_z and Cu - O(1) dpσ bands varies because of a reduction in the c-axis dimension? The latter may be achieved by pressure[13] or, alternatively, by substituting smaller cations for Nd^{3+}. For this purpose, two relatively small (compared to Sr^{2+}) divalent cations, Mg^{2+} and Ca^{2+}, were chosen to substitute for Nd^{3+} in the following Ce-codoped systems: $Nd_{2-x-y}Mg_xCe_yCuO_{4-\delta}$ (system A, y = x - 0.20), $Nd_{2-x-y}Ca_xCe_yCuO_{4-\delta}$ (system B, y = x - 0.20), and $Nd_{1.80-y}Ca_{0.20}Ce_yCuO_{4-\delta}$ (system C, y = 0.05 ~ 0.25). Systems A and B were prepared in an attempt to achieve an isoelectronic substitution (see later discussion). In system C, the effect on superconductivity was investigated by reducing the oxidation state of the copper cation through the substitution of trivalent Nd^{3+} cations with tetravalent Ce^{4+} cations.

All magnesium and calcium of Ce-codoped compounds adopted the Nd_2CuO_4 structure (T´-phase) according to the XRD patterns. In Figure 2 a typical XRD pattern of Mg-doped $Nd_{2.00-x}Mg_xCuO_{4-\delta}$ (x = 0.20 and δ = 0.24) is shown and compared with that of T´-Nd_2CuO_4. The XRD patterns of the latter compound match with that of the

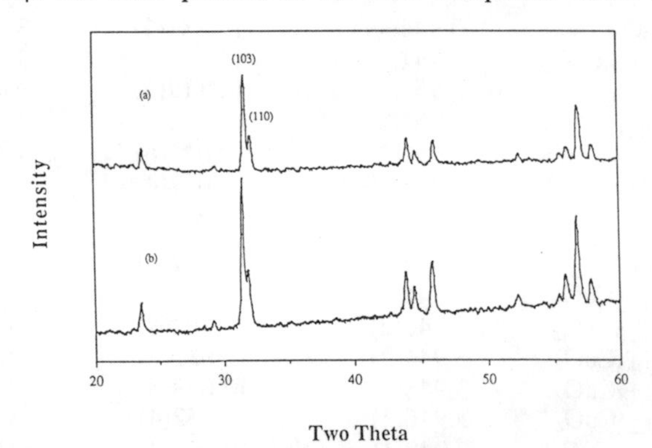

Figure 2. Powder x-ray diffraction patterns of (a) Nd_2CuO_4 (T´-phase) and (b) T´-$Nd_{1.80}Mg_{0.20}CuO_{3.76}$.

previously reported T´-Nd_2CuO_4.[14] The 103 (or 013) and 110 reflections, that show a very small separation around 32° in 2θ, are indicative peaks of T´-phase in XRD. No evidence shows that the divalent cations, Mg and Ca in this case, are ordered.

As reported in Table I, since the size of the Nd^{3+} cation is different from that of Mg^{2+}, Ca^{2+} and Ce^{4+}, a change in the cell parameters is expected. The Shannon[15] crystal radii for eight coordinated Nd^{3+}, Ce^{4+}, Ca^{2+} and Mg^{2+} ions are 1.249Å, 1.11Å, 1.26 Å and 1.03Å, respectively. The calculated magnesium radius is comparable with that observed in synthetic pyrope ($Mg_3Al_2Si_3O_{12}$).[16] This cation size effect can be readily seen from the merging of the 103 (or 013) and 110 reflections, as shown in Figure 3. For example, these two diffraction peaks in the $Nd_{2-x-y}Mg_xCe_yCuO_4$ XRD patterns move closer toward each other as the total substituents (x + y) increases from 0.2 to 0.6. A closer examination of the peak positions revealed that the (103) peak moves to higher angles and the (110) peak to slightly lower angles indicating a contraction of the c-axis and an expansion of the a-axis

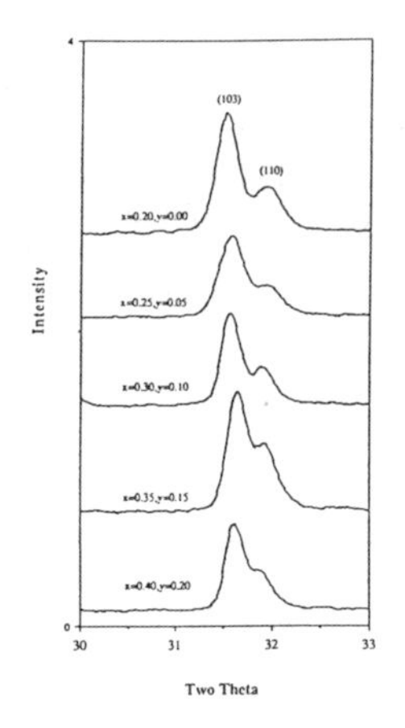

Figure 3. Sections of x-ray diffraction patterns showing the (103) and the (110) reflections of the $Nd_{2-x-y}Mg_xCe_yCuO_{4-\delta}$.

as the concentration of the dopant increases. For both magnesium and calcium doped systems, the c/a ratios range from 3.05 to 3.09 (see Table I), which is also a characteristic of the T´-phase.

In Figure 4, a linear relationship between c and (x + y) is shown to indicate that the cation substitution in compounds $Nd_{2-x-y}A_xCe_yCuO_4$ is successful. The solubility limits

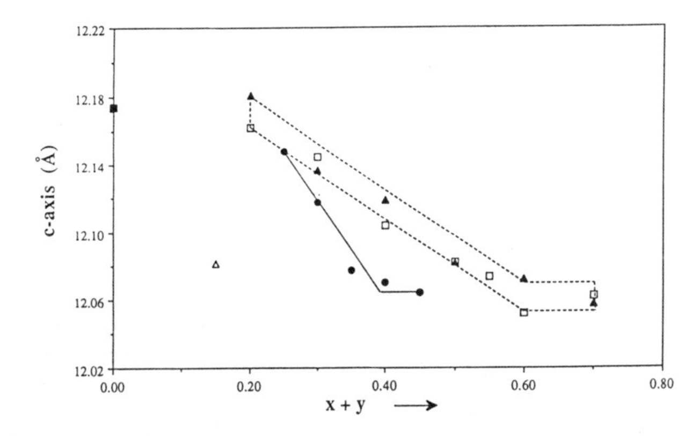

Figure 4. Values of lattice dimensions vs total amount of dopant (x + y) of the $Nd_{2-x-y}A_xCe_yCuO_{4-\delta}$: system A ($A^{II}$ = Mg, ---▲---), system B (A^{II} = Ca,---☐---), system C (—●—), and, for comparison, lattice parameters of Nd_2CuO_4 (■) and $Nd_{1.85}Ce_{0.15}CuO_4$ (Δ).

(x_c) for systems A (Mg-doped), B (Ca-doped) and C are 0.60, 0.60 and 0.35, respectively, which are defined by Vegard's Law. These values are all significantly higher than that of Ce-doped $Nd_{2-x}Ce_xCuO_4$ ($x_c = 0.2$).[17] The c lattice parameter decreases as the concentration of the dopant increases. Consequently, the unit cell volume decreases with increasing (x + y). The contraction in c is thought to be primarily due to the presence of the smaller Ce^{4+} cation. This effect is more profoundly shown in system C. However, the change in the cell dimensions is rather complicated. It is more likely attributed to the combination of the size of the substituent, the oxidation state of the copper atom, and the oxygen deficiency (δ).

Let us first look at the so called "isoelectronic" systems A and B, where the copper oxidation state presumably stays the same if $\delta = 0$. For each divalent magnesium (Mg^{2+}) or calcium (Ca^{2+}) cation substituted for a trivalent neodymium (Nd^{3+}), a tetravalent cerium (Ce^{4+}) cation was co-substituted to keep the charge balanced. The resultant c lattice is complicated by the large amount of oxygen deficiency (see later discussion). In $Nd_{1.80}A_{0.20}CuO_z$ systems (A = Mg and Ca), the c-axis of Mg^{2+}-doped compounds is larger than that of Nd_2CuO_4, *e.g.*, 12.181(2)Å and 12.174(2)Å, respectively, while that of Ca^{2+}-doped is smaller, *e.g.*, 12.162(4)Å. This comparison in c, *e.g.*, Nd(Mg) > Nd > Nd(Ca), seems to be opposite than what one would expect in a decreasing cation size, *e.g.*, Ca > Nd > Mg. Further studies show that the trend is consistent with the oxygen stoichiometry (z), *e.g.*, Ca (4.06) > Nd (3.95) > Mg (3.76). That is to say that the Mg-doped sample has the greatest oxygen deficiency. While the charge of electropositive cations (Nd^{3+}, Ca^{2+} and Mg^{2+}) remains unchanged, the reduction (or oxidation) due to the change in z may occur on the copper site. The reduction of copper causes extra electrons, presumably filling the Cu-O(1) dpσ antibonding orbitals, to give rise to the longer c-axis.

The size effect, due to the substitution of a smaller cation, Ce, is evident, *i.e.*, with Ce-codoping, the lattice constant decreases as shown in Figure 4. The c-parameters of two known compounds, *e.g.*, Nd_2CuO_4 and $Nd_{1.85}Ce_{0.15}CuO_{4-\delta}$, are also included in the figure for comparison. It is clear that the Ce^{4+} cation is largely responsible for the lattice contraction.

The thermogravimetric analysis (TGA) reveals that these compounds are highly oxygen deficient. In $Nd_{1.50}Mg_{0.35}Ce_{0.15}CuO_{3.67}$, for example, the average oxidation state of copper is as low as 1.54. Based only on powder XRD there is no structural evidence to show that the oxygen vacancies are ordered, however more complete characterization with HREM is required. It is also noted that $Nd_{1.8}Ca_{0.2}CuO_4$ shows little oxygen non-stoichiometry (within the experimental error by TGA) while $Nd_{1.8}Mg_{0.2}CuO_{4-\delta}$ ($\delta = 0.24$) shows a significant oxygen deficiency. A large oxygen deficiency is observed in both systems when doped with the smaller Ce^{4+} cations. This may be attributed to the decrease in the coordination number around small cations.

The results from system C are rather interesting according to TGA. All the Ce-doped compounds have an approximately equal oxygen deficiency, *e.g.*, $\delta \sim 0.09$-0.13 (see Table I). Thus the overall oxidation state of copper is about the same throughout the solubility range. (The formal oxidation state ranges from 1.82 to 1.89.) In increasing the Ce-dopant level, systematic trends are again shown by the decreasing c-lattice parameter. This suggests that the oxidation state of copper is not a determining factor, but rather the size of the cations, accompanied with the oxygen vacancies, for the c axis dimension. Moreover, the relatively low oxygen deficiency, even with high concentration of smaller Ce^{4+} cations, is believed to be attributed to the net electrostatic effect. The higher oxidation state, 4+ for Ce, requires more anionic charge to balance the oxidation state and in turn more oxygen is required.

Figure 5 shows the temperature dependent resistivity curves for $Nd_{2-x-y}Mg_xCe_yCuO_4$. Although the undoped compound Nd_2CuO_4 is a typical semiconductor,[11,18] doping with Mg^{2+} and Ce^{4+} ions increases the conductivity remarkably. The quenched samples (from ca. 1050 °C in air down to room temperature) show semiconducting behavior down to liquid helium temperature. Post oxygen annealing in an ambient atmosphere does not increase the oxygen content nor change the bulk conductivity significantly. While the results are preliminary, our findings do reflect two points: First, doping with smaller cations improves the conductivity by a factor of 1000 and second, high pressure oxygen annealing may help in decreasing oxygen vacancies, possibly favoring superconductivity. It is known that high pressure oxygen produces bulk superconductivity in some Sr-doped phases.[10,19]

All of these newly prepared compounds, $Nd_{2-x-y}M_xCe_yCuO_{4-\delta}$, (M = Mg and Ca) crystallize in the T´-structure as opposed to Sr-doped compounds which form the T*-phase[19]. This research demonstrates that the size limitation on A-site cations is responsible for retaining the T´-phase. It is not known, however, whether the lack of superconductivity is associated with the difference in the crystal structures (T´ vs T*) or to the extended oxygen vacancies. Nevertheless, codoping with both divalent and tetravalent cations extends the solubility limit (0.60 for the Mg^{2+} and Ca^{2+}-doped compounds), and, moreover, significantly improves the conductivity of the T´-phase. The small cations affect on the resistivity coupled with the large oxygen deficiency indicates that high pressure oxygen annealing studies need further investigation in order to understand the T_c with respect to the carrier concentration and crystal structure.

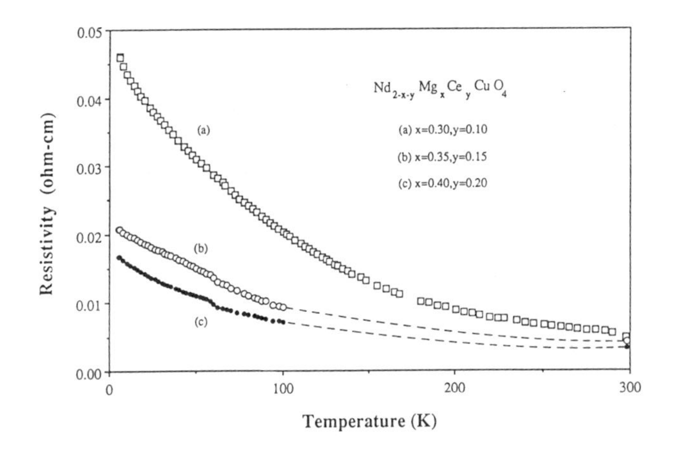

Figure 5. Resistivity of $Nd_{2-x-y}Mg_xCe_yCuO_{4-\delta}$ as a function of temperature.

4. Acknowledgements

This work was supported by a Rice University startup grant (S.-J. Hwu) and by the Science and Technology Center for Superconductivity (Grant NSF-DMR-8809854).

5. References and Notes:

1. Y. Tokura, H. Takagi and S. Uchida, *Nature* **337** (1989) 345.
2. V. J. Emery, *Nature* **337** (1989) 306.
3. A. C. W. P. James, S. M. Zahurak and D. W. Murphy, *Nature* **338** (1989) 240.
4. J. T. Markert and M. B. Maple, *Solid State Commun.* **70** (1989) 145.
5. J. T. Markert, E. A. Early, T. Bjørnholm, S. Ghamaty, B. W. Lee, J. J. Neumeier, R. D. Price, C. L. Seaman and M. B. Maple, *Physica C.* **158** (1989) 178.
6. A. Fujimori, Y. Tokura, H. Eisake, H. Takagi, S. Uchida and E. Takayama-Muromachi, preprint.
7. E. E. Alp, S. M. Mini, M. Ramanathan, B. Dabrowski, D. R. Richards and D. G. Hinks, *Phys. Rev. B.* **40(4)** (1989) 2617.
8. S. Massidda, N. Hamada, J. Yu and A. J. Freeman, *Physica C.* **157** (1989) 571.
9. J. Akimitsu, S. Suzuki, M. Watanabe, and H. Sawa, *Jpn. J. Appl. Phys.* **27** (1988) L1859.
10. E. Takayama-Muromachi, Y. Matsui, Y. Uchida, F. Izumi, M. Onoda and K. Kato, *Jpn. J. Appl. Phys.* **27** (1988) L2283.
11. H. Sawa, S. Suzuki, M. Watanabe, J. Akimitsu, H. Matsubara, H. Watabe, S.-I. Uchida, K. Kohusho, H. Asano, F. Izumi and E. Takayama-Muromachi, *Nature* **337** (1989) 347.
12. LATT: F. Takusagawa, Ames Laboratory, Iowa State University, Ames, Iowa, unpublished research, 1981.
13. Pressure-dependent electrical resistivity measurements reveal an increase of T_c with applied pressure for $Nd_{1.85}Ce_{0.15}CuO_{4-\delta}$ ($\delta = 0.02$) at a rate of ~ $dT_c/d_p \approx 0.025$ $K/_{kbar}$. For details, see the following reference: J. T. Markert, E. A. Early, T. Bjørnholm, S. Ghamaty, W. B. Lee, J. J. Neumeier, R. D. Price, C. L. Seaman, and M. B. Maple, *Physica C.* **158** (1989) 178.
14. No. 24-777, Joint Committee on Powder Diffraction Standards, Swarthmore, PA.
15. R. D. Shannon, *Acta Crystallogr. Sect. A.* **32** (1976) 751.
16. (a) G. V. Gibbs and J. V. Smith, *Am. Mineral* **50** (1965) 2023.

    (b) Note: The Mg-O bond distances in a distorted MgO_8 cube are 2.198Å and 2.343Å. The eight coordinated Mg radii, resulting from subtracting the eight coordinated O^{2-} crystal radius (1.28Å), are 0.92Å and 1.06Å.
17. T. C. Huang, E. Moran, A. I. Nazzal and J. B. Torrance, *Physica C.* **158** (1989) 148.
18. P. Ganguly and C. N. R. Rao, *Mat. Res. Bull.* **8** (1973) 405.
19. S.-W. Cheong, Z. Fisk, J. D. Thompson and R. B. Schwarz, *Physica C.* **159** (1989) 407.

EFFECT OF ALKALI ELEMENT DOPING ON PROPERTIES OF $Bi_2Sr_2CaCu_2O_8$

S.X. Dou, W.M. Wu, H.K. Liu, W.X. Wang, and C.C. Sorrell

School of Materials Science and Engineering
University of New South Wales
Kensington, NSW 2033, Australia

W.M. Bian and C.L. Ji

Northeast University of Technology
Shenyang, PR China

ABSTRACT
The effect of alkali element doping on the superconducting properties of the Bi-Sr-Ca-Cu-O system has been investigated. It was found that a T_c of 94 K and T_o of 90 K in alkali element doped $Bi_2Sr_2CaCu_2O_8$ were achieved through the use of a melt-processing technique. XRD, SEM, and EDS examinations showed that the samples did not contain the high-T_c phase $Bi_2Sr_2Ca_2Cu_3O_{10}$. Alkali elements were found to be excellent sintering aids for processing Bi-based superconductors since they enhance the T_c and promote grain growth and grain alignment in the bulk materials. Two modulated structures with vectors 4.75b and 2b were observed to coexist in the alkali element doped samples.

INTRODUCTION

The superconducting phase $Bi_2Sr_2CaCu_2O_8$ (2212) has been extensively studied. The T_c for this phase varies largely depending on the processing conditions [1-3]. A T_c up to 92 K has been reported by treating the sample in low oxygen partial pressure or quenching [3]. Recently, an improvement in T_c by alkali doping has been reported by Kawai and co-workers [4]. They found that Li doping raised the T_c of 2212 to 95K while K doping depressed the T_c significantly; Na did not affect the T_c.

In the present paper, we report the effect of alkali element doping on the properties of the 2212 phase. It has been found that Li,

Na, and K doping all raised the T_c to 90 K without the need for low oxygen partial pressure treatment or quenching. Alkali element doping also has a strong effect on the microstructures and the formation of superlattice .

EXPERIMENTAL PROCEDURE

Mixtures of Bi_2O_3 (99.9%), $SrCO_3$ (99.5%), A_2CO_3 (99%) [where A = Li, Na, or K], $CaCO_3$ (99%) and CuO (99.9%) in appropriate proportions were ground with a mortar and pestle and calcined at 720°C for 12 h and 780°C for 10 h, with intermediate grinding. The calcined powders were then pressed into pellets and sintered or partially melted in Ag boats at 740° to 890°C for 5 min. to 15 h.

The electrical resistivity was measured by the standard four-probe d.c. technique. The a.c. magnetic susceptibility was measured by a mutual inductance method at a frequency of 83 Hz. Microstructural and compositional analyses were performed with a JEOL JSM-840 scanning electron microscope (SEM) and JEOL FX2000 transmission electron microscope (TEM), both of which were equipped with a Link Systems AN10000 energy dispersive spectrometer (EDS). X-ray powder diffraction patterns were obtained with a Philips type PW1140/00 powder diffractometer using CuKα radiation and scanning rate of $2\theta = 1^o$/min. Chemical analyses were carried out with a Labtam International Plasmalab inductively coupled plasma atomic emission spectrometer.

RESULTS AND DISCUSSION

Figure 1 shows resistivity-temperature curves for samples with the cation ratio Bi:Sr:Ca:Cu:A = 2.2:1.8:1.05:(2.15-x):x, where A = Li, x = 0.7; A = Na, x = 0.6; and A = K, x = 0.7. The heat treatment conditions for these samples were as follows. The Li-doped sample was sintered at 830°C for 1 h in nitrogen and cooled at a rate of 10°C/h to 790°C and then at 60°C/h to room temperature. The Na-doped sample was treated at 885°C for 5 min. in air and cooled at a rate of 60°C/h to room temperature. The K-doped sample was treated at 850°C for 10 h and cooled at a rate of 60°C/h to room temperature. It can be seen that Na raised the T_c to 94 K and T_o to 90 K, while K and Li raised the T_c the same degree, where the T_c = 92 K and T_o = 88 K. The undoped standard sample, which was partially melted at $890^{o}C$ for 1 h with cooling rate of $60^{o}C/h$, showed the T_c = 84 K and T_o = 80 K. These results are further rate confirmed by the a.c. magnetic susceptibility measurements, as shown in Figure 2. It is clear that Na, K, and Li doping increased the T_c of the 2212 phase relative to the undoped sample.

EDS analysis and X-ray examination show that the alkali element doped samples consisted of 2212 as the major phase. No high T_c phase (2223) was detected, indicating that the enhancement in the T_c for the alkali element doped samples cannot be attributed to the presence of 2223.

At sintering temperature between $810^{o}C$ to $885^{o}C$, the samples were partially melted owing to the formation of eutectics between alkali elements oxides and Bi_2O_3. These liquid phases act as fluxes, which provide a fast diffusion path, which accelerates the formation of the superconducting phase. Furthermore, the substitution of alkali element in the superconducting phase may give rise to a reducing effect on the oxygen content, similar to the effect of low oxygen partial pressure treatment, which leads to enhancement of the T_c.

The melting process through alkali element doping not only promotes the formation of the superconducting phase but also affects the microstructure. Figure 3 shows the electron diffraction patterns on the (001) plane for Na-doped Bi-Sr-Ca-Cu-O. A new modulated structure along the b axis is evident, as shown by the satellite spots. The modulation vector becomes 2 b in addition to the 4.75 b, which has been commonly reported in the literature [4,5]. The change in the modulation periodicity in the Na-doped sample may be attributed to the change in oxygen content of the superconducting phase. The modulation structure with a vector of 4.75 is also observed in the alkali element doped samples.

Figure 4 shows a micrograph of the arrays of dislocations in the Na-doped sample, indicated by the arrows. It is seen that the lattice fringes are shifted systematically.

The melt processing of alkali element doped samples promotes a high degree of grain alignment, as indicated by the micrograph in

Figure 5. It is possible that a continuous process for texture growth can be achieved with a directional temperature gradient.

Alkali element doping also affects the impurity phase Sr-Ca-Cu-O solid solution, which is commonly found in the Bi-Sr-Ca-Cu-O system. The lattice parameters for this phase have been reported to a = 1.328 nm, b = 1.277 nm, and c = 0.389 nm, with face-centered orthorhombic symmetry (6). In the Na-doped samples, the lattice parameters are changed to a = 1.628 nm, b = 1.174 nm, and c = 0.410 nm, although the face-centered symmetry remains unchanged. An incommensurate modulated structure was also observed, which has a modulation vector of 2.4 c.

In summary, alkali element doping in the Bi-Sr-Ca-Cu-O system enhances the T_c, promotes the grain alignment, and affects the mmodulation structure, defects, and microstructure.

ACKNOWLEDGEMENTS

The authors gratefully acknowledge the financial support of Metal Manufactures Ltd., The Commonwealth Department of Industry, Technolog and Commerce.

REFERENCES

1. S.X. Dou, H.K. Liu, A.J. Bourdillon, N.X. Tan, N. Savvides, J.P. Zhou, and C.C. Sorrell, Supercond. Sci. Technol, 1 (1988) 178.
2. T. Kawai, T. Horiuchi, K. Mitsui, K. Ogura, S. Takagi, and S. Kawai, Physica C, 161 (1989) 561.
3. J.L. Tallen, R.G. Buckley, P.W. Gilberd, M.R. Presland, I.W.M. Brown, M.E. Bowden, L.A. Christian, and R. Goguel, Nature 333 (1988) 153.
4. E.A. Hewat, M. Dupuy, P. Bordet, J.J. Capponi, C.Chaillout, J.L. Hodeau, and M. Marajio, Nature 333 (1988) 53.
5. T.M. Shaw, S.A. Shivashankar, S.J. La Plica, J.J. Luomo, T.R. McGuire, R.A. Roy, K.H. Kelleher, and D.S. Yee, Phys. Rev. B, 37 (1988) 9856.

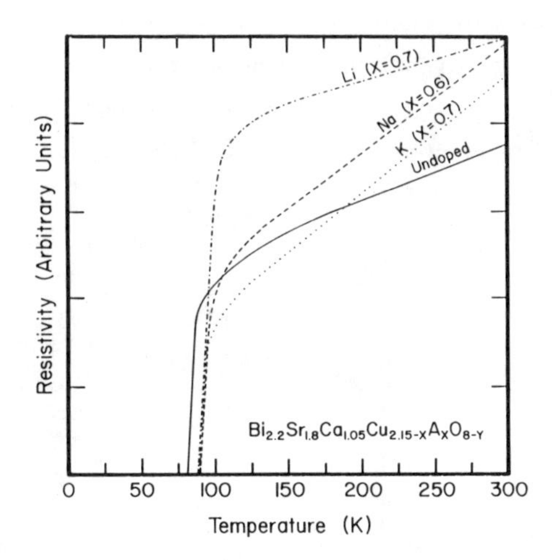

Figure 1. Effect on the transition of alkali doping: Undoped, 890°C for 1h in air; Li-doped, 850°C for 1h in air; Na-doped, 885°C for 5 min. in air; and K-doped, 850°C for 10h in air.

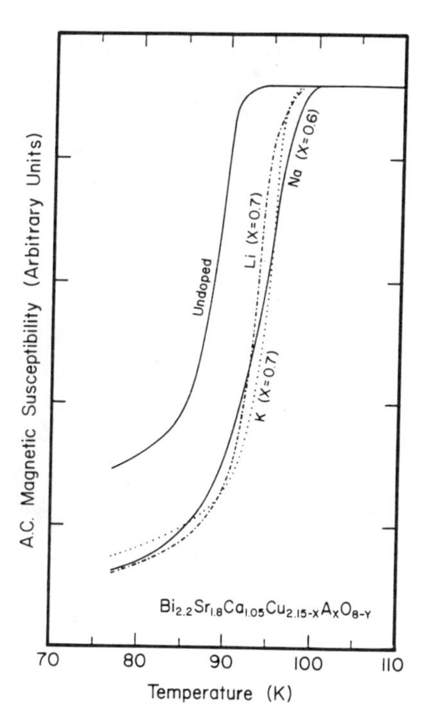

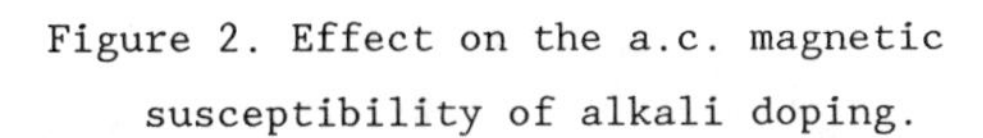

Figure 2. Effect on the a.c. magnetic susceptibility of alkali doping.

Undoped, 890°C for 1h in air
Li-doped, 830°C for 1h in air
Na-doped, 885°C for 5 min. in air
K -doped, 850°C for 10h in air

Figure 3. Electron diffraction patterns on (001) plane showing new microstructures of (a) Li-doped and (b) Na-doped samples.

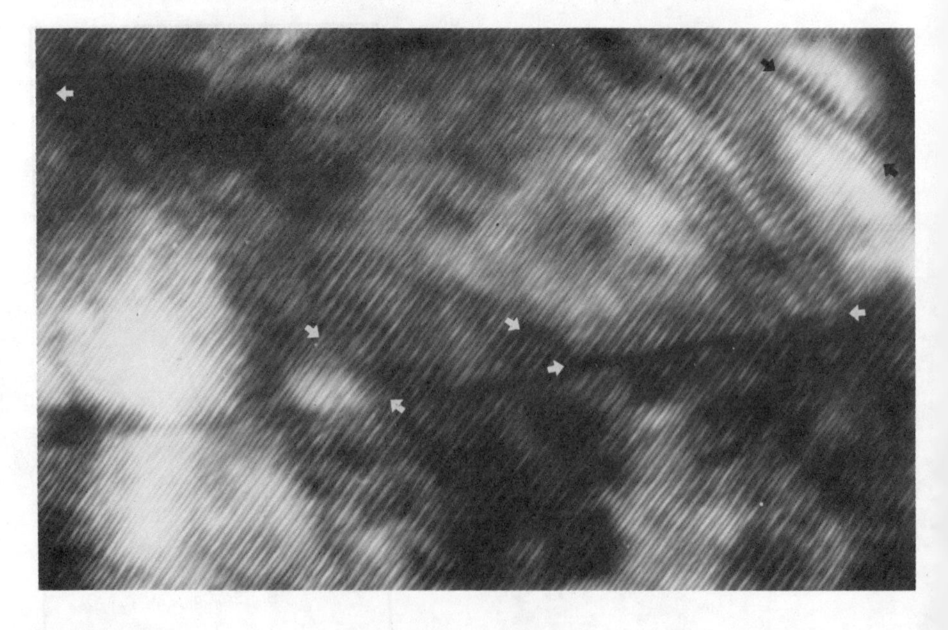

Figure 4. Micrograph of Na-doped Bi-Sr-Ca-Cu-O showing the systematic shift of lattice fringes due to dislocations.

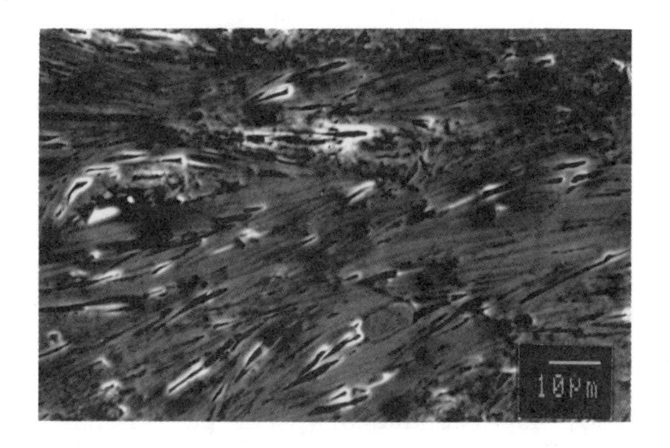

Figure 5. Micrograph of a Na-doped sample showing the grain alignment through the melt processing.

THE CRITICAL TEMPERATURE OF NEW CLASS AND OLD TYPE SUPERCONDUCTORS

Om P. Sinha
Clark Atlanta University, Atlanta, GA 30314, USA

ABSTRACT

The basic processes correlated with the critical temperature of superconductors are being examined. In the old type of superconductors, the pair bondage energy Δ is reduced to zero at the critical temperature. We examine the possibilities of interactions different from the B.C.S. type of phonon interaction for this bondage energy. The critical temperature may also be correlated with the Bose-Einstein condesation temperature for the Bosons. We show that the B.E. condesation temperature is the critical temperature of the doped type of the new class of superconductors.

INTRODUCTION

Electrons in superconductors are in pairs as Bosons with spin zero. This theoretical idea was proposed and fairly examined by several authors[1-4], but could not lead to the explanation of critical temperatures and other theoretical predictions. However, the pairing was identified experimentally[5] and theoretically[6] through the examination of the magnetic fux quantization in superconductors. The most helpful and fairly complete explanation came through the B.C.S. Theory[7] in which the coupling has been correlated to the phonon interaction related bondage at the ground state of the carriers of the superconductors. This has been expected even for very low critical temperatures, most probably through the correlation of the phonons with Heisenberg uncertainty principle.

Our basic idea is that Heitler-London type of covalent bondage calculated through wave functions in Hilbert space can acquire results somewhat similar to the B.C.S. theoretical results which has been worked out in Fock space with Many-Body techniques. Sometimes it is not only easier but also more useful to have calculations without the Boson-Fermion interaction related Many-Body technique. In such cases the more practical paramenters like dielectric constant etc. can be used which are not obtainable with experimental agreement using the Fock space related Many Body technique.

Our purpose is to go through the Heitler-London theoretical results and examine the possibility of its extension to the crystalline superconductor solids, then compare these results with the B.C.S. theoretical results.

HEITLER LONDON COVALENCY RELATED PAIR BONDAGE

In Heitler-London calculation of the ground state

of H_2 molecule, three types of energy levels are obtained[8]. In the normal case the energy eigenvalue is given by:

$$W_N = \iint u_{1s}(r_1-A)\,u_{1s}(r_2-B)\,\mathcal{H}\,u_{1s}(r_1-A)\,u_{1s}(r_2-B)\,d^3r_1\,d^3r_2 \quad (1)$$

where r_1, r_2, A and B are coordinates of the electrons and nuclei respectively. The wave function U_{1s} is the ground level wave function of H atom. Obviously the energy level is dependent upon r_{AB}, the distance between the two nuclei.

This normal energy W_N is only slightly below $-2R_y$ for a long range of r_{AB}, where R_y is the ground state energy depth $\simeq$ 13.6 e.v. for a single H atom.

The two other energy levels belong to the spin wise triplet and singlet states. In these cases an antisymmetric and symmetric combinations of the wave functions are used. For these combinations the used expectation values are $H_{I,I}$ and $H_{I,II}$, where $H_{I,I}$ is W_N and $H_{I,II}$ is given by:

$$H_{I,II} = \iint u_{1s}(r_1-A)\,u_{1s}(r_2-B)\,\mathcal{H}\,u_{1s}(r_1-B)\,u_{1s}(r_2-A)\,d^3r_1\,d^3r_2 \quad (2)$$

For the combination of the two valence electrons in spinwise triplet case, the energy level is,

$$W_A = (H_{I,I} - H_{I,II})/(1-N_2^2) \quad (3)$$

where N_2 used for the normalization is,

$$N_2^2 = \iint u_{1s}(r_1-A)\,u_{1s}(r_1-B)\,u_{1s}(r_2-B)\,u_{1s}(r_2-A)\,d^3r_1\,d^3r_2 \quad (4)$$

The energy level for the triplet case, as a function of r_{AB}, and antisymmetric wave function combination, is much higher than $-2R_y$. On the other hand, the energy level for the singlet state is:

$$W_S = (H_{I,I} + H_{I,II})/(1+N_2^2) \quad (5)$$

This energy level W_s for the singlet state of the electron pair goes through a minimum energy value

$\simeq$ 2.25 R_y at $r_{AB} \simeq$ 1.5a_B, where a_B is the Bohr radius. Beyond r_{AB} = 1.5a_B, the energy comes back closer and closer to $-2R_y$.

The lowest ground energy level value for the singlet state is a little bit above this minimum energy level at $r_{AB} \simeq$ 1.5 a_B. This ground level E_g has the value:

$$E_g = E_{min} + \hbar\omega/2 \qquad (6)$$

where,

$$\omega^2 = \left(\frac{2}{M}\frac{\partial^2 E}{\partial r_{AB}^2}\right) \qquad (7)$$

M is the single nuclear mass. This ground state is a discrete energy level as expected below $-2R_y$. This addition to E_{min} is correlated with Heisenberg uncertainly principle by which r_{AB} cannot have particularly single value in these continuum values of the calculated negative energies. The other discrete energy eigen values are:

$$E_n = E_{min} + (n+1/2)\hbar\omega \qquad (8)$$

These discrete energy levels have to lie below W_N (and below $-2R_y$). These are the discrete eigenstates of the pair of electrons with zero spin and symmetric combination of wave functions. These discrete levels come from the oscillation related variation of r_{AB} due to the uncertainty principle. We regard these states belonging to the spin zero related Boson pairs of electrons. If these states were not localized by having large distances between different molecules, they would show the Boson property related multiple occupation of a single eigen state through Mott transition for the multiple nuclei in a solid. These Boson type pairs of electrons may still have a significant average expected pair distance of 1.5 a_B in H_2 molecules.

It may be better to represent the ground state Boson related pairs through Boson operators in Fock space rather than Fermion pair operators.

COVALENCY RELATED PROCESS FOR ELECTRON PAIRING IN SUPERCONDUCTORS

We suppose there are two valence electrons available in superconducting type crystalline elements. There are two possibilities for these Boson related electron pairings.

(a) Both valence electrons of each atom may combine as in the He atom. The ground state of He atom is a spin-wise singlet state. The wave functions are symmetric whereas spin-wise they have to be antisymmetric for being Fermions. However, there is a fair chance that for the singlet pair of electrons as in He atom there is a pair bondage which brings the electrons to the Boson related wave function $\psi_{g\uparrow}(r_1)\psi_{g\downarrow}(r_2) + \psi_{g\downarrow}(r_1)\psi_{g\uparrow}(r_2)$ rather than the Fermion related antisymmetric combination. For a singlet pair of electrons the wave function related symmetry or antisymmetry is easily observed by the calculation of energy eigen-value. The spin wise symmetry or antisymmetry can be observed only through the difference in singlet and triplet related multiplicity. There is a fair chance that in the ground state of the bivalence atom, the electron pairs may be occupying a Boson related pair bondage. These pairs of valence electrons in a crystal will have Bloch related momentum with hopping as Boson pairs; as long as they are not excited to decoupling through phonons at higher temperatures.

(b) The second possibility is that the two valence electrons of each atom in a crystal may have Heitler London type of bondage with the two nearest neighbors along one axis. This may provide a chain related cluster of electrons which will be Bosons with spin zero. At higher level of energy provided by lattice vibration relation phonons they will still have Fermion related interhopping along that axis as $c_{i\uparrow}^{+} c_{j\uparrow} c_{j\downarrow}^{+} c_{i\downarrow}$. The lattices i and j are the nearest neighbors along that axis. In other directions there will be pair related hopping as $c_{i\uparrow}^{+} c_{m\uparrow} c_{i\downarrow}^{+} c_{m\downarrow}$. Most probably this type of Boson related bondage occurs in doped type of new superconductors whereas in the metallic type of old superconductors the other type bondage as mentioned above is more likely. This pair bondage is correlated to the energy gap between the superconducting and the normal conducting states of the electrons.

SOME DERIVATIONS OF B.C.S. TYPES OF RESULTS

It is difficult to find theoretically the exact pair bondage in crystalline structures. The energy gap, on the other hand, can be found experimentally by several methods.[9] These experiements are correlated with the determination of specific heat or thermal conductivity and microwave absorption and tunnelling etc. Let us assume that the transfer through this energy gap Δ occurs with lattice related phonon contribution.

At critical temperature T_c, the crystalline lattice will provide enough phonon energy to raise the Bosons above the gap Δ. At any temperature T, the energy E_T provided by the phonons is[10]:

$$E_T = \frac{9NkT^4}{T_D^3}\int_0^{\hbar\omega_m/kT} \frac{x^3\,dx}{e^x - 1} \tag{9}$$

where $\hbar\omega_m$ is the highest level of phonon energy. N is the number of lattices, T_D the Debye temperature and $X = \hbar\omega/kT$. For ordinary temperatures, $\hbar\omega_m/kT >> 1$ and therefore the upper limit changed to ∞ does not make any difference. For the lower limit of integration, it may be better to replace zero by Δ/kT for each phonon to provide enough energy to raise a Boson pair through energy Δ at the critical temperature.

The total energy needed for transferring the N pairs of electrons from the superconducting state should normally be expected to be $N\Delta$. However, when the temperature is raised to T_c, the specific heat or phonon contribution comes not only from the lattices but also from the electrons. In other words the total energy needed by the N electron pairs will be less than $N\Delta$ by N times the average energy per electron pair. Using Maxwell Boltzman statistics the average value per electron pair for kinetic energy will be $(3/2)kT$. For Bosons, below the Bose-Einstein condesation temperature, the chemical potential is at the lowest energy level and the average kinetic energy is as follows:

$$\begin{aligned} E_K &= (3kT/2)[\zeta(5/2)/\zeta(3/2) \\ &\simeq 0.77\,kT \end{aligned} \qquad (10)$$

Hence the critical temperature T_c at which the lattice phonons raise up all Boson pairs through the energy gap is given by the following equation:

$$N(\Delta - 0.77)kT_c) = \frac{9NkT_c^4}{T_D^3}\int_{\Delta/kT}^{\infty}\frac{x^3\,dx}{e^x - 1} \qquad (11)$$

The critical temperature T_c can be calculated from the above equation if the energy levels Δ and kT_D are known. However, the Debye temperature T_D, dependent upon the acoustic velocity and elasticity does not have the same value at much lower temperatures. On the other hand, the isotope effect can be easily seen from the above equation. The Debye temperature T_D is correlated with the acoustic velocity C_s through:

$$kT_D = \hbar C_s (6\pi^2 N/V)^{1/3} \qquad (11b)$$

Since C_s is proportional to $1/\rho^{1/2}$ where ρ is the density of the material; it is clear that both C_s and T_D are dependent upon the isotope effect. They should be proportional to $1/M^{1/2}$ where M is the atomic mass of the particular isotope element.

From Eq (11):

$$(\Delta/kT_D)\{1 - 0.77kT_c/\Delta + 1.8\Delta^2 kT_c/(kT_D)^3\} = 58.445\,(kT_c/kT_D)^4 \quad (12)$$

Hence,

$$(\Delta/kT_D)^{1/4}\left\{1 - \frac{0.1925\,kT_c}{\Delta} + \frac{0.45\,\Delta^2 kT_c}{(kT_D)^3}\right\} \simeq 2.765\,(kT_c/kT_D) \quad (13)$$

This gives:

$$\frac{T_c}{T_D} = \frac{0.36167\,(\Delta/kT_D)^{1/4}}{\{1 + 0.6962\,(kT_D/\Delta)^{3/4} - 0.1627(\Delta/kT_D)^{2.5}\}} \quad (14)$$

This may not be agreeable with the experimental values of T_c and Δ the energy gap, unless the Debye temperature is shown to have much lower value at temperatures far below the room temperature. This may be true because of the lowering of elasticity and the acoustic speed C_s. However, one can assume the following relation from Eq (14):

$$kT_c \propto \Delta^{1/4}\,(kT_D)^{3/4} \quad (15)$$

Regardless of any change in T_D, it should satisfy the following relation at all temperatures from the isotope effect:

$$T_D M^{1/2} = \text{constant} \quad (16)$$

where M is the atomic mass for different isotopes of the same element. From Eqs (15) and (16), it is evident:

$$T_c M^{3/8} = \text{constant} \quad (17)$$

According to B.C.S. theory, $T_c M^{1/2}$ is expected to be constant. However from the experimental results[11],

$$T_c M^{0.5(1-\zeta)} = \text{constant} \quad (18)$$

where ζ lies between 0.1 and 0.3. Hence the isotope

effect related Eq (17) is much closer to the experimental results expressed by Eq (18) as compared to the B.C.S. theoretical result. The other possibility is that if the Debye temperature gets significaly lowered at temperatures much below the room temperature; then Eq (14) may numerically agree with the relations between T_C, T_D and Δ . That will also show that the higher value of T_C for any superconductor is correlated with its higher value of T_D and elasticity.

THE DOPED TYPE OF THE NEW CLASS SUPERCONDUCTORS

There are three basic conditions required for the doped type superconductors.

(a) The doped superconductors have molecular formulas like $A_{1-x} B_x C_n D_{n'}$ or $A_1 B_2 C_3 D_{n'+x}$ etc., where A, B, C, D are the elements of the compound. The normal number of atoms like n, n etc. are integers as expected in the molecular structures. The symbol x is representing the fractional doping with $X \leq 1$.

(b) The doped materials are basically the semiconductor type in which the excess valency of the doped element is ± 2 or ± 1, which provides the donor or acceptor type localized states below the conduction band. These localized states of the carriers are like in H or He atoms. The Boson related pairing is acquired in the spin-wise singlet state as in He atom or H_2 molecule. The singlet state acquires Boson property not only with spin being zero but also having the symmetry in the combination of the wave functions.

(c) The doping density should be heavy enough to provide only a single discrete energy related ground state below the conduction band. With this heavy doping density this single discrete energy state is shared by all electron pairs as Bosons through semi-Mott transition.

The observed critical temperature in the doped type superconductors is the Bose Einstein Condesation temperature rather than the critical temperature associated with the transfer of Boson pairs to the dissociated Fermions in the normal case of the old type superconductors.

The B.E. condensation temperature (T_B) correlated with the phase transition of Boson free particles and the critical temperature T_C associated with the transition to the Fermion eigenstates, have different comparative relations in these two types of superconductors. In the doped type of superconductors,

$$kT_B < \Delta_0 \leq kT_c \tag{19}$$

where Δ_c is the energy gap which is the depth of the ground state from the conduction band in these doped type superconductors. In the other type superconductors,

$$kT_c \leq \Delta_0 << kT_B \tag{20}$$

Thus the observed critical temperature is much below the B.E. condesation temperature in this usual metallic type superconductors. The change in the number of Boson carriers with temperature occurs in this case through the transfer from Boson to Fermion eigenstates.

The B.E. condensation related observed critical temperature has slight differences in the isotropic and anisotropic type of doped superconductors.

ISOTROPIC TYPE DOPED SUPERCONDUCTORS

In the isotropic type doped superconductors, the doped atoms, providing the electron (or hole) pairs, have almost similar distances along the different lattice axes. The Boson pairs occupying the conduction band have three dimensional momentums available. The occupied energy levels satisfy the following equation for B.E. statistics:

$$N = \frac{1}{[e^{(E_g-\mu)/kT} - 1]} + \frac{V}{(2\pi)^3}\int \frac{d^3k}{[e^{(E_k-\mu)/kT} - 1]} \tag{21}$$

Where N is the total number of pair of carriers, V the volume and μ the chemical potential. E_g and E_k are the energies of the single ground state and the momentum states respectively. We are going to use the infinite momentum related sphere rather than the Brillouin zone as an approximation. Below the B.E. condensation temperature a significant fraction of N particles are at the ground state and the chemical potential comes close to the value E_g. For example if at the temperature T_n, there are N/n particles at the ground state, where $2 \leq n \leq 10$, then the chemical potential μ_n is given by:

$$\mu_n = E_g - nkT/N \tag{22}$$

We define the B.E. condensation temperature as the temperature at which $N/5$ particles are at the ground state and naturally $4N/5$ are in the continuum states. The chemical potential μ and the B.E. condensation temperature T_D can be evaluated from the following two equations for this new definition:

$$\mu = E_g - 5kT/N \tag{23}$$

$$4N/5 = V/(2\pi)^3 \int \frac{d^3k}{[e^{(E_k + |E_g|)/kT} - 1]} \tag{24}$$

Hence from Eq (24),

$$(8\pi^{3/2}/5)(1/a_d^3) = \left(\frac{2mE_g}{\hbar^2}\right)^{3/2} \sum_{n=1}^{\infty} e^{-n\alpha}/(n\alpha)^{3/2} \tag{25}$$

Where M is the mass of the electron pair, $a_d^3 = V/N$ and $\alpha = E_g/kT_B$. Obviously α or kT_B can be calculated from Eq (25) if Eg and the doping density $1/a_d^3$ are known. For the energy depth Eg, we define the Bohr type radius a_B with the following equation:

$$E_g = -\hbar^2/(2ma_B^2) \tag{26}$$

For having a single discrete energy level Eg, the average distance a_d between the doped atoms has to satisfy the following relation:

$$8a_B > a_d > 2a_B \tag{27}$$

This explains how the superconductivity starts in the doped type superconductors with a certain minimum doping density needed for a single discrete energy level. Above a certain level of doping the value of Eg for this single discrete energy level goes to zero and the critical temperature also drops down to zero level to disappear.

ANISOTROPIC DOPED SUPERCONDUCTORS

In these type of doped superconductors, the average distance between the doped atoms is larger along some axis than in other directions. This can be seen from the crystalline structure of the new type doped superconductors. Let us suppose that along that particular axis the average distance between the doped atoms is large enough to have a localized eigenstate. On the other hand, in the plane perpendicular to this axis, the average distance between the doped atoms is small enough to have hopping related Bloch momentum states. Let us suppose that the large average distance between the doped atoms is along x axis. This attracting one dimensional potential V(x) will have a localized energy eigenvalue $-E_x$. Once again we suppose that this potential V(x) is weak enough to have a single value of negative energy E_x. In the perpendicular directions the energy eigenvalues are $(\hbar^2/2m)(k_y^2+k_z^2)$. At the ground level the energy eigenvalue is E_x because K_y and K_z will have the value zero. At excited levels, E_x is replaced by $\hbar^2 k_x^2/2m$. From the conditions, Eq (21) is changed in the following way:

$$N = \frac{1}{[e^{(E_x-\mu)/kT}-1]} + \frac{V}{(2\pi)^3}\iiint \frac{dk_x\, dk_y\, dk_z}{[e^{(E_k-\mu)/kT}-1]} \cdot \quad (28)$$

Once again we consider the infinite momentum related sphere (or ellipse) replacing the Brillouin zone for the Bloch momentum space. Hence the result similar to Eq (25) is:

$$(8\pi^{3/2}/5)\left[\frac{1}{a_{dx}}\cdot\frac{1}{a_d^2}\right] = \left[\frac{2mEg}{\hbar^2}\right]^{3/2}\sum_{n=1}^{\infty} e^{-n\alpha}/(n\alpha)^{3/2} \qquad (29)$$

where a_{dx} is the average distance between the doped atoms along the x axis. Thus the only difference between this equation and Eq (25) is correlated with replacing Eg by E_x and a_d by $(a_{dx}\cdot a_{dy}\cdot a_{dz})^{1/3}$. The value of E_x is given by,

$$E_x = 2m a_{dx}^2 V_0^2/4\hbar^2 \qquad (30)$$

where V_o is the maximum depth of the exponential type of the potential V(x) with its range of about $a_{dx}/2$. This energy depth and V_o have to be greater than kT_B, otherwise this temperature becomes the critical temperature of the transfer from Boson to Fermion eigenstates. For having a single discrete eigenstate the potential depth V_o and its range $a_{dx}/2$ should satisfy the following relation:

$$\hbar^2/(2ma_{dx}^2) > V_0/4 \qquad (31)$$

Eq (31) shows the lowest doping density needed to satisfy this equation to start superconductivity with having a single discrete energy state.

The doped superconductor $Y_1Ba_2Cu_3O_{7-\delta}$ is the antisotropic type. It has different resistivity along Cu-O planes and perpendicular to these planes. Along CU-O planes, the resistivity $\rho_{a\text{-}b}$ is a little lower and varies with temperature above T_c as in a metal[13]. This is so because the number of carriers occupying the momentums K_y and K_z along this plane remain constant above this critical temperature T_c. Hence the resistivity increases with temperature from scattering. On the other hand, the resistivity ρ_c in perpendicular direction has semiconducting property because of the ground energy state E_x. The number of carriers depends upon temperature and therefore the resistivity ρ_c decreases with the increase of temperature[13].

The supercurrent in the doped type superconductors comes below the critical temperature from the systematic excitation of the Boson pairs lying at the single discrete ground state.

Acknowledgment

The author wishes to express his thanks to Professors Morrel H. Cohen, H. S. Hayre and C. Brown for their help and suggestions etc. Thanks to Ms. Rose Hill for help in the preparation of the manuscript.

References

1. W. Heisenberg, Z. Naturforsch, 2a, 185 (1947).
2. M. R. Schafroth and J. M. Blatt, Nuovo Cimento 4, 786 (1955).
3. M. R. Shafroth, Phys. Rev. 100, 463 (1955).
4. John M. Blatt, Theory of Superconductivity, (Academic Press Publication, 1964), Chapters II, III and VII.
5. B. Deaver and Wm. Fairbank, Phys. Rev. Letters, 7, 43 (1961).
6. L. Onsager, Phys. Rev. Letters 43, 50 (1961).
7. J. Bardeen, L. N. Cooper and J. R. Schrieffer, Phys. Rev. 108, 1175 (1957).
8. Linus Pauling and E. B. Wilson, Introduction to Quantum Mechanics with Application to Chemistry, (McGraw Hill Publication 1935), Chapter XII.
9. See for example, Charles G. Kuper, An Introduction to the Theory of Superconductivity (Oxford University Press, 1968) p. 28.
10. See for example, J. Richard Christman, Fundamentals of Solid State Physics (John Wiley & Sons Publication, 1988) p. 219.
11. J. W. Garland, Phys. Rev. Letters 11, 114 (1963).
12. See for example, D. ter Haar, Selected Problems in Quantum Mechanics, (Academic Press Inc., NY 1964) Chapter I.
13. S. W. Tozer et. al. Phys. Rev. Letters 59, 1768 (1987).

LEVITATION AND INTERACTION IN A MAGNET-SUPERCONDUCTOR SYSTEM

Z.J. Yang[1], T.H. Johansen[1], H. Bratsberg[1], G. Helgesen[1], A.T. Skjeltorp[2]

[1]: Department of Physics, University of Oslo, Oslo, NORWAY

[2]: Institute for Energy Technology, N-2007 Kjeller, NORWAY

Abstract

The ceramic high temperature superconductor (HTSC) materials show great promise for applications in superconducting levitation. At very low magetic fields the flux lines penetrate into the bulk HTSCs, and determine the properties of the magnet-HTSC system. Various techniques are used today to study levitation and flux line motion in HTSC materials including vibration reed, several kinds of balances etc. We are using a magnet on a mechanical pendulum to investigate the interaction between a magnet and a HTSC for the static and dynamic behavior. The experiments on a bulk $Y_1Ba_2Cu_3O_{7-\delta}$ *sample show a lateral restoring force on the magnet and the effect of transport currents on the collective motion of flux lines in the HTSC.*

PACS Numbers: 06.70.Dn, 41.10.Dq, 74.60.Ge, **74.60.-w**, **74.90.+n** and 85.25.Ly

I. Introduction

Before the discovery of the high temperature oxide superconductors (HTSC), major efforts were made to apply superconductors to magnetic bearing systems

(Homer et al. 1977 and Brandt 1989). However, the expensive helium cooling systems for the conventional superconductors hence have limited the applications in many fields. For instance, the levitation train (Maglev) project employing conventional superconductors has for a large part been abandoned. The United States shut down their Maglev project more than 10 years ago. West Germany has not used the superconducting system in their ongoing Maglev project either (Moon 1984, Brandt 1989, and references therein). The historical discovery (Bednorz et al. 1986) of HTSCs has stimulated more and more people to reconsider the possible applications of magnet-superconductor systems, in particular for levitation trains (Maglev) (Carlsen 1990). With the possibility of using liquid nitrogen for cooling, the building of cost-efficient superconducting levitation trains has come closer to realization.

While some theoretical analyses have been reported on the levitation and lateral forces, most papers have focused on experimental investigations.

In most cases, fundamental research on superconducting levitation has involved studies of systems formed by a small magnet and a large superconductor. Besides this, other systems are considered and investigated, such as a ring-shaped magnet with a small HTSC body (Kitaguchi et al. 1989, Terentiev 1990, and Brandt 1990 a) and a ring-shaped superconductor with a small cylindrical magnet (Wojtowicz 1990).

There has been considerable progress in engineering designs of levitation systems. For example, high-speed rotation of magnets on HTSC bearings have been investgated by some groups (Moon et al. 1990, Weinberger et al. 1990 and Martini et al. 1990), and a micro superconducting actuator has been fabricated and tested (Kim et al. 1990 a, 1990 b). A superconducting magnetic levitation system for the transport of light payloads has also been made and investigated (Wolfshtein et al. 1989).

It is generally accepted that the HTSCs are extreme type-II superconductors and the magnetic properties of the bulk HTSCs are dominated by the weak links among the granulars at low external fields (Tinkham et al. 1989 and Clem 1990). The lower critical field of intergranulars is $B_{c_1 J} << B_{c_1}$ (the lower critical field of the intragranulars). Tinkham and Lobb have estimated $B_{c_1 J} \sim 0.5 * 10^{-4}$ T based on a three-dimensional, cubic array weak-link model. Clem has obtained the result $B_{c_1 J} \sim 1 * 10^{-4}$ T from the weak Josephson coupling parameters. Due to the field penetration into the HTSCs in a magnet-HTSC levitation system, it is of crucial importance to study the interaction between a magnet and a HTSC.

In addition, if the magnet is moved by an external force, it is also very important to study the dynamic response of flux lines for the applications of levitation.

In this paper, after briefly surveying the activities of levitation studies for the magnet-HTSC systems, we present the experimental studies on the static and dynamic behavior of a magnet-HTSC system using a mechanical pendulum in sections II-ii and III-i. By extending the utilities of the pendulum, we have studied the effect of transport currents on the collective motion of flux lines in a bulk HTSC in section III-ii. A short summary is given in section IV.

II. Static Forces in a Magnet-Superconductor System

It may be shown (Yang et al. 1990 d) that the field produced around a homogeneously polarized spherical permanent magnet (SPM) with moment $\mathbf{M}$ is the same as that created by a point dipole with moment $\mathbf{M}$ located at the center of the sphere. Consider a SPM with moment $\mathbf{M}$ over an infinite flat superconductor as shown in Fig. 1. For type-I superconductors, (or when $B < B_{c_1}$ for conventional type-II superconductors), we may use a dipole-dipole model and the image method to describe the interactions. The potential can thus be expressed as (SI units hereafter):

$$V = \frac{\mu_0}{8\pi} \frac{|\mathbf{M}||\mathbf{M}'|}{(2z)^3} [\cos(\alpha + \alpha') + 3 \sin\alpha \sin\alpha'] \ , \tag{1}$$

where $\alpha' = \alpha$ is the vertical inclining angle, and $\mathbf{M}'$ is the image dipole of the dipole $\mathbf{M}$ and $|\mathbf{M}|=|\mathbf{M}'|$.

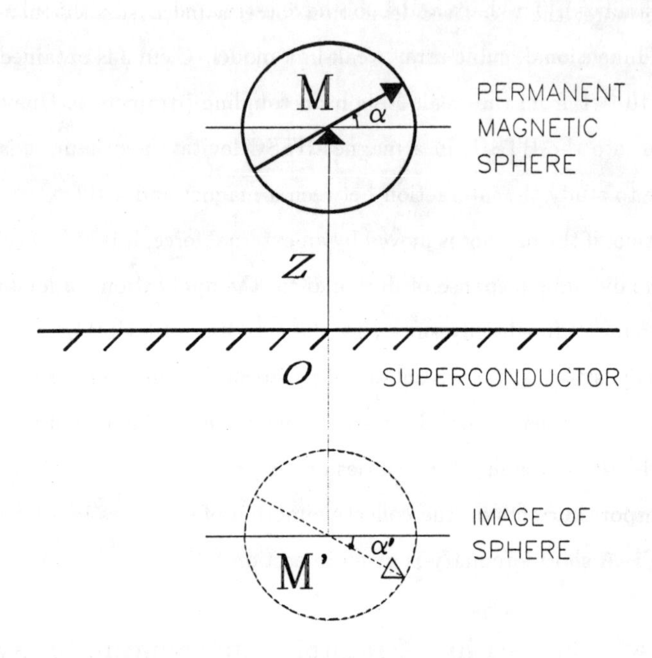

Fig. 1. *The configuration for a dipole with moment* $\mathbf{M}$ *placed above an infinite flat type-I superconductor (or when* $B < B_{c_1}$ *for conventional type-II superconductors). The equivalent description for the effect of the superconductor is the image dipole with moment* $\mathbf{M}'$ *acting on the dipole as discussed in the text.*

The potential can then be written

$$V = \frac{\mu_0}{64\pi} \frac{|\mathbf{M}||\mathbf{M}'|}{z^3}(1+\sin^2\alpha) \ , \tag{2}$$

with minimum $\alpha = 0$, which is the equilibrium position. At the equilibrium position, the only force on the magnet is then given by:

$$\mathbf{F} = -\nabla V = \frac{3\mu_0}{64\pi} \frac{|\mathbf{M}||\mathbf{M}'|}{z^4}\mathbf{k} \ , \tag{3}$$

where $\mathbf{k}$ is the unit vector in the $z-$direction. No lateral force and twist torque can exist in the dipole-dipole model, but experiments show that there are a significant lateral force and a pronounced twist torque in the magnet-HTSC system (Moon et al. 1988, Williams et al. 1988, Braun et al. 1990, and Johansen et al. 1990 b). Furthermore, the measured levitation force is much smaller than the value evaluated by using the dipole-dipole model ($\sim 1/5 - 1/2$) (Johansen et al. 1990 a). This means that the dipole-dipole model can not describe the magnet-HTSC system due to the penetration at very low field, and a new model is required to interpret the observed phenomena.

II–i. Levitation (Vertical) Force in a Magnet-HTSC System

Most activities on the force measurements in the magnet-HTSC systems have focused on the levitation (vertical) forces (Harter et al. 1988, Hellman et al. 1988, Moon et al. 1988, Marshall et al. 1989 and 1990, Jin et al. 1989, Weeks 1989, and Johansen et al. 1990 a).

Hellman and co-workers have suggested a complete penetration model to calculate the levitation force (Hellman et al.'s 1988). Considering that the energy cost of flux lines' penetration in the HTSC contributes to the origin of levitation, they showed that the levitation force is $F \propto z^{-2}$, where z is the distance between the center of the SPM and the upper surface of the HTSC. However, no other experiments have provided proof of this model.

Moon and co-workers (Moon et al. 1989 and 1990 b) have measured the levitation force and its stiffness as a function of seperation. After a series of experiments, they have introduced an empirical exponential formula to express the relationship between the levitation force and the separation:

$$F = F_0 \exp(-\alpha z) \ , \tag{4}$$

and a power-law formula to describe the stiffness (spring constant) as a function of the levitation force:

$$\kappa = C F^{\alpha'} \ . \tag{5}$$

However, so far they have not given the physical arguments for their empirical formulas.

The levitation force and lateral force measurements show hysteresis (Brandt 1988, Moon et al. 1988, and Johansen et al. 1990 a) due to the hysteresis in magnetization of the HTSCs. It is now clear that the magnetization hysteresis loop depends on the pinnings of the flux lines in superconductors. A stronger pinning results in a larger area of the hysteresis loop and higher critical current density J_c. By using the melted-quenched preparation method, it has been possible to increase the J_c of ceramic (bulk) HTSCs, and the levitation force has been enhanced significantly (Moon et al. 1990 b and Murakami et al. 1990).

Similar to Moon et al.'s measurements Weinberger and co-workers (Weinberger et al. 1990) have also measured the levitation force and its stiffness as functions of the separation using an analytic balance. They have also measured the vertical mechanical resonances of a freely levitated magnet above a HTSC sample. By measuring the levitation force as a function of temperature, they have proved that, at a weak field condition ($50*10^{-4}$ T), the force is correlated with the magnetization data measured by a SQUID.

Following the first reports of the suspension force in HTSCs (Peters et al. 1988 and Shapira et al. 1989), Adler and co-worker (Adler et al. 1988, 1990) have observed that a small piece of the conventional superconductor Nb_3Sn could also be suspended below a magnet. These results indicate that pinned flux lines must play a crucial role in type-II superconductors.

It seems therefore reasonable that studies of flux penetration is a good start point to interpret the results found for the magnet-HTSC systems.

We have therefore measured the levitation forces on a SPM placed above several $Y_1Ba_2Cu_3O_{7-\delta}$ (Y123) disks (Johansen et al. 1990 a and 1990 d), and been able to confirm Moon et al.'s empirical formula, Eq. (4), for both configurations: the magnetic moment of the SPM normal (N-state) and parallel to (P-state) the upper surface of the HTSC sample. Moreover, we have observed the crossover behavior of two exponential coefficients when the maximum field on the HTSC reaches $8*10^{-3}$ T for both configurations. This was interpreted on the basis of a formulation given in the following.

First, we calculate the number of fluxoids in the HTSC disk penetrated by the SPM. The magnetic field around a SPM with moment $\mathbf{m}$ is

$$\mathbf{B} = \frac{\mu_0}{4\pi}\left(\frac{3\mathbf{m}\cdot\mathbf{r}}{r^5}\mathbf{r} - \frac{\mathbf{m}}{r^3}\right),$$

where $\mathbf{r}$ is the spacial vector from the center of the SPM. For the N-state, the total flux on the HTSC disk with radius a at distance z is

$$\begin{aligned}\Phi_\perp &= \frac{\mu_0}{4\pi}2\pi m\int_0^{\rho_1}\rho d\rho\left[\frac{3z^2}{(\rho^2+z^2)^{5/2}} - \frac{1}{(\rho^2+z^2)^{3/2}}\right] \\ &\quad -\frac{\mu_0}{4\pi}2\pi m\int_{\rho_1}^{a}\rho d\rho\left[\frac{3z^2}{(\rho^2+z^2)^{5/2}} - \frac{1}{(\rho^2+z^2)^{3/2}}\right] \\ &= \frac{\mu_0}{4\pi}\left[\frac{8\pi m}{\sqrt{27}}\frac{1}{z} - \frac{2\pi m a^2}{(a^2+z^2)^{3/2}}\right], \end{aligned} \tag{6}$$

where $m = |\mathbf{m}|$. Here, we use $\rho_1^2 = 2z^2$ to take into account the sign change of the flux lines when $\rho_1 < a$. One may use the approximation $\frac{a^2}{(a^2+z^2)^{3/2}} \sim 0$ when $a >> z$. The number of fluxoids is thus found to be:

$$n_\perp = \frac{\Phi_\perp}{\Phi_0} \sim \frac{\mu_0}{4\pi}\frac{8\pi m}{\sqrt{27}\Phi_0}\frac{1}{z}. \tag{7}$$

where $\Phi_0 = h/2e = 2.07*10^{-15}$ Wb, the quantum fluxoid.

Similarly, for the P-state, the total flux is:

$$\begin{aligned}\Phi_\parallel &= \frac{\mu_0}{4\pi}2m\int_0^{a}\rho d\rho\int_{-\pi/2}^{\pi/2}d\theta\frac{3z\rho\cos\theta}{(\rho^2+z^2)^{5/2}} \\ &= \frac{\mu_0}{4\pi}\frac{4m}{z}\frac{a^3}{(a^2+z^2)^{3/2}}. \end{aligned} \tag{8}$$

One may use the approximation $\frac{a^3}{(a^2+z^2)^{3/2}} \sim 1$ when $a >> z$. The number of fluxoids is thus obtained:

$$n_\parallel = \frac{\Phi_\parallel}{\Phi_0} \sim \frac{\mu_0}{4\pi}\frac{4m}{\Phi_0}\frac{1}{z}, \tag{9}$$

Second, we postulate that the *effective interaction energy*, E_{eff}, is a function of the number of fluxoids (n) in the sample, and obeyes the following differential equation:

$$\frac{\partial E_{eff}}{\partial n} = C\frac{1}{n}\frac{E_{eff}}{n} , \tag{10}$$

i.e. the relevant increase in the effective energy is proportional to the energy per fluxoid and the density of fluxoids due to the correlation among the fluxoids. Here, C is a sample- and magnet-dependent constant, and also temperature-dependent. (E_{eff} is certainly also temperature-dependent, but we do not take this into consideration here.)

By solving Eq. (10), one obtains the following solution:

$$E_{eff} = E_0 \exp(-\frac{C}{n}) , \tag{11}$$

where the prefactor E_0 is a sample-dependent constant, i.e. a function of $H_{c_1 J}$ and the thickness of the sample.

This is qualitatively consistent with the interaction energy of the London vortices (Brandt et al. 1979 and Brandt 1990 b). As we know that, for $B < 0.25B_{c_2}$ (this is satisfied by all levitation studies), the vortex cores do not overlap and the London theory is valid for all temperature $0 < T < T_c$. For isotropic superconductors the interaction energy between two straight vortices tilted against each other by an angle ϕ and with separation d is given by London theory:

$$U_{int}(d,\phi) = \Phi_0^2\frac{\cot\phi}{2\mu_0\lambda}\exp(-\frac{d}{\lambda}) , \tag{12}$$

where λ is the London penetration depth.

Let us consider a system formed by regularly distributed *vortex-lattices* with the interaction energies described by Eq. (12). The exponent part is thus proportional to the *lattice constant d*. The exponent part of Eq. (11) is proportional to the *lattice density* $1/n$, which also gives the *lattice constant* argument.

The levitation forces are found directly by taking the derivative of the effective interaction energy: $\mathbf{f} = -\nabla E_{eff}$. For the two configurations of a magnet sphere-HTSC disc pair, we obtain:

$$
\begin{aligned}
f_{\perp} &= -E_{\perp 0}\frac{\partial}{\partial z}\exp(-\frac{C}{n_{\perp}}) \\
&= \frac{4\pi}{\mu_0}\frac{\sqrt{27}E_{\perp 0}C\Phi_0}{8\pi m}\exp(-\frac{4\pi}{\mu_0}\frac{\sqrt{27}\Phi_0 C}{8\pi m}z) \ ,
\end{aligned} \tag{13}
$$

and

$$
\begin{aligned}
f_{\parallel} &= -E_{\parallel 0}\frac{\partial}{\partial z}\exp(-\frac{C}{n_{\parallel}}) \\
&= \frac{4\pi}{\mu_0}\frac{E_{\parallel 0}C\Phi_0}{4m}\exp(-\frac{4\pi}{\mu_0}\frac{C\Phi_0}{4m}z) \ .
\end{aligned} \tag{14}
$$

If we rewrite the formulas in the form: $f = f_0\exp(-\alpha z)$, we have the result:

$$
\frac{\alpha_{\perp}}{\alpha_{\parallel}} = \frac{\sqrt{27}}{2\pi} = 0.827 \ , \tag{15}
$$

which is a sample- and magnet-independent constant. If we take into account that the crossover point corresponds to H_{c_1} of the sample, the only modifications in the effective energy and the force formulas are that the constants C and E_0 are changed to C' and E_0', respectively. This means that the ratio between the α's is not affected!

The high resolution measurements on the levitation forces for thick Y123 disc samples are in excellent agreement with this model. The experiments also showed that, for the thin Y123 samples, the emprical exponential formulae were not valid, and the force behavior was thickness dependent (Johansen et al. 1990 d).

It should be pointed out that, since the constant C is magnet-dependent, there is not a linear dependent relationship between α and $1/m$ for different magnet spheres.

II–ii. Lateral (Horizontal) Force in a Magnet-HTSC System

Due to the fact that the most important aspect of levitation applications is the ability of magnet-HTSC system to lift "heavy" bodies, most studies have been focused on investigating levitation forces. There are therefore only a few published papers reporting lateral force studies. Williams and Matey (Williams et al. 1988) were the first to measure the lateral force on a small magnet floating above a HTSC disk using a resonance method. Moon and co-workers measured the hysteresis behavior of the lateral force (Moon et al. 1988).

We have previously studied the vibrations of a free SPM levitated above a superconductor (Yang et al. 1989). By extending this idea, we designed a mechanical pendulum to investigate the interaction in a magnet-HTSC system (Yang et al. 1990 b). The mechanical pendulum, shown in Fig. 2, has been used to study both static and dynamic behavior of the magnet-HTSC device.

The vital components in the horizontal force detection system is shown in Fig. 2 (a), (b) and (c).

A planar disk of high-T_c superconductor material is placed on a block of aluminum in a polystyrene container filled with liquid nitrogen. The top surface of the sample is aligned with the horizontal plane. The container rests on an electrically movable table. A step motor can move the sample 1.25 μm per step in a lateral direction. A manual micrometer, 0.5 mm/rev., is used to set the vertical position.

A Nd-Fe-B permanent magnet, M, is glued at the bottom end of a 4 mm diameter tubular quartz rod. At the other end, the 54 cm long rod is attached to a metal frame that permits the rod to swing as a vertical planar pendulum. The horizontal axis of rotation is provided by a razor blade fixed to a support not shown in the figure. The projected motion of the magnet onto the horizontal plane is parallel to the lateral displacement of the sample, the x-direction.

When the sample is in the superconducting state, we observe that as it is moved, the magnet will try to follow. This is because a part of the flux from the magnet gets pinned in the superconductor, and a friction-like interaction arises between the two bodies.

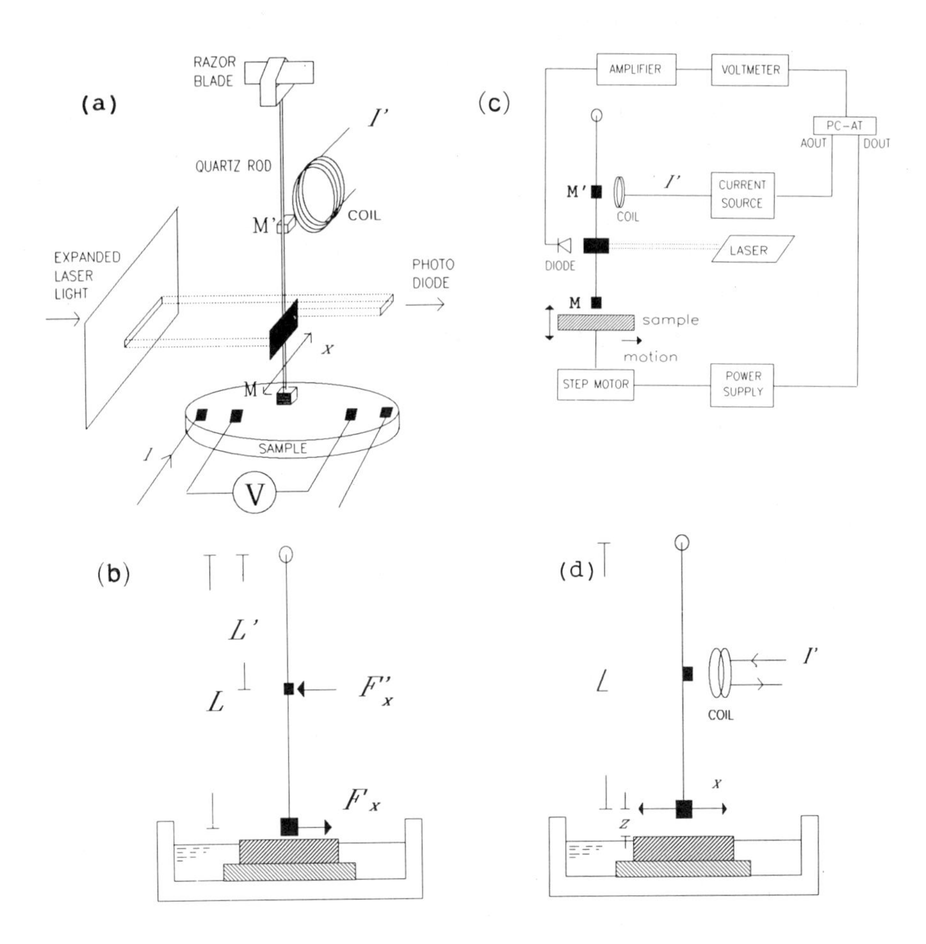

Fig. 2. *Schematic diagrams of the experimental setup: (a) perspective view, (b) and (d) front view, and (c) block diagram of the instrument connections. (b) shows the operation mode for lateral force measurements, and (d) shows for dynamical measurements. The horizontal motion of the sample was controlled by the PC/AT, but the vertical motion was performed manually. The function of the small magnet and coil with current I' is to keep the pendlum vertical for lateral force measurements, and to start the oscillations for the dynamical measurements as discussed in the text.*

The horizontal motion of the magnet is detected by a photodiode, which collects an amount of light that is linearly related to the position of the magnet. The light comes from a HeNe laser, and the beam is expanded to fill homogeneously a 2×15 mm^2 hole in a stopper plate. The rectangular beam is partly intersected by the edge of a razor blade glued onto the quartz rod. The passing light is then focused on the active area of the diode.

With this arrangement it is possible to follow the interesting dynamic behaviour of the pendulum during oscillations as described below. However, for studies of the static lateral interaction, the system is operated in a different mode. Instead of observing an oscillatory diode signal we now insist on keeping the signal constant, i.e. fix the magnet's position while the sample moves. This requires that the pendulum is acted on by an additional controllable force. The central point then is that from the magnitude of this extra force we can determine directly the strength of the lateral interaction under conditions of constant vertical separation.

The additional balancing force, F'_x, is generated by applying a field gradient to a second small permanent magnet, M', attached near the middle-point of the rod. The gradient is produced by a 3 cm diameter coil mounted on a support about 1 cm away from the magnet. With the pendulum in the equilibrium position, the current, I', supplied to the coil is proportional to the force, $F'_x = \alpha I'$. Since the torques of the two forces now must be equal, one has

$$F_x = \frac{L'}{L} \alpha I'$$

for the lateral force on the magnet close to the sample.

Calibration of the system is quite easy by simply putting a dead load on one side of the horizontal part of the metal frame (the supporting structure at the top of the pendulum) so that a torque of known magnitude is produced. This forces the pendulum away from equilibrium, and by measuring the current required to pull it back to equilibrium position the parameter α is determined. The ratio L'/L is 0.66.

We performed the experiments on a sample made of sintered Y123 sample, which was shaped as a planar circular disk of thickness 7 mm and diameter 70 mm. The superconducting material, which had a density of 90% of the theoretical value and $T_c = 90K$, was provided by Kali-Chemie Aktiengesellschaft, Hannover, W. Germany.

The magnet close to the sample is 9 mm long and has a square cross-section of 2×2 mm^2. The magnet is magnetized parallel to the x-direction indicated on the figure. Its strength was measured with a Hall probe, which gave $1.3*10^{-3}$ T for the field intensity at a point along the x-oriented symmetry axis 1 mm away from the poles. According to the infinite dipole-bar model (Eq. (19) below), the maximum tangential field on the HTSC sample is then $B_{max} = 1.3 * 10^{-3}$ T for our measurements.

The experimental results are shown in Fig. 3. Compared with Williams et al.'s result (Williams et al. 1988) and Moon et al.'s result (Moon et al. 1988), our experimental data show a higher resolution. A very important result is that the lateral stiffness is independence of the lateral displacement as shown in Fig. 3 (c).

It should be pointed out that we have not observed the **S**-shaped hysteresis curve as reported in Moon et al.'s paper. The size ratio between the magnet and the HTSC disc in our experiments is much smaller than that in Moon et al.'s measurements. This means that the edge effect in Moon et al.'s experiments should be more pronounced than ours, and the **S**-shaped hysteresis curve may be due to the edge effect.

Refering to Williams and Matey's work, Davis (Davis 1989, 1990) has suggested a model to calculate the lateral force based on Bean's critical state model. Davis has obtained the following expression for the force:

$$F_x(x_0) = -\frac{\mu_0}{4J_c}\{H_0^3(0) - \int_{-\infty}^{0} dx[H_0(x - x_0)\frac{\partial H_0^2(x)}{\partial x} + H_0^2(x - x_0)\frac{\partial H_0(x)}{\partial x}]\}, \quad (16)$$

where $H_0(x)$ is the field profile produced by the magnet at the lateral displacement x. Davis evaluated the force numerically using the field profile of an infinite long current carrying wire

$$H_0(x) = \frac{H_0(0)h^2}{h^2 + x^2}, \quad (17)$$

where x is the lateral displacement, and h is the height of the field source from the superconductor. In fact, this can be calculated analytically (Yang et al. 1990 c):

$$\frac{-F_x(x)}{\mu_0 H_0^3(0)/4J_c} = \frac{4 + 7(x/h)^2 + (x/h)^4}{[4 + (x/h)^2]^2} + \frac{(x/h)[20 + (x/h)^2]}{[4 + (x/h)^2]^3}(\pi - \arctan\frac{x}{h}) - \frac{4[4 + 5(x/h)^2]}{(x/h)^2[4 + (x/h)^2]^3}\ln[1 + (x/h)^2] . \quad (18)$$

However, the field profile of Eq. (17) can not describe the field created by any practical permanent magnet. The field in the x-direction created by a permanent magnetic bar infinitely long in the y-direction and $x \times z$ dimensions of $2l \times 2w$, can be expressed as:

$$H_x = 2m_v(\arctan\frac{h - w}{x + l} + \arctan\frac{h + w}{x - l} - \arctan\frac{h - w}{x - l} - \arctan\frac{h + w}{x + l}), \quad (19)$$

where m_v is the dipole-moment volume density and h is the distance from the center of the bar magnet to the superconductor surface. When the vertical distance (h) is larger than $2w$, the field can be approximated by a dipole-line model (polarized along the x-direction and transverse to the line direction) and expressed as:

$$H_x = 2m_l\frac{x^2 - h^2}{(x^2 + h^2)^2} \quad (20)$$

where m_l is the dipole-moment linear density (per unit length), x is the lateral displacement, and h is the height of the dipole-line above the superconductor.

Since Davis' model is not valid for the alternative fields (Davis 1990), we have tried several positive field profiles to evaluate the forces, such as fractions and Gaussians, we have not found one in agreement with the experimental results. The fractional field profile $H_0(x) = H_0(0)/[1 + (x/h)^2]^3$ is a good approximation for the infinite dipole-line model given by Eq. (20). The comparable results are shown in Fig. 3 (a).

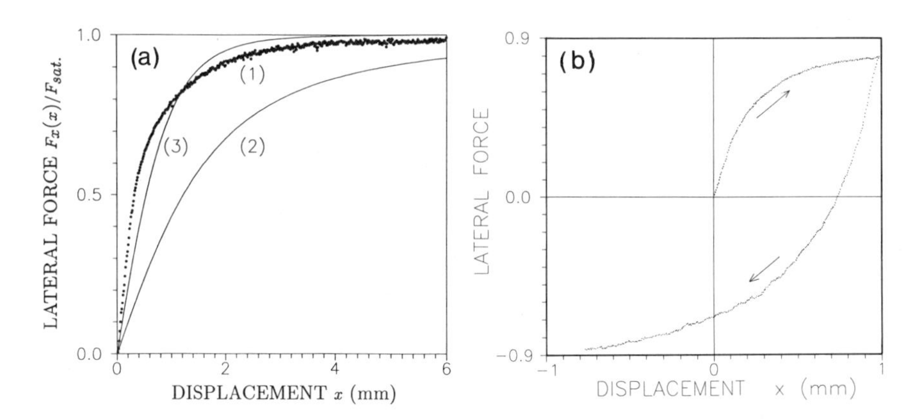

Fig. 3. *(a) Normalized lateral force $F_x(x)/F_{saturation}$ as function of horizontal displacement: (1) experimental, (2) analytical solution Eq. (18) with field profile $H_0(x) = \frac{H_0(0)}{1+(x/h)^2}$, and (3) numerical result with field profile $H_0(x) = \frac{H_0(0)}{(1+(x/h)^2)^3}$. (b) Hysteresis behaviour of the lateral force. (c) The stiffness of lateral force F_x measured at various lateral displacements. The normalization of the plot corresponds to 5.1 mN for the saturation force.*

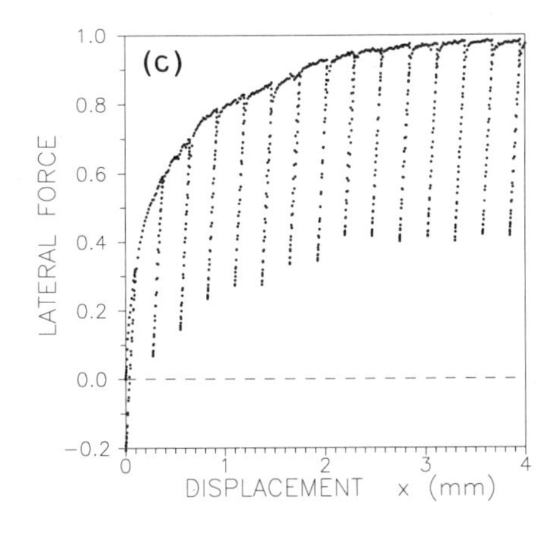

III. Dynamic Behavior in a Magnet-Superconductor System

When a permanent magnet placed above a HTSC is moved by an external force, such as the graviational force, the flux lines in the HTSC will follow the motion. If the flux lines are in the vertex-liquid phase, the magnet will be influenced by the viscous force of the flux lines, but if the flux lines are in the vertex-glass phase, the magnet will be influenced by the pinning force of flux lines in the HTSC. These aspects are very important in levitation applications with motion between the magnet and the HTSC.

Figs. 2 (a) (c) and (d) show schematically the vital components for studying the dynamic behavior of the magnet-HTSC system. A cube-like Nd-Fe-B permanent magnet (M) (with dimensions $6.5 * 6.5 * 7.5 \mathrm{mm}^3$) with rounded edges was used for these measurements. The magnet was magnetized along the longest side. The maximum field on the surface of the magnet was measured to be 0.4 T with a Hall-probe.

The magnet could be set in oscillations above the Y123 sample. (This sample was also used for the lateral force measurements discussed in section II-ii.) The sample was immersed in liquid nitrogen and the distance z between the sample and permanent magnet could be adjusted by elevating the sample/nitrogen container. The optical system employed to monitor the time dependence of the "horizontal displacement" of the oscillations was described in Section II-ii. A controlled and reproducible initiation of the oscillation was obtained by using the small permanent magnet, M' and the adjacent coil. The rod could thus be pulled out to a fixed position ($< 0.4°$ from vertical) by supplying current to the coil. By switching off the current, the oscillation started.

In the experiments, the direction of the magnetic moment **M** could be made either normal (N configuration) or parallel (P configuration) with the direction of oscillations of the pendulum.

The performance of the pendulum was also checked with the sample in the normal state. It was found that the free-damping Γ_0 was very small ($\Gamma_0 \sim 0.011$ Sec^{-1}) and negligible compared to the damping resulting from the superconducting sample. Also, no change in the oscillation frequency ν_0 ($\nu_0 = 0.798$ Hz) of the free pendulum was observed. Thus, by a relatively rapid sampling rate (188 Sec^{-1}) of the amplitude signal (pendulum position), we could directly extract information about the magnet-HTSC interactions from frequency and damping changes.

III–i. "Effective Interaction" Studied by the Mechanical Pendulum

Fig. 4 shows the typical response curves: horizontal displacements versus time for two configurations with different distances z between the magnet and the superconductor. The frequency shift and damping change can clearly be seen. In order to affirm that this is a specific interaction, normal metals and alloys (cooper, aluminium, brass, etc.) were used as a substitute for the superconductor. Damping was observed for these, but frequency changes were never observed.

The present results were confirmed by measuring other Y123 samples with various sizes made by different companies.

By fitting parabolas to the extreme regions of the response curves, the positions and values of the peaks were accurately determined, and the frequency ν and damping factor Γ were calculated precisely. The experiments show that, at a given distance, as the amplitude decreases with time, the frequency initially increases slightly and then saturates, meanwhile the damping factor decreases slightly in the beginning and then reaches a constant level.

The frequency shift, $\nu - \nu_0$, due to the interaction for decreasing distance, z, can be seen clearly in Fig. 5. The experiments show that: (i) for the N-configuration, the frequency initially decreases slightly (Fig. 5 (c)), and increases drastically with decreasing z; (ii) for the P-configuration, the frequency increases monotonically with decreasing z; and (iii) for both configurations, the frequency shifts show hysteresis for a cycling measurement (first decreasing z and then increasing z). The differences between the two configurations will be discussed below.

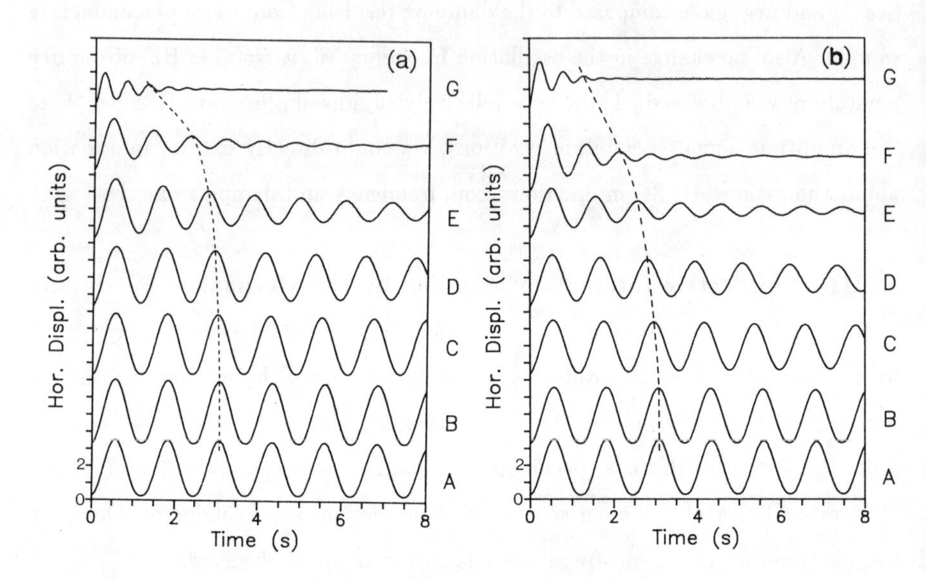

Fig. 4. *Typical measurements of the horizontal displacements (proportional to the diode signals) versus time for decreasing distances between the magnet and the superconductor: (a) for the N-configuration, and (b) for the P-configuration. Curves A show the free oscillations. The distances are* $z = 37$ *mm (B); 25 mm (C); 21 mm (D); 17 mm (E); 13 mm (F); and 9 mm (G). The dashed lines indicate the frequence shift and increasing damping, diminishing* z *from A to G. A sampling rate of 188* sec^{-1} *was used.*

From Fig. 6, one may see that the quantities $\nu^2 - \nu_0^2$ show very different behavior for decreasing z (virgin curves) and increasing z (except for first some points for the N-configuration). The virgin curves show exponential z-dependence, but after the magnetized, the squared frequency shifts show power law-like dependence with increasing z. The data are in excellet agreement with the fitted functions over three decades and these will be discussed below.

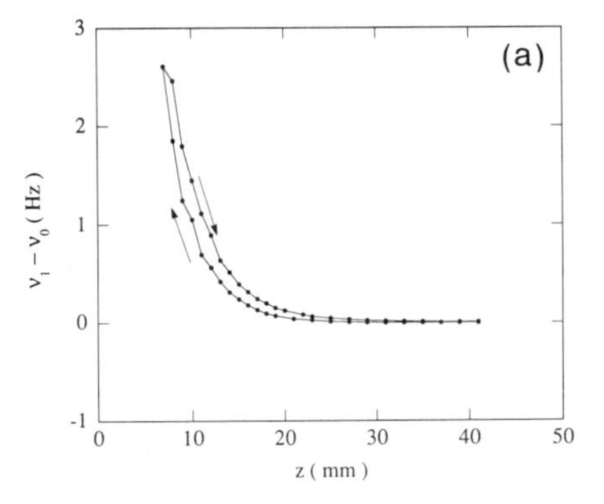

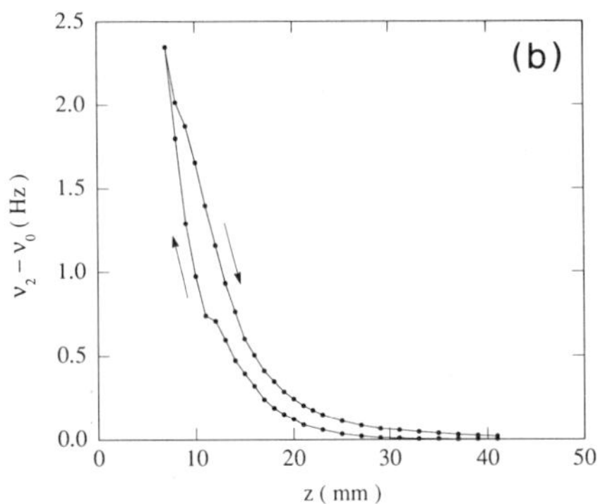

Fig. 5. *The frequency shift, $\nu - \nu_0$, versus vertical distance (z) for the data shown in Fig. 4 (ν_0 is the free pendulum frequency): (a) for the N-configuration, and (b) for the P-configuration. (c) shows the detail of (a) for the range 41 $\rightarrow$ 27 mm for the decreasing z in the virgin curve (the corresponding fields at the center of the HTSC were estimated to be $2.2 * 10^{-4} \rightarrow 7.7 * 10^{-4}$ T).*

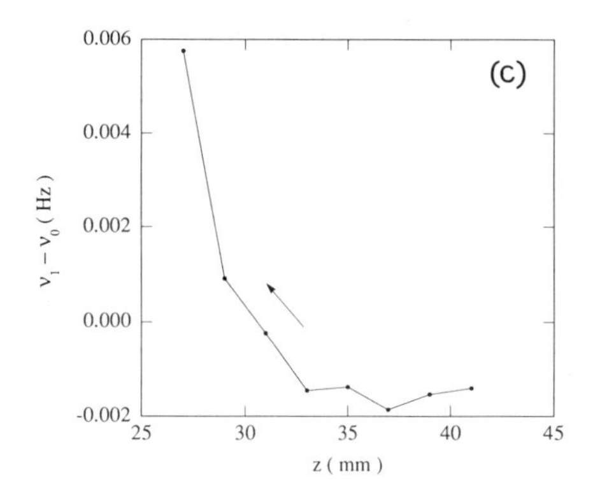

In a low order approximation, we may use a sphere approximation instead of the cube-like magnet to evaluate the field on the HTSC disc. The maximum field on the surface of a SPM with moment $\mathbf{M}$ and radius r_0 is given by $B_{max} = \frac{\mu_0|\mathbf{M}|}{2\pi r_0^3}$. Let us consider a SPM placed above the center of a large HTSC disc with the moment $\mathbf{M}$ parallel with the upper surface of the superconductor. We define a critical separation, $z_1 = (\frac{B_{max}}{2B_{c_1}})^{1/3} r_0$, for which $B = B_{c_1}$ at the center of the superconductor surface. (For this configuration, one may show that the maximum field on the superconductor is $\frac{\mu_0|\mathbf{M}|}{4\pi z^3}$). We may estimate the field range for our experiments with the parameters chosen: $B_{max} = 0.4$ T, $B_{c_1} = 0.01$ T and $r_0 = 4.2$ mm ($r_0 = (3 * 6.5 * 6.5 * 7.5/4\pi)^{1/3}$ is the equivalent radius of a sphere with the same volume as the cube-like magnet), then $z_1 = 11.4$ mm.

As discussed in the previous sections, one may see that if the superconducting disc is in the perfect diamagnetic state, the repelling force due to the Meissner effect, expressed by the dipole-dipole model, should only cause a decrease in frequency. Using the dipole-dipole model, at the critical distance z_1, the vertical repelling force was estimated to be ~2.5 g or about 10% of the weight of the pendulum. This should reduce the pure inertia frequency ν_0 by approximately 5%. But the present experiments show that, within experimental resolution, the frequency increases even for separations much larger than the critical distance z_1. From Fig. 6, using the sphere approximation, we may choose $z_0 = 30.0$ mm (for which the frequency shift is $\nu_1 - \nu_0 = 0$ for the N-configuration). The corresponding field at this separation was estimated to be $B(z_0) = 5.6 * 10^{-4}$ T $<< B_{c_1}$. The present experiments further affirm that the dipole-image model cannot describe the magnet-HTSC system.

The frequency increase is due to some kinds of attractive forces with lateral or vertical components. As discussed above, the physical origins of frequency increase are the flux lines penetrating the HTSC via the inter-granulars or grainboundaries. The pinned flux lines attract the magnet and contribute to the frequency increase. (Unfortunately, we cannot perform this experiment with the small single crystals available at the present time).

The present experimental results support the weak link model of HTSCs (Tinkham et al. 1989 and Clem 1990). Since the vertical repelling force only contributes to the reduction of frequencies, the penetration takes place even for $z > z_0 =$ 30.0 mm ($B(z_0) = 5.6 * 10^{-4}$ T), i.e. $B_{c_{1J}} < B(z_0)$. As the separation, z, decreases, the physical processes are: (i) the vertical force between the magnet and the HTSC causes the frequency, ν, to decrease when $z > z_{1J}$ (defined by $B(z_{1J}) = B_{c_{1J}}$); (ii) the penetration of the flux lines starts from $z = z_{1J}$ and the pinned fluxoids at intergranulars result in a lateral restoring force on the magnet which contributes to the frequency increase, and the effect of the lateral force increases more quickly than that of the vertical repelling force; (iii) the effect of the lateral force on the frequency is equal to the effect of the vertical force when $z = z_0$, and is larger when $z < z_0$; and (iv) when $z = z_1$ the penetration of the flux lines occurs via intragranulars.

This penetration could also be used to interpret the fractional (incomplete) Meissner effect of the HTSC (Phillips 1989). Due to the penetration of weak fields ($<< B_{c_1}$), one cannot observe the perfect Meissner effect even for very dense bulk samples. (To our knowledge, there have been no reports on bulk sample showing perfect Meissner effect in the whole the field range of $B < B_{c_1}$. Due to limited single crystal size at present, it has not been clarified whether the fractional Meissner effect is an intrinsic property).

All our measurements were performed with an initial amplitude less than 1 mm ($< 0.4°$ from vertical). For this case, the lateral restoring force on the long permanent magnet bar (Section II-ii) is linear with the displacement. These direct lateral force measurements thus provide the experimental basis for us to employ a linear approximation to analyse the present results.

Without the superconductor, the equation of motion of the mechanical compound pendulum can be expressed as:

$$\ddot{\theta} + \Gamma_0\dot{\theta} + 4\pi^2\nu_0^2\theta = 0 \ , \tag{21}$$

where θ is the deflection angle, Γ_0 is the mechanical damping factor, and the intrinsic frequency of inertia is $\nu_0 = (1/2\pi)(lgM_{eff}/G)^{1/2}$ (g is the graviational constant). Here, the moment of inertia, G, can be measured or calculated, M_{eff} is the effective mass of the inertia compound pendulum, and l is the length of the pendulum rod.

The solution to Eq. (21) is

$$\theta(t) \ = \ \theta_0 \exp(-\Gamma_0 t/2)\cos(2\pi\nu_{0,meas}t + \delta_0) \ , \tag{22}$$

where θ_0 is the initial angle, δ_0 is the initial phase, and the measured frequency is $\nu_{0,meas} = \sqrt{\nu_0^2 - (\Gamma_0/4\pi)^2} \simeq \nu_0$ for $4\pi\nu_0 >> \Gamma_0$, which is satisfied for the present measurements.

In the linear approximation ($\theta_{max} < 0.4°$) the motion of the compound pendulum can be described by the following equations:

$$\begin{aligned} \ddot{\theta} + \Gamma_n\dot{\theta} + 4\pi^2\nu_n^2\theta &= 0 \qquad \text{for N-configuration,} \\ \ddot{\theta} + \Gamma_p\dot{\theta} + 4\pi^2\nu_p^2\theta &= 0 \qquad \text{for P-configuration.} \end{aligned} \tag{23}$$

where Γ_i $(i = n,\ p)$ is the damping factor, and ν_i $(i = n,\ p)$ is the "intrinsic oscillation frequency" due to the interaction.

Similarly to Eq. (21), the solutions to Eq. (23) are

$$\theta_i(t) = \theta_{i,0}\exp(-\Gamma_i t/2)\cos(2\pi\nu_{i,meas}t + \delta_i) , \tag{24}$$

where $\theta_{i,0}$ $(i = n,\ p)$ is the initial angle, δ_i $(i = n,\ p)$ is the initial phase, and the measured frequency is $\nu_{i,meas} = \sqrt{\nu_i^2 - (\Gamma_i/4\pi)^2}$.

The quantities $\nu_i^2 - \nu_0^2$ and $\Gamma_i - \Gamma_0$ express the contributions of the interaction to the elastic and dissipative parts, respectively. Classically, the parameter $\nu_i^2 - \nu_0^2$, which is proportional to the change of the stiffness constant, is a measure of the energy changes from the pure inertia pendulum. The kinetic energy, for the system described by Eq. (24), is $T_i \sim 2\pi^2\nu_i^2\exp(-\Gamma_i t)$, and the "potential" is $V_i \sim 2\pi^2\nu_i^2\exp(-\Gamma_i t)$. The total energy is $E_i \sim 4\pi^2\nu_i^2\exp(-\Gamma_i t)$, and the relevant energy loss per period is $\exp(-\Gamma_i/\nu_i)$. Since $\Gamma_i/\nu_i \leq 0.1 << 1$ for most cases in our measurements, the interaction between the magnet and the superconductor mainly contributes to the elastic part. It is then reasonable to assume that the parameter $\nu_i^2 - \nu_0^2$ is a measure of the interaction in this system.

It was shown in Section II-ii that the lateral stiffness is independence of lateral displacement, and may only depend on the vertical separation for spacial coordinates, and by extending the postulated idea described in Section II-i, we intuitively introduce the exponential functions to express the interaction energies for the virgin processes:

$$
\begin{aligned}
E_N &\propto \nu_n^2 - \nu_0^2 = \delta\nu_n \exp(-\alpha_n z) \\
E_P &\propto \nu_p^2 - \nu_0^2 = \delta\nu_p \exp(-\alpha_p z) ,
\end{aligned}
\tag{25}
$$

where the constants α_n and α_p are functions of temperature, and sample- and magnet-dependent. As may been seen from Fig. 6 (a) the data are in excellent agreement with the exponential functions over three decades for both configurations. Based on the discussion in section II-i, we argue that the ratio between α_p and α_n should be a sample- and magnet-independent parameter. The present experimental data show that

$$\alpha_p/\alpha_n = 0.7 . \tag{26}$$

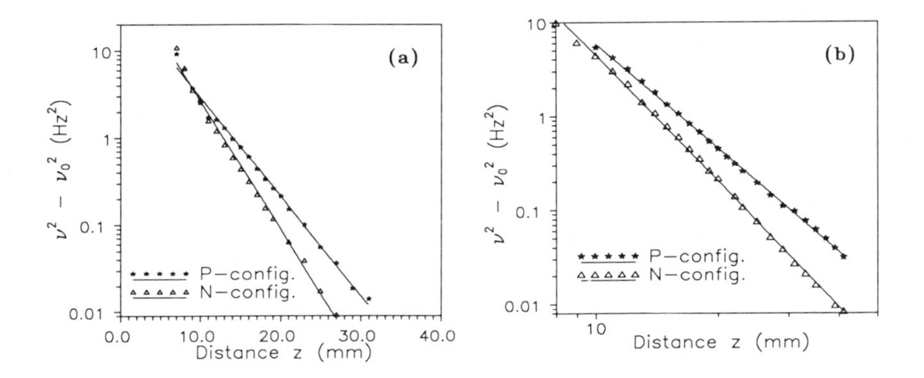

Fig. 6. *The quantity* $(\nu^2 - \nu_0^2)$ *against the distance* z*: (a) for decreasing* z *(virgin curves), and (b) for increasing* z*. The least-squares fits of exponetial functions,* $(\nu_i^2 - \nu_0^2) = \delta\ \nu_i \exp(-\alpha_i\ z)$*, to the curves in (a) show* $\delta\nu_p = 27.7$, $\delta\nu_n = 79.8$, $\alpha_p = 0.24$ *and* $\alpha_n = 0.34$*. The least-squares fits of power law functions,* $\nu_i^2 - \nu_0^2 = \delta\nu_i z^{-\beta_i}$*, to the curves in (b) show* $\delta\nu_p = 2.78 * 10^4$, $\delta\nu_n = 1.27 * 10^5$, $\beta_p = 3.68$ *and* $\beta_n = 4.45$.

As the magnet is moved away from the sample, the interactions are very different from the virgin processes due to the pinning of flux lines in the sample. The interactions may thus be express as power laws:

$$\begin{aligned} E_N &\propto \nu_n^2 - \nu_0^2 = \delta\nu_n z^{-\beta_n} \\ E_P &\propto \nu_p^2 - \nu_0^2 = \delta\nu_p z^{-\beta_p} . \end{aligned} \tag{27}$$

The exponents can be obtained by fitting the data in Fig. 6 (b): $\beta_n = 4.45$ and $\beta_p = 3.68$. The ratio is $\beta_p/\beta_n = 0.83$.

It is interesting to look at the ratios between the vertical stiffness exponent and the lateral stiffness exponent in terms of the levitation force in the Moon and co-workers' results (P-.Z. Chang et al. 1990 and Moon et al. 1990 b). Their experimental data for various samples, such as melted-quenched and sintered samples, show the ratios are changed from 0.69 to 0.80. Somehow, the ratios in Moon et al.'s paper showed different meaning from the present results.

It should be pointed out that: (i) The present measurements were performed in the field range around B_{c_1} as limited by the experimental resolution, but we believe that the present method could also be applied to the higher field range, and in turn provide the basis for the superconducting bearing and similar applications. (ii) Since the frequency response is very sensitive, it is believed that an improved pendulum could also be applied to investigate the dynamic behavior of isolated fluxoids in HTSC films. The characteristic frequency (Dew-Hughes 1988) of collective vibrations of flux-lines in HTSCs could possibly be measured by the use of forced oscillations and this is presently being tried out in our laboratory.

III–ii. Collective Motion of Flux Lines Investigated by the Pendulum

Flux line (fluxoid) dynamics in type-II superconductors reveals information about the motion of vortices as influenced by various pinning effects from physical defects like inhomogenieties, strains, and vacancies. These effects are important for the depinning critical current and possible applications of high T_c superconductors (HTSC). Traditional a.c. susceptibility and magnetization measurements have revealed giant thermal flux creep in these materials (Müller et al. 1987, Yeshurum et al. 1988, Malozemoff et al. 1988, Palstra et al. 1988, Tinkham 1988, and Gupta et al. 1989)

Related to the dynamics of fluxoids in HTSCs, it is also of interest to find the relationship between the transport currents and the pinning barrier. Several experimental (Zeldov et al. 1990) and theoretical (Feigel'man et al. 1989, Griessen 1990,and Kes et al. 1989) papers have been published in journals and proceedings. All of these reports have focused on the behavior of the intragranulars, i.e. the physical properties of single crystals and highly orientated thin films. This research is very important for the applications in microelectronics. The application of the bulk HTSC materials is equally important in other fields of science and technology.

The lower critical field of the intergranulars ($H_{c_{1J}} \sim 1$ Oe) is much lower than the lower critical field of the intragranulars ($H_{c_1} \sim 100$ Oe for the Y-Ba-Cu-O and 20 Oe for the Bi(Pb)-Sr-Ca-Cu-O). When a relatively weak external field H in the range $H_{c_{1J}} < H < H_{c_1}$ is applied to a bulk HTSC sample, the penetration takes place via intergranulars. The pinning properties of the flux lines in this field range have not been understood yet.

By extending the pendulum experiments described in the previous sections, we have studied the effect of the transport currents on the dynamic responses of the flux lines in a bulk Y123 HTSC in the field range of $H_{c_{1J}} < H < H_{c_1}$.

A standard four-lead setup with indium contacts was used to supply current (I) to the sample and monitor the superconducting state of the sample as shown in Fig. 2 (a). In all the experiments, the sample current direction was parallel with the light beam, and perpendicular to the direction of the motion of the magnet. Thus, the current could be in one direction (positive) or reversed (negative).

As stated above, the interaction between a free-swinging permanent magnet and a superconducting disk leads to frequency change and strong damping. As shown below, this may be related to the flux bundle activation barrier U (Anderson 1962 and Feigel'man et al. 1989). Specifically, we find that the predicted power-law dependence: $U(I) \propto I^{-\alpha}$ describes the data well for a wide range of currents.

The measurements of the oscillation frequency ν and damping factor Γ of the pendulum with sample currents from 0A to 5A in steps of 0.5A were performed with various distances between the magnet and the superconductor. Some typical pendulum oscillations are shown for increasing sample currents I =0A−5 A in Figs. 7 (a)-7 (b) for the N and P configurations, respectively. The distance from the center of the magnet to the surface of the superconductor was 14 mm. In this case the magnetic field at the center of the sample was estimated to be $54*10^{-4}$ T using a sphere approximation. As may be seen, there are significant changes in both frequency and damping with currents.

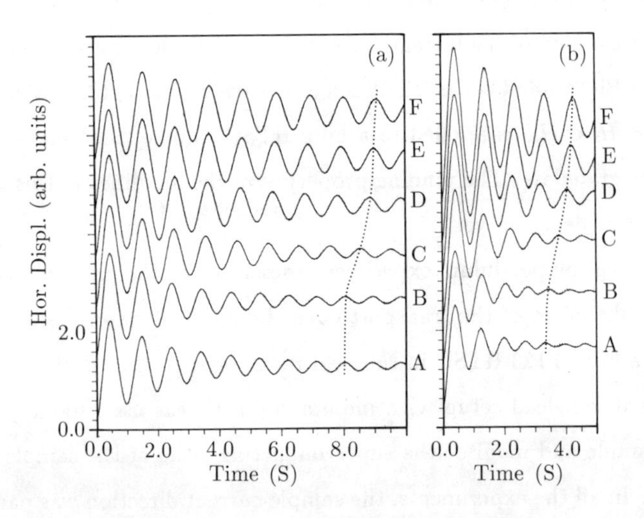

Fig. 7. *Horizontal displacements of the magnet versus time for various sample currents: (a) for the N configuration, and (b) for P configuration. Here the separation d = 14 mm measured from the center of the magnet to the surface of the sample. A: I = 0A, B: I =1.0A, C: I =2.0A, D: I =3.0A, E: I = 4.0A and F: I = 5.0A. The dashed lines show the frequency shift and decreasing damping, increasing I from A to F.*

Fig. 8 shows typical experimental results for the quantity $\nu^2 - \nu_0^2$ against the sample current I in a log − log plot (the reason for this mode of presentation will become clear below).

A characteristic feature of the frequency change with sample current I was a cross-over effect: for low I, ν was almost independent of I, followed by a decrease in ν for increasing I. This effect reflects the reduced interaction strength due to sample currents as discussed below.

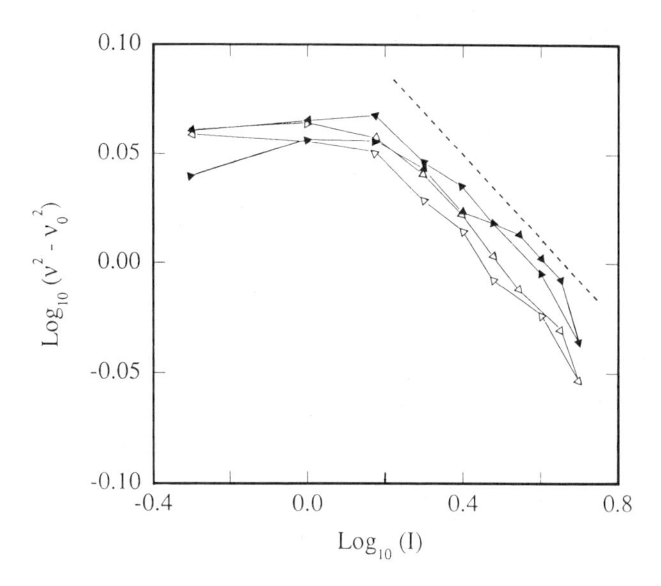

Fig. 8. *Typical* log – log *plot of the quantity* $\nu^2 - \nu_0^2$ *versus sample current* I *in the P configuration cycled between 0 A and 5 A (up-down marked by arrows) at* $d = 18$ *mm (the maximum field was estimated to be* $B_{max} = 25.4 * 10^{-4}$ T *on the sample).* ▲ *is for positive and* △ *for reversed currents. The dashed line (shifted upwards for clarity) represents the average slope* $\alpha \simeq 0.19$ *in the high current region for all the measurements.*

The experiments also show that, for increasing field (reduced distance), there is a shift in cross-over region to the left corresponding to smaller sample currents.

When the magnet is moved by the graviational force, the fluxoids will follow the magnet and perform a collective motion (hopping). The collective motion needs to overcome the pinnings at randomly distributed defects such as the interfaces of the intergranulars. The pinned fluxoids attract the magnet, and as a consequence, the frequency ν increases due to the collective pinning of weak disorder. The frequency shift is thus a measure of the collective pinnings. From a classical point of view, $4\pi^2(\nu^2-\nu_0^2)$ expresses the energy gain of the pendulum due to the collective pinnings in the HTSC sample. Therefore, we argue that the quantity $\delta\nu^2 = \nu^2 - \nu_0^2$ is proportional to the bundle activation energy U.

The collective pinning barrier U is of the order of the elastic energy of hopping flux bundles (Anderson 1962 and Feigel'man et al. 1989). When a current is passed through the sample, the Lorentz force $\mathbf{F} = \mathbf{I} \times \mathbf{B}$ enhances the collective flux motion, which is equivalent to reducing the collective pinning barrier U. As a result of the reduced magnet-HTSC interaction, the "*effective stiffness*" of the pendulum is also reduced.

Feigel'man et al. have studied the flux-creep phenomena in the case of collective pinning by weak randomly distributed defects. Based on the Anderson concept of flux bundle, they have shown that the bundle activation barrier U has a power-law dependence on current I: $U(I) \propto I^{-\alpha}$. For different regimes of currents, α changes from 1/7 to 7/9. Since the physical basis for this theory is the pinning of weak disordered defects, the results should also be applicable for analysing the present experimental data although they obtained the theory based on the consideration of single crystals. The physical reason for the present results is also the randomly distributed pinning centers.

The transport current in the superconductor creates a magnetic field parallel with the direction of the pendulum motion. For the P configuration, the direction of the current has no effect on the barrier exponent α. For the N configuration α is affected by the direction of the current when the field is relative weak. The relative effect on the different current directions decreases with increasing field.

The field dependence of the barrier exponent α has also been investiged and this is shown in Fig. 9, which is in qualitative agreement with Feigel'man et al.'s theory.

We found that the effect of sample currents on the frequency and damping is very different for weak fields ($H \sim H_{c_{1J}}$ corresponding to $\nu \sim \nu_0$ without sample current and only a few penetrating flux lines present) and strong fields ($H \gg H_{c_{1J}}$ corresponding to $\nu \gg \nu_0$ without sample current and with a certain amount of flux lines). It appears that when $H \succeq H_{c_{1J}}$, the frequency and damping factors are almost independent of sample currents. It seems to be a very small increase in frequency with an increase in sample current, but this is on the level of our experimental resolution. However, when the field is much higher than $H_{c_{1J}}$, the effect of sample currents is significant.

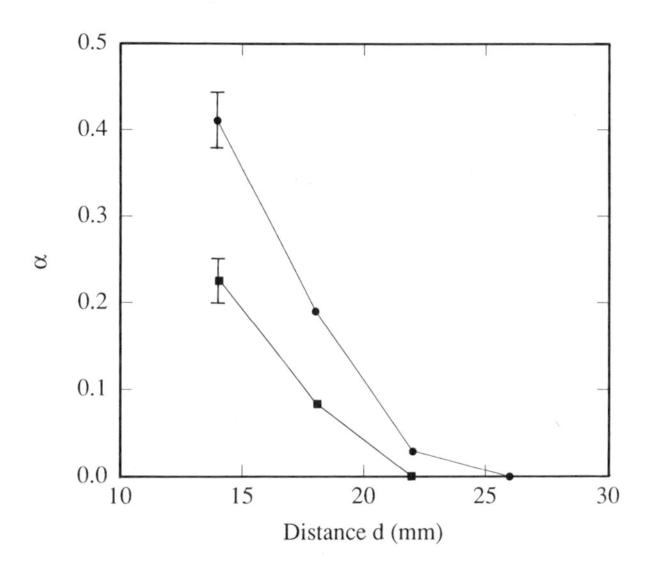

Fig. 9. *The variation of the slope α with magnet-superconductor separation d for the P configuration (•), and the N configuration (□). The maximum fields on the HTSC sample were estimated to be 8.4*10^{-4} T (d = 26 mm), 13.9*10^{-4} T (d = 22 mm), 25.4*10^{-4} T (d = 18 mm), and 54*10^{-4} T (d = 14 mm) using a sphere approximation.*

Kes and co-workers (Kes et al. 1989) have shown that the activation barrier U is independent of the field for thermally assisted flux flow (TAFF) at small driving forces. Kes et al.'s theory is a good description for isolated fluxoids. For the TAFF, the hopping of fluxoids is governed by the convensional linear diffusion equation. In this case, the fluxoids relax individually, and the correlation of the fluxoids does not play an important role. The present experiments at low fields are consistent with Kes et al.'s theory. In the case of low fields, the fluxoid-fluxoid interaction is very weak since only a few fluxoids are present in the sample. The motion of fluxoids is similar to the TAFF in this case. However, when the fields are high enough, the correlation of fluxoids is more pronounced. As a consequence, the motion of fluxoids becomes collective, i.e. the flux-"creep" is the case of collective pinnings by weak disorders. This is beyond the description of Kes et al.'s TAFF theory. As a result of an increase in fluxoid-fluxoid correlation, the barrier exponent α increases monotonically with increasing fields.

The experiments also show that for the P and N configurations α has different field-dependence, which is consistent with the current-free experiments. Phenomenologically, the parallel motion has an extra contribution of the magnetic moment $\mathbf{M}$ compared with the normal motion. Therefore, the former needs more energy to overcome the barrier.

IV. Summary

The new HTSC materials are clearly promising for bearing and levitation applications. The updated experimental results outlined in this paper are encouraging. Enhanced performance is clearly achievable through optimisation of materials and component designs. However, before practical application can be realized, further theoretical and experimental investigations must be undertaken to understand the extreme type-II superconductive behavior of the HTSCs.

Acknowledgement

The research was supported in part by the Norwegian Research Council for Science and the Humanities (NAVF). One (ZJY) of the authors would like to thank Dr. R. Hilfer for a helpful discussion, and the authors acknowledge the assistance for the experiments from P. M. Hatlestad, K. Nilsen and J. Yao.

References

R.J. Adler and W.W. Anderson (1988), Appl. Phys. Lett. **53** 2346.

R.J. Adler and W.W. Anderson (1990), J. Appl. Phys. **68** 695.

P.W. Anderson (1962), Phys. Rev. Lett. **9** 309.

J.G. Bednorz and K.A. Müller (1986), Z. Phys. **64** 189.

E.H. Brandt, J.R. Clem, and D.G. Walmsley (1979), J. Low. Temp. Phys. **37** 43.

E.H. Brandt (1988), Appl. Phys. Lett. **53** 1554.

E.H. Brandt (1989), Science, **243** 349.

E.H. Brandt (1990 a), Proceedings of the NASA-Conference "*Adv. in Mater. Sci. and Appl. of High Temp. Superconc.*", April 2-6, Greenbelt, MD., USA.

E.H. Brandt (1990 b), **LT 19** (Physica **B**) Invited paper.

M. Braun, P. Buszka, T. Motylewski, W. Przydróźny, and C. Śliwa (1990), submitted to Physica **C**.

R.C. Budhani, D.O. Welch, M. Suenaga, and R.L. Sabatini (1990), Phys. Rev. Lett. **64** 1666.

L. Carlson (1990), Superc. Industry, Vol.3 No.2 19.

P.-Z. Chang, F.C. Moon, J.R. Hull and T.M. Mulcahy (1990), accepted by Jpn. J. Appl.Phys.

J.R. Clem (1990), in "*Physics and Materials Science of High-Temperature Superconductors*", Eds. R. Kossowsky, S. Methfessel, and D. Wohlleben, (Kluwer Academic Publishers. Dordrecht) p 79.

L.C. Davis (1990), J. Appl. Phys. **67** 2631.

L.C. Davis, E.M. Logothetis and R.E. Soltis (1988), J. Appl. Phys., **64** 4212.

D. Dew-Hughes (1988), Cryogenics **28** 674.

M.V. Feigel'man, V.B. Geshkenbein, A.I. Larkin and V.M. Vinokur (1989), Phys. Rev. Lett. **63** 2303.

R. Griessen (1990), Phys. Rev. Lett. **64** 1674.

A. Gupta, P. Esquinazi, H.F. Braun and H.-W. Neumüller (1989), Phys. Rev. Lett. **63** 1869.

W.G. Harter, A.M. Hermann and Z.Z. Sheng (1988), Appl. Phys. Lett. **53** 1119.

F. Hellman, E.M.Gyorgy, D.W. Johnson, Jr., H.M. O'Bryan and R.C. Sherwood (1988), J. Appl. Phys. **63** 447.

G.J. Homer, T.C. Randle, C.R. Walters, M.N. Wilson, and M.K. Bevir (1977), J. Phys. **D 10** 879.

S. Jin, R.C. Sherwood, E.M. Gyorgy, T.H. Tiefel, R.B. van Dover, S. Nakahara, L.F. Schneemeyer, R.A. Fastnacht, and M.E. Davis (1989), Appl. Phys. Lett. **54** 584.

T.H. Johansen, H. Bratsberg, Z.J. Yang and G. Helgesen (1990 a), Poster, Pres. at EPS 10th Conf. Cond. Matt. Phys., Lisbon, April.

T.H. Johansen, H. Bratsberg, Z.J. Yang, G. Helgesen, and A.T. Skjeltorp (1990 b), Rev. Sci. Instrum. in press.

T.H. Johansen, Z.J. Yang, H. Bratsberg, G. Helgesen and A.T. Skjeltorp (1990 c), to be published.

T.H. Johansen, Z.J. Yang, H. Bratsberg, G. Helgesen and A.T. Skjeltorp (1990 d), to be published.

P.H. Kes, J. Aarts, J.van den Berg, C.J. van der Beek, J.A. Mydosh (1989),

Supercon. Sci. and Tech. **1** 242.

Y.K. Kim, M. Katsurai, and H. Fujita (1990 a), Sensors and Actuators, **20** 33.

Y.K. Kim, M. Katsurai, and H. Fujita (1990 b), Proceedings of 9th Sensors Syposium, 121.

H. Kitaguchi, J. Takada, K. Oda, A. Osaka and Y. Miura (1989), Physica **C** **157** 267.

A.P. Malozemoff, L. Krusin-Elbaum, D.C. Cronemyer, Y. Yeshurum and F. Holtzberg (1988), Phys. Rev. **B38** 6940.

D.B. Marshall, R.E. DeWames, P.E.D. Morgan and J.J. Ratto (1989), Appl. Phys. **A** **48** 87.

D.B. Marshall, R.E. DeWames, P.E.D. Morgan and J.J. Ratto (1990), Appl. Phys. **A** **50** 445.

G. Martini, A. Rivetti, and F. Pavese (1990), Adv. Cryog. Eng. **35** 639.

F.C. Moon (1984), "*Magneto-Solid Mechanics*", (John Wiley and Sons, New York)

F.C. Moon and P.-Z. Chang (1990 a), Appl. Phys. Lett. **56** 397.

F.C. Moon and P.-Z. Chang, H. Hojaji, A. Barkatt, and A.N. Thorpe (1990 b), accepted by Jpn. J. Appl. Phys. Lett.

F.C. Moon, K.-C. Weng and P.-Z. Chang (1989), J. Appl. Phys. **66** 5643.

F.C. Moon, M. M. Yanoviak and R. Ware (1988), Appl. Phys. Lett. **52** 1534.

M. Murakami, H. Fujimoto, T. Oyama, S. Gotoh, Y. Shiohara, N. Koshizuka, and S. Tanaka (1990), pres. at ICMC'90, High-Temperature Superc. May 9, Garmisch-Partenkirchen, FRG.

K.A. Müller, M. Takashige and J.G. Bednorz (1987), Phys. Rev. Lett. **58** 1143.

T.T.M. Palstra, B. Batlogg, L.F. Schneemeyer and J.V. Waszczak (1988), Phys. Rev. Lett. **60** 1662.

P.N. Peters, R.C. Sisk, E.W. Urban, C.Y. Huang, M.K. Wu (1988), Appl. Phys. Lett. **52** 2066.

J.C. Phillips (1989), "*Physics of High-T_c Superconductors*", (Academic Press Inc., San Diego) p 272, and references therein.

Y. Shapira, C.Y. Huang, E.J. Mcniff Jr, P.N. Peters, B.B. Schwartz and M.K. Wu (1989), J. Magn. Magn. Mat. **78** 19.

A.N. Terentiev (1990), Physca **C 166** 71.

M. Tinkham (1988), Phys. Rev. Lett. **60** 1658.

M. Tinkham and C.J. Lobb (1989), Solid State Physics, **42** 91.

D. E. Weeks (1989), Appl. Phys. Lett. **55** 2784.

B.R. Weinbergert, L. Lynds, and J.R. Hull (1990), Supercond. Sci. Technol. **3** 381.

R. Williams and J.R. Matey (1988), Appl. Phys. Lett. **52** 751.

P.J. Wojtowicz (1990), J. Appl. Phys. **67** 7154.

D. Wolfshtein, T.E. Seidel, D.W. Johnson, Jr., and W.W. Rhodes (1989), **2** 211.

Z.J. Yang, H. Bratsberg, T.H. Johansen, G. Helgesen and A.T. Skjeltorp (1990 a), to be published.

Z.J. Yang, T. H. Johansen, H. Bratsberg, G. Helgesen and A.T. Skjeltorp (1989), Physica **C160** 461.

Z.J. Yang, T.H. Johansen, H. Bratsberg, G. Helgesen and A.T. Skjeltorp (1990 b), Physica **C165** 397.

Z.J. Yang, T.H. Johansen, H. Bratsberg, G. Helgesen and A.T. Skjeltorp (1990 c), J. Appl. Phys. **67** in press.

Z.J. Yang, T.H. Johansen, H. Bratsberg, G. Helgesen and A.T. Skjeltorp (1990 d), to be published.

Y. Yeshurum and A.P. Malozemoff (1988), Phys. Rev. Lett. **60** 2202.

E. Zeldov, N.M. Amer, G. Koren, A. Gupta, M.W. McElfresh, and R.J. Gambino (1990), Appl. Phys. Lett. **56** 680.

7. MICROWAVES AND SQUIDS

An Improved Sensitivity Configuration for the Dielectric Probe Technique of Measuring Microwave Surface Resistance of Superconductors

S.J. Fiedziuszko, J.A. Curtis, and P.D. Heidmann
Ford Aerospace, Space Systems Division
3825 Fabian Way, Palo Alto, California

D.W. Hoffman and D.J. Kubinski
Ford Scientific Research Laboratories
20,000 Rotunda Drive, Dearborn, Michigan

ABSTRACT

The measurement of the microwave properties of HTSC materials, both bulk and thin film, is very important to the potential users of these materials. Several techniques have been developed for the measurement of HTSC microwave surface resistance including cavity end wall replacement, disc resonator, and resonant transmission line structures. The post dielectric resonator used as a probe offers a simple technique for measuring the microwave surface resistance of superconductors. Advantages of this technique include its simplicity and the ability to measure superconductor samples of small surface area, or nondestructively measure selected areas of larger samples. A new physical configuration for the dielectric probe which offers improved sensitivity is described, and selected measurement results are given.

1. INTRODUCTION

Since the discovery of high temperature superconductivity in 1987, there has been considerable effort focused toward developing the microwave applications of high temperature superconductors (HTSC). In conjunction with this effort, several techniques have been proposed and developed for evaluating the microwave surface resistance (R_S) of superconductors. Many of these methods have been successfully used to evaluate microwave performance of superconductors, however, all of them suffer from certain drawbacks. The dielectric resonator probe measurement technique offers a number of attractive advantages over other measurement techniques, particularly as a simple, quick, reliable method of screening superconductor samples for microwave applications.

In this paper, we will present a brief overview of some of the more popular microwave surface resistance measurement techniques along with some of their properties. We will also discuss the desirable characteristics of an R_S measurement method used for HTSC materials. Next, we will present the dielectric probe in an improved sensitivity configuration as an alternative measurement technique and discuss its advantages over the other methods. Some measurement results will also be presented.

2. OVERVIEW OF POPULAR MICROWAVE SURFACE RESISTANCE MEASUREMENT TECHNIQUES

Nearly all of the methods used to evaluate the microwave properties of superconductors involve measuring the quality factor (Q) of some resonant microwave structure which in some way incorporates the superconductor sample under test. The more popular microwave surface resistance measurement techniques include cavity end cap replacement, cavity perturbation, parallel plate resonator, and a variety of resonant microstrip structures. Some properties of these techniques are briefly discussed below.

2.1 TE011 Mode Cavity End Cap Replacement

In the cavity end cap method, one of the end caps of a TE011 mode resonant cavity is replaced by the superconductor test sample. The contribution of the superconducting end cap to the measured cavity Q can easily be calculated to determine the surface resistance of the sample. This method is illustrated in Figure 1, and a cavity end cap test fixture is shown in Figure 2. This measurement technique is successfully used to measure microwave surface resistance at a number of laboratories, however, it does

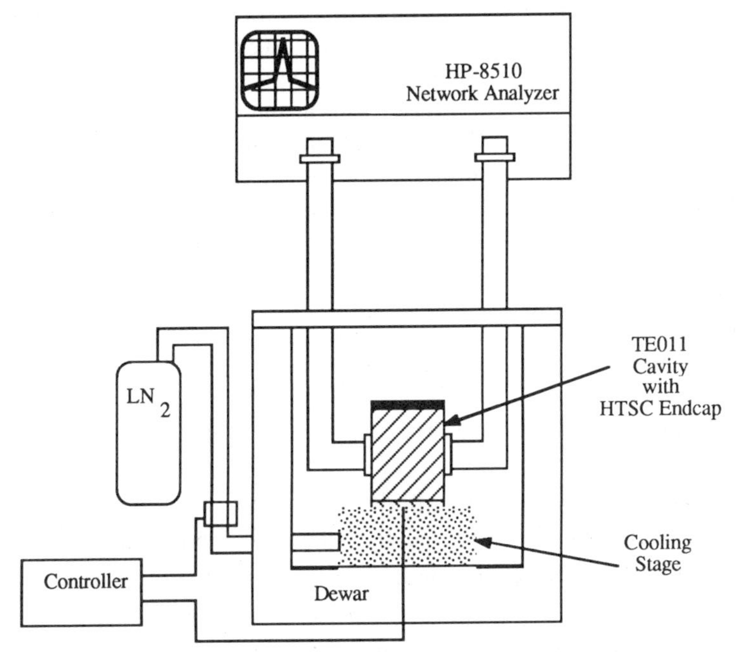

Figure 1 Measurement set-up for the TE011 mode cavity end wall replacement technique.

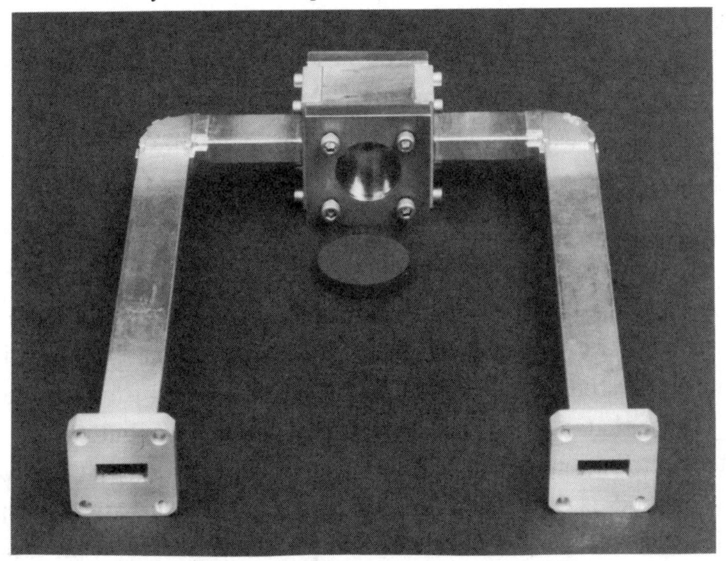

Figure 2 Test fixture for cavity end wall replacement measurements at 20 GHz.

have a few drawbacks. The most serious drawback is that it requires test samples of relatively large surface area to make measurements at lower microwave frequencies (greater than 5 cm^2 for frequencies below 20 GHz) where the materials are most likely to be used and where Q measurements are easier to make. Also, the sensitivity and accuracy of the measurements are decreased since the losses in the metal cavity dominate the Q value, particularly for measurements above the critical temperatures of low temperature superconductors. Cavities made from HTSC materials may be an option for increased sensitivity at higher temperatures in the future, but are currently not practical because of the relatively poor microwave performance of bulk materials.

2.2 Disk or Parallel Plate Resonator Method

The disk or parallel plate resonator technique proposed by Belohoubek ct. al. [1] and used by Taber [2] incorporates two parallel plates or disks of superconductor separated by a thin dielectric spacer as shown in Figure 3. An average surface resistance value for the two test samples can be calculated from the measured Q. One drawback of this technique is that the measured R_S value is an average of two samples. This is a particular problem when measuring at higher temperatures where niobium can't be used for one of the plates of the resonator. A second drawback is that it is limited to one specific sample size and geometry for a given test fixture.

2.3 Cavity Perturbation Techniques

A variety of cavity perturbation techniques have been proposed and are being used for R_S measurements. These involve incorporating the superconductor test sample within a resonant cavity such that it perturbs the cavity field distribution in a way that its contribution to the cavity Q can be calculated. An example of this method is illustrated in Figure 4 [3]. Perturbation techniques typically require very small test samples and suffer from decreased sensitivity at temperatures where niobium cavities can't be used.

Disk/ Parallel Plate Resonator Measurement Technique

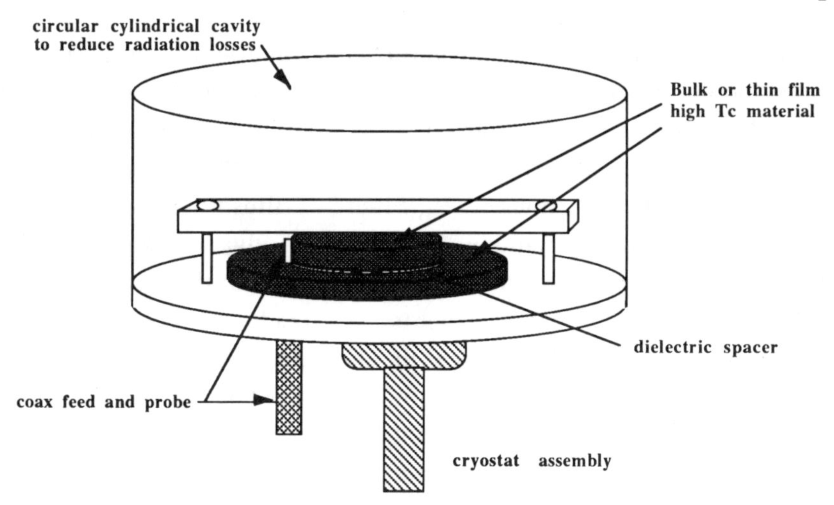

Figure 3 Illustration of the measurement set-up for the disk resonator technique. Taber later modified this technique in the parallel plate resonator [1,2].

Cavity Perturbation Technique

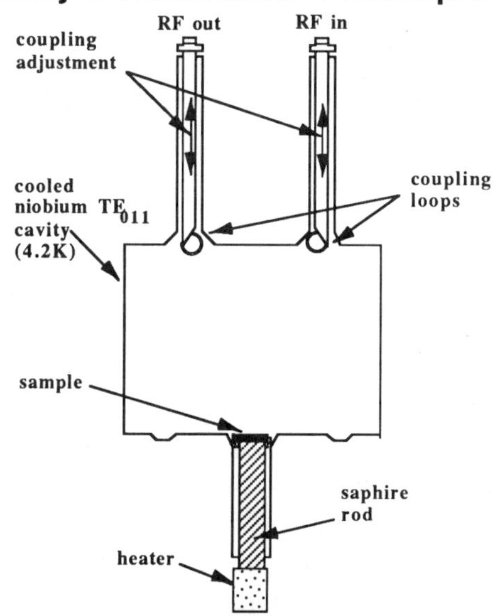

Figure 4 Illustration of the cavity perturbation technique from Moffat et. al. [3]. The sample temperature is controlled independently from the cavity temperature which is at 4.2K.

2.4 Resonant Microstrip Structures

Another technique for evaluating the microwave properties of superconductors is to measure the Q value of resonant structures of superconducting microstrip transmission line. A wide variety of transmission line structures have been used including ring resonators and meanderlines [4,5]. Figure 5 illustrates two of these structures [5,6]. This technique has distinct advantages over other methods since it measures the film properties on the substrate side of the film, and the measurements are made on structures that closely resemble the actual usage of the films in many microwave applications. However, microstrip measurement techniques have drawbacks in that they measure only a small portion of the film surface, they are limited to a specific film size for a given test fixture and pattern, and most importantly, the films can't be used for any other purpose after being tested.

3.0 CHARACTERISTICS OF A GOOD MICROWAVE SURFACE RESISTANCE MEASUREMENT TECHNIQUE

All of the measurement techniques outlined above have desirable characteristics, yet they also have important drawbacks. The desirable characteristics for a good R_S measurement technique include the following.

- Accuracy, Repeatability, Sensitivity
- Simplicity, and ability to perform measurements quickly
- Ability to use the sample after testing (nondestructive)
- Ability to test small samples and flexibility in sample dimensions
- Ability to test at more than one frequency and at reasonably low frequencies
- Maintained sensitivity at temperatures above low T_C

Each of the measurement methods outlined in the previous section has many of these characteristics, but none of them has all of these properties. In this paper, the dielectric resonator probe in

Microstrip Resonator Techniques

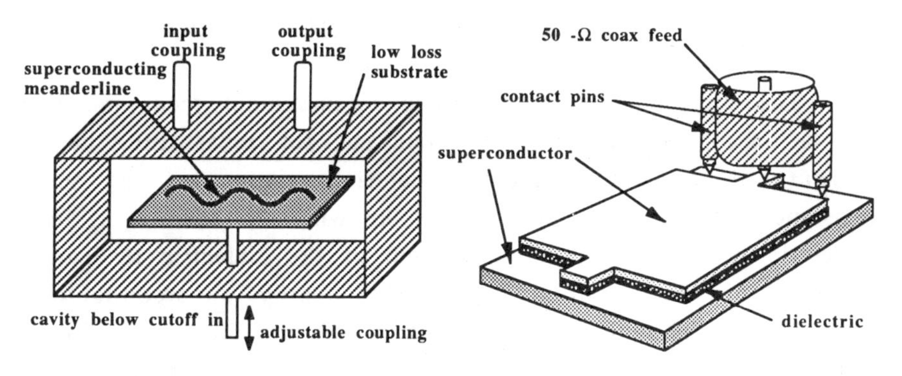

Figure 5 An illustration of two types of resonant microstrip structures used for HTSC Rs measurements [5,6]. A variety of other microstrip patterns are also used.

Dielectric Probe Technique

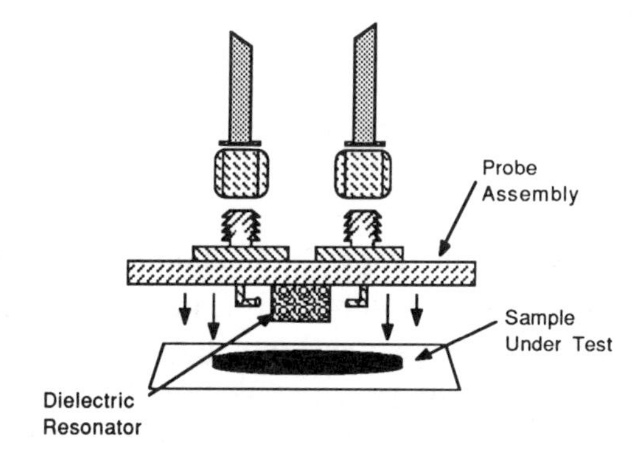

Figure 6 Illustration of the post resonator configuration for the dielectric probe. An improved sensitivity configuration for the probe is shown in Figures 12 and 13.

an improved sensitivity configuration is presented as an alternative for measuring the microwave surface resistance of high temperature superconductors.

4.0 THE DIELECTRIC RESONATOR PROBE TECHNIQUE

The dielectric resonator probe measurement technique is an adaptation of a well known technique for measuring the loss tangents of dielectric resonator materials given that the surface resistance of nearby conductors is known [7,8]. However, in the dielectric probe case, the loss tangent of the dielectric resonator is known and the conductivity of the nearby superconductor is unknown. The general configuration of the dielectric probe is illustrated in Figures 6 and 7 [7]. In these figures, a weakly coupled circular dielectric resonator is sandwiched between two conductive plates, one of these plates being the superconductor sample under test. As in the case of the other measurement techniques discussed in this paper, R_S is calculated from a measured Q value. The TE011 mode is used for measurements since it is easily identified, relatively insensitive to small gaps between the dielectric and the test sample, and has no axial currents across any possible discontinuities in the fixture (similar to the TE011 mode used in the metal cavity end cap replacement technique). The field configuration for the TE011 mode is illustrated in Figure 8.

4.1 Calculation of Absolute Rs Values From Measured Q Values

For the structure of Figures 6 and 7, the absolute surface resistance of a sample under test can be calculated from a measured Q value using the formulas below after substituting the known values for the dielectric loss tangent and the surface resistance of the copper fixture [7].

Figure 7 Photograph of the dielectric probe in the post configuration. The dielectric resonator is sandwiched between two conductive plates. Energy is coupled to and from the resonator using the coupling probes shown [7].

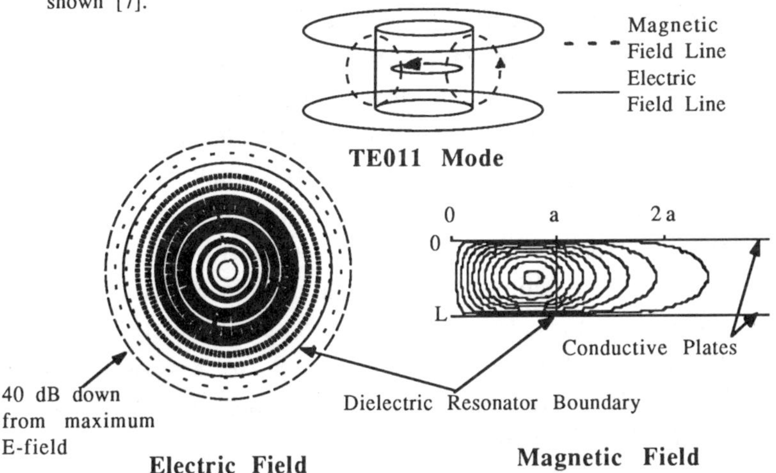

Figure 8 Illustrations of the field configurations in the dielectric probe. The top drawing shows the general TE011 mode configuration. The bottom two drawings are computer generated plots of the electric and magnetic fields stored within the resonant circuit. These plots illustrate the confinement within the resonator.

$$R_{smeas}=(A/Q_{meas} - \tan\delta)/B - R_{sfixt}$$

where 9];

$$A=1 + W/\varepsilon, \qquad B=(\lambda_o/2L)^3 * (1 + W)/(60\pi^2\varepsilon)$$

$$W = \frac{J1^2(\zeta)\,[K0(\zeta)K^2(\zeta)-K1^2(\zeta)]}{K1^2(\zeta)\,[J1^2(\zeta)-J0(\zeta)J2(\zeta)]}$$

and, R_{meas} - surface resistance of the sample
R_{sfixt} - surface resistance of the fixture
L - length of the dielectric resonator
$\tan\delta$ - loss tangent of the dielectric resonator
ε - dielectric constant of the dielectric resonator
λ_o - wavelength of measurement
Jo, J1, J2, K0, K1, K2, - regular and modified Bessel functions

$$\zeta^2 = (2\pi/\lambda_o)^2 - (\pi/L)^2$$

$$\zeta^2 = (\pi/L)^2 - (2\pi/\lambda_o)^2$$

A key to the sensitivity of the dielectric probe technique is that the electric and magnetic fields of the resonant circuit are largely confined within the low loss, high dielectric constant material. Hence, the measured Q value is determined largely by the losses from the portion of the conductive plates directly under the dielectric resonator. The computer generated plots in Figure 8 illustrate this field confinement for one particular dielectric probe configuration. The filling factor which is defined as the ratio of the energy stored within the dielectric material to the total energy stored in the circuit is plotted in Figure 9 for a variety of dielectric resonators. The filling factor shows the extent to which the measured Q value is determined by the conductivity of the end plates directly under the resonator and is indirectly a measure of the probe sensitivity. Figure 9 illustrates that over 95% of the stored energy is within the dielectric resonator region for most

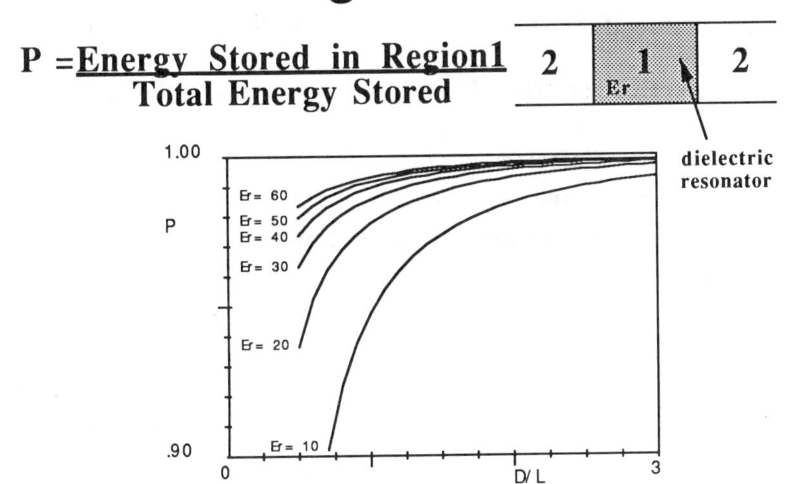

Figure 9 Plot of filling factor verses diameter to length ratio for dielectric resonators with various permitivities. Filling factor is indirectly a measure of dielectric probe sensitivity.

Figure 10 Photograph of dielectric resonators similar to those used in the dielectric probe. The variety of dielectric resonators available adds the advantage of frequency flexibility to a given probe fixture.

practical probe configurations. Hence, over 95% of the conductor losses contributing to the measured Q are from the portion of the conductors directly under the ends of the dielectric resonator.

4.2 Properties of the Dielectric Resonators Used

Since the dielectric resonator is the central element in the dielectric probe fixture, its properties largely determine the performance of the measurement technique. Dielectric resonators made from a variety of materials with relative dielectric constants between 25 and 80 are commonly available in a number of standard sizes. A variety of these are shown in Figure 10. The small resonator size for a given resonant frequency, the high degree of field confinement as illustrated in Figure 8, and the high Q performance of these resonators, particularly at cryogenic temperatures where measurements are performed, are key elements to the dielectric probe approach.

Since most of the energy is stored within the dielectric resonator, the losses from the dielectric material must be extremely small compared to the losses from the test sample in order to ensure a sensitive measurement. Figure 11 is a plot of the unloaded Q of a dielectric resonator verses temperature [10]. At 77K where R_S measurements are commonly made, unloaded dielectric resonator Q values of over 100,000 are common. This extremely low loss performance ensures high sensitivity for dielectric probe measurements at 77K.

An additional benefit to the dielectric probe technique can be derived from the variety of high Q dielectric materials available. This variety makes it possible to make measurements at a number of different frequencies with the same test fixture simply by interchanging the dielectric resonator with a new resonator of the same dimensions but having a different dielectric constant. Measurements at selected frequencies over a broad band are possible from this option.

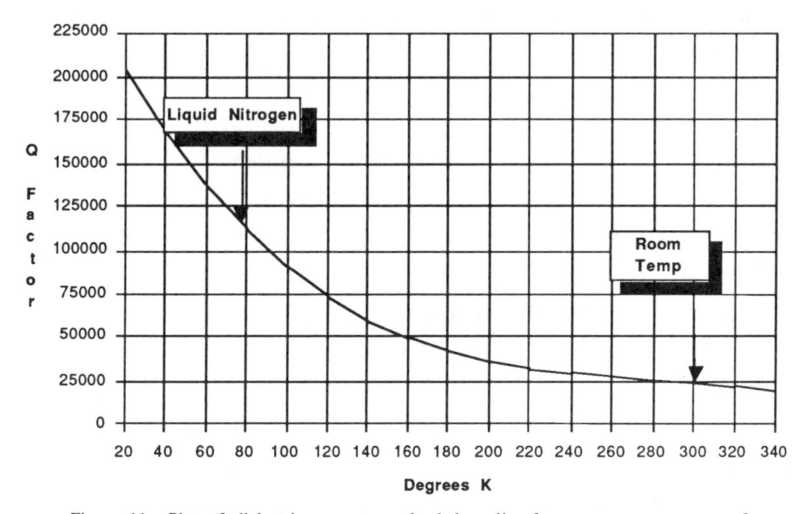

Figure 11 Plot of dielectric resonator unloaded quality factor verses temperature for a selected dielectric resonator. This high Q performance is a key to the sensitivity of the dielectric probe measurement technique.

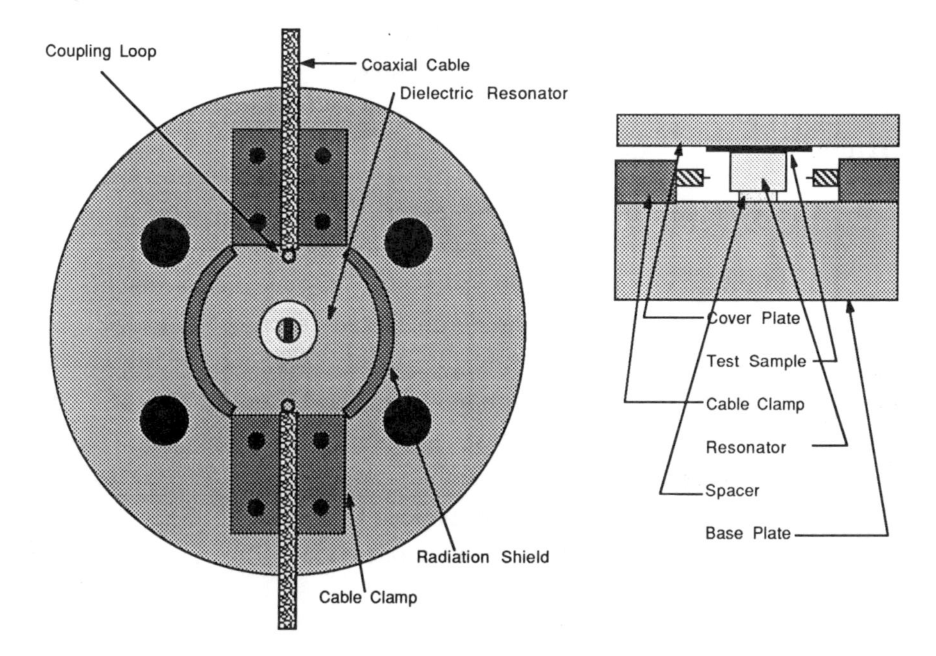

Figure 12 Drawings of the improved sensitivity configuration for the dielectric probe. Top view left, and cross sectional view right.

4.3 Improved Sensitivity Configuration for the Dielectric Probe

The dielectric probe configuration illustrated in Figures 6 and 7 employs a dielectric resonator sandwiched between two conductive metal plates in a post resonator configuration. As discussed earlier, the measured Q from this configuration is dominated by the losses from the conductive plates directly above and below the dielectric resonator. In this configuration, the Q value is determined by the combination of the test sample at one end of the dielectric resonator and the copper plate at the opposite end of the resonator. The improved sensitivity configuration for the probe shown in Figure 12 alleviates this problem by separating the dielectric resonator from the copper plate using a low loss plastic spacer. The physical separation of the probe from the copper plate significantly decreases the contribution of the copper to the measured Q. In this configuration, the dielectric probe maintains the advantages of the post resonator configuration, but has the advantage of significantly increased sensitivity. Figure 13 is a photograph of an improved sensitivity dielectric probe test fixture.

4.4 Analysis of Dielectric Probe Q Measurements

Absolute R_S values from dielectric probe measurements can be calculated by the technique discussed earlier, but for most measurement purposes absolute R_S values are not required, and measurements relative to a common standard are sufficient. The dielectric probe represents a fast and simple method to determine the surface resistance of HTSC samples relative to a copper or other normal metal standard. The measured Q value is determined by contributions from the following components as illustrated in Figure 14.

Figure 13 Photogragh of the dielectric resonator probe in the improved sensitivity configuration.

$$\frac{1}{Q_m} = \frac{1}{Q_d} + \frac{1}{Q_{t\,1}} + \frac{1}{Q_{t\,2}} + \frac{1}{Q_b} + \frac{1}{Q_r} + \frac{1}{Q_s}$$

$$\frac{1}{Q_{t\,1}} = \frac{1}{Q_m} - \frac{1}{Q_{fix}}$$

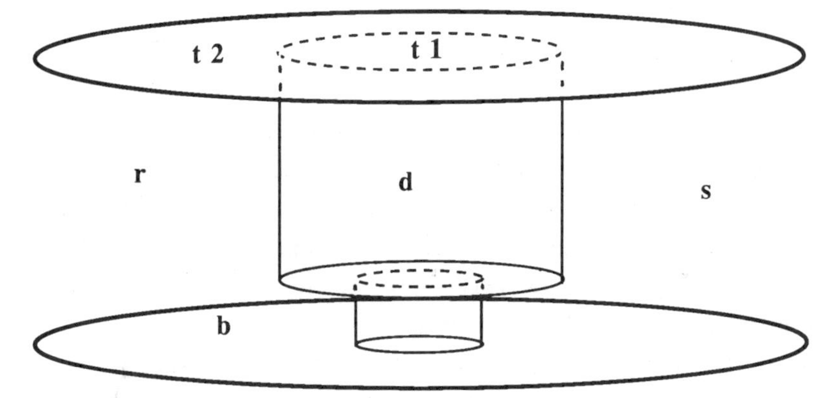

Figure 14 Illustration of the improved sensitivity configuration of the dielectric probe showing contributions to the measured Q value.

$$1/Q_m = 1/Q_d + 1/Q_{t1} + 1/Q_{t2} + 1/Q_b + 1/Q_r + 1/Q_s$$

where; Q_m - measured quality factor
Q_d - Q contribution from the dielectric and spacer combined
Q_{t1} - Q contribution from the test sample
Q_{t2} - Q contribution from the top plate excluding the test sample
Q_b - Q contribution from the bottom plate
Q_r - Q contribution from radiation losses
Q_s - Q contribution from the side walls

By defining Q_{fix} as the Q of the fixture excluding the test sample and solving for Q_{t1}, the Q of the test sample, we get

$$1/Q_{t1} = 1/Q_m - 1/Q_{fix}$$

By measuring Q_m for two samples of known conductivity, gold and copper for example, and substituting into the above equation, we get a value for Q_{fix}. Once Q_{fix}, the Q contribution of the fixture, is established, R_s measurements relative to the copper and gold calibration pieces can easily be calculated from the above equation.

4.5 Measured Results Using the Dielectric Probe

The dielectric resonator probe shown in Figure 13 has been used to measure over 60 HTSC thin film samples from a variety of sources. Measurements have been found to be repeatable, fast, simple to perform, and accurate. The best films measured have had R_s values less than 1/20th that of copper, and there is no indication that this is near the limit of the fixture sensitivity. Figures 15 through 18 show measured results from the dielectric probe. Figure 15 shows a relatively broad band view of the frequency response of the probe illustrating the easily identified TE011 resonant mode. Figure 16 shows the response of the probe with a YBCO test sample at both room temperature and at 77K. Figure 17 shows the response of the probe with a copper calibration sample,

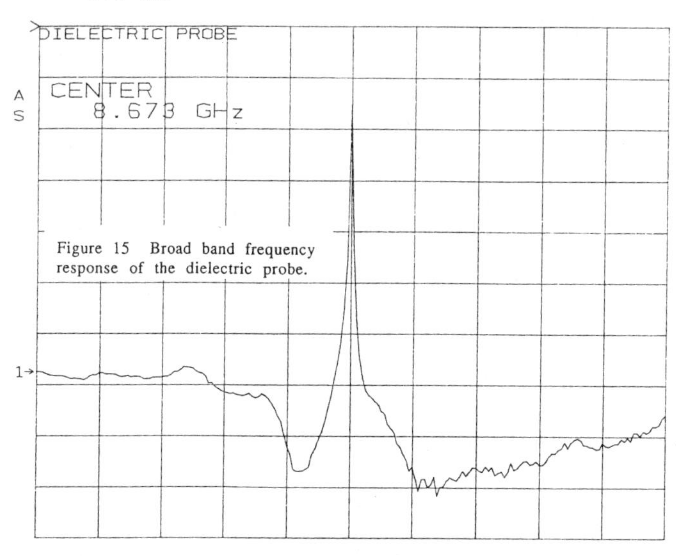

Figure 15 Broad band frequency response of the dielectric probe.

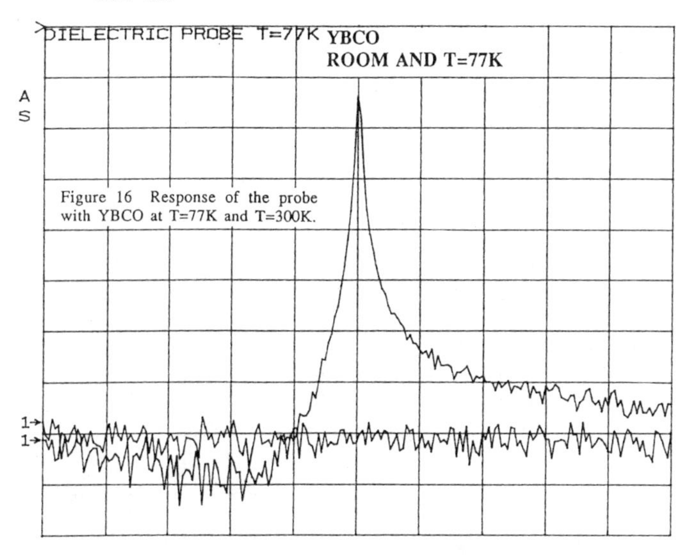

Figure 16 Response of the probe with YBCO at T=77K and T=300K.

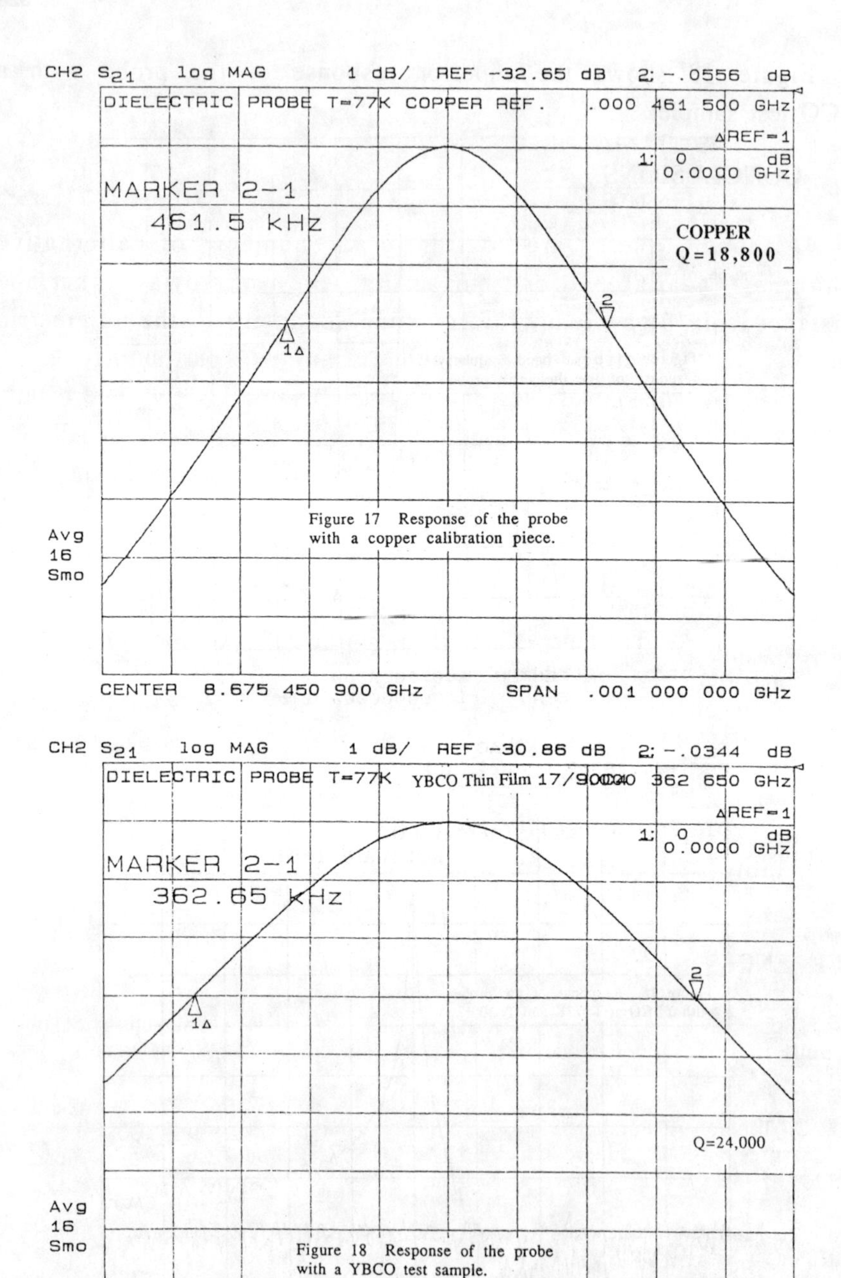

Figure 17 Response of the probe with a copper calibration piece.

Figure 18 Response of the probe with a YBCO test sample.

and Figure 18 shows the superior response of the probe with a YBCO test sample.

5. CONCLUSIONS

Although there are currently a number of alternative techniques available for measuring the microwave surface resistance of high temperature superconductors, the dielectric probe is an attractive measurement technique, particularly as a quick, simple means of screening HTSC samples for microwave applications. Some of the properties of the improved sensitivity configuration of the dielectric resonator probe technique include the following.

- Accuracy, Sensitivity, and Repeatability
- Ability to measure small samples at reasonably low microwave frequencies or to nondestructively measure selected areas of larger samples
- Ability to measure at more than one frequency with the same test fixture
- Simple to perform tests quickly
- Maintained sensitivity at temperatures above T_C
- Ability to obtain either absolute or relative R_S values
- Ability to measure either bulk or thin film samples.

REFERENCES

1. E. Belohoubek, A. Fathy, and D. Kalokitis, "Microwave Characteristics of Bulk High Tc Superconductors", IEEE MTT-S, pp. 445-448, New York, June, (1988).

2. R.C. Taber, "A Parallel Plate Resonator Technique for Microwave Loss Measurements on Superconductors", Rev. Sci. Instrum. 61 (8), 2200 (1990).

3. D. L. Moffat, K. Green, J. Gruschus, J. Kirchgessner, H. Padamsee, D.L. Rubin, J. Sears, Q-S Shu, T-W Noh, R. Buhrman, S. Russek, and D. Lathrop, Materials Research Society Spring Meeting, April 5-9, (1988).

4. A.J. DiNardo, J.G. Smith, and F.R. Arams, "Superconducting Microstrip High-Q Microwave Resonators", Journal of Applied Physics, vol. 42, no. 1, (1971).

5. D. Kalokitis, A. Fathy, V. Pendrick, R. Brown, B. Brycki, E. Belohoubek, L. Nazar, B. Wilkens, T. Venkatesan, A. Inam, and X.D. Wu "Measurement of Microwave Surface Resistance of Patterned Superconducting Thin Films", Jnl. of Elect. Mat., 19, (1990).

6. J.S. Martens and J.B. Beyer, "Measure RF Losses of Superconductors in Planar Circuits", Microwaves and RF, vol. 27, no. 9, September, (1988).

7. S.J. Fiedziuszko and P.D. Heidmann, "Dielectric Resonator Used as a Probe For High Tc Superconductor Measurements", IEEE MTT-S, Long Beach, (1989).

8. B.W. Hakki and P.D. Coleman, "A Dielectric Resonator Method of Measuring Inductive Capacities in the Millimeter Range", IRE Trans. Microwave Theory and Tech., vol. MTT-8, PP. 402-410, July, (1960).

9. Y. Kobayashi, M. Katoh, "Microwave Measurement of Dielectric Properties of Low-Loss Materials by the Dielectric Rod Resonator Method", IEEE MTT Trans., vol. MTT-33, pp. 586-592, July, (1987).

10. Y. Kobayashi, Y. Kabe, Y. Kogami, and T. Yamagishi, "Frequency and Low-Temperature Characteristics of High-Q Dielectric Resonators", IEEE MTT-S, pp. 1239-1242, Long Beach, (1989).

FREQUENCY MEASUREMENT in MILLIMETER and SUBMILLIMETER ELECTROMAGNETIC WAVE BANDS USING JOSEPHSON JUNCTIONS

Denisov A.G., Larkin S.J., Obolonsky V.A., Khabaev P.V.

"Saturn",Research-Production Association,

252148, Kiev,USSR

Abstract

The paper is concerned with the problem of Josephson junctions use for frequency measurement in millimeter- and submillimeter- wave bands. The results obtained for breadboard models point to a compatability of the measuring systems, realizing the described methods, have been presented.

Frequency measurement in submillimeter-wave and the shortest part of millimeter-wave bands is connected with certain problems. First, this part of frequency spectrum is critical for a number of semiconductive elements and so the effect of frequency spectrum transfer of the tested signal by heterodyning becomes problematic. Second, the application of the measurement methods dealt with spectrum conversion as well as the application of resonance methods of frequecy measurement presume fairly large signal levels (about 10^{-3} W), necessary for the start that essentially limits practical implementation of these measurements in the mentioned frequency bands.

At the same time it is known that the quantum nature of the superconductive elements using Josephson effect allows to use such elements up to the infrared region of the spectrum of electomagnetic waves [1)]. One of the consequences of the AC Josephson effect is the advent of the Josephson oscillation [2)]. At this the frequency of the oscillation being produced is directly proportional to the voltage across the junction:

$$\omega = 2eU/\hbar \quad (1)$$

where: ω - oscillation frequency; $2e$ - the charge of the Coper pair; $\hbar$ - Planck constant; U - constant voltage across the junction.

When a Josephson junction with an external microwave signal being irradiated one can observe some peculiarities on its current-voltage characteristic the so-called, "Shapiro steps" ocurring at synchronizing of the free oscillation of the junction with the external signal. Thus, the voltage in the region of "Shapiro steps" is proportional to the frequency of the signal irradiating the junction [2]. The latter allows to use the above peculiarities of the current-voltage characteristic of the Josephson junction for the frequency measurement of microwave signal in the millimeter- and submillimeter-wave bands.

The use of the Josephson junction as a measuring element has a number of merits as compared to traditional ones. We may point out the following:

- broadband performance; the order of magnitude is attributed to the applicability of Josephson junctions for frequency measurement in the upper part of radiofrequency spectrum, and the cut off frequency for Josephson effect is determined by the relaxation time of the superconductive parameter of which approximately gives the frequency about 7500 GHz [3];
- sensitivity; the fundamental limit will be determined by the parameters which define the sensitivity of aselective Josephson detector. Referring to [2),4)], the value of the threshold sensitivity (NEP) will be not less than 10^{-13}W/Hz;
- fundamentality; the frequency of the tested

microwave signal is linearly connected through the fundamental constans with the constant voltage being measured. This simplifies the methodological approaches and consequently widens possibilities of automatic control of the measuring process.

It should be pointed out that due to such unique frequency features of the Josephson junction there is no need to use special heterodynes, the apriori knowledge of the approximate band of the measured frequencies is not necessary, and one can make measurements in a broad band of frequencies using only one function unit. Besides, the use of the peculiarities of the current-voltage characteristic is also possible for the measurement of microwave-signal power attributed to the change of the critical current caused by the Bessel function, and the synthesis of frequency and power measurements and application of Gilbert spatial conversion to the current-voltage characteristic study, promote the possibility of carrying out the highly sensitive spectral analysis in the submillimeter-wave band [5].

Generalizing the results of our research of the last few years we may recommend some methodological approaches to the frequency measurement with the help of Josephson effect.

The first approach is characterized by the use, as a bias source, of the working point along the current-voltage characteristic of the junction, of a voltage source with linearly increasing amplitude (Fig. 1).

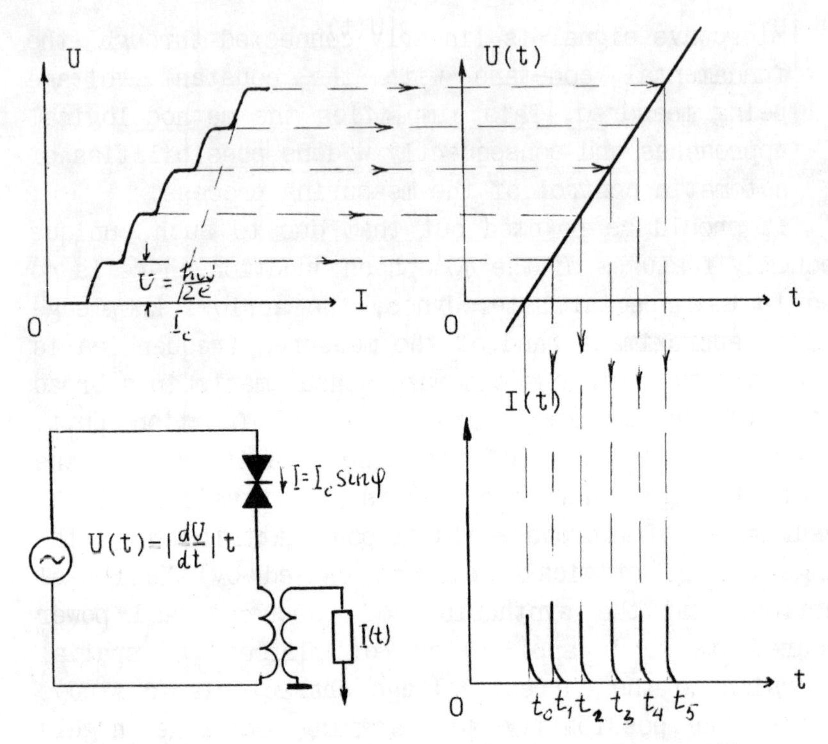

Fig. 1. The method using as the base source the source of a voltage with linearly-rising amplitude U(t).

$$U=|du/dt|*t \qquad (2).$$

Because of the fact, that the current-voltage characteristic of the irradiated junction acquires' a step-like form at the moment of time t_0, t_1, t_2, then the output unit will register current steps $I_{out}(t)$. The time interval $\Delta t=t_2-t_1$ and the step voltage of the step U_{st} we may present as follows:

$$\Delta t=U_{st}*(|du/dt|)^{-1} \qquad (3),$$

and substituting then in (1) will obtain the expression for the desired frequency as.

$$\omega=(2e|dU/dt|\Delta t)/\hbar \qquad (4)$$

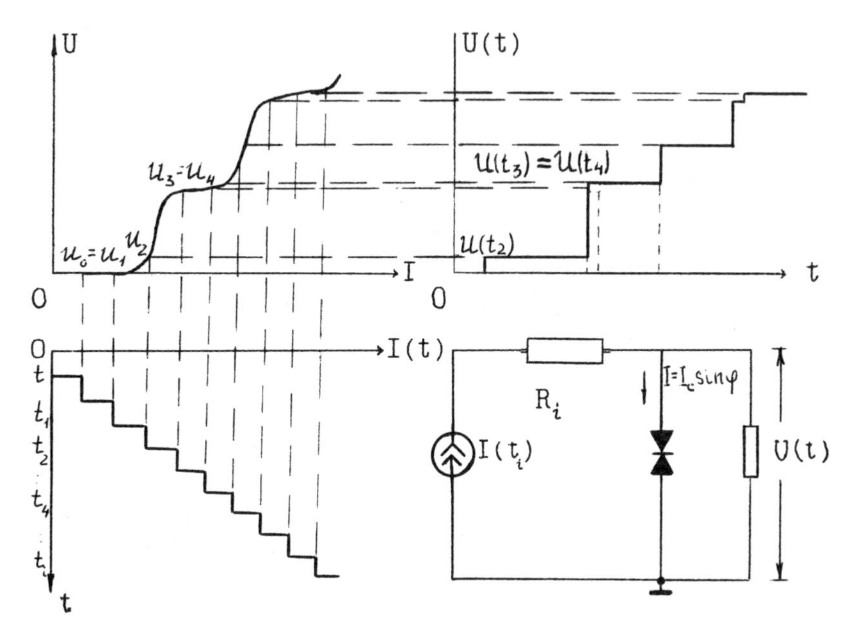

Fig. 2.The method using as the base source the source of current with dicretely rising amplitude I(t).

The second method is based on the implementation of the current source with discretely increasing signal amplitude $I(t_N)$ for the working point to shift along the current-voltage characteristic of the Josephson junction (Fig. 2). At this, the voltage ocurring across the junction $U(t_N)$ is stored by the output unit and in every-measurement cycle it is compared with the voltage value $U(t_{N-1})$ obtained in $N-1$ cycle:

$$\Delta U=(U(t_N)-U(t_{N-1})) \qquad (5).$$

It is seen from Fig.2 that at the moments of time (t_3, t_4) the difference $\Delta U=0$ because $U(t_3)=U(t_4)$. As far as this measurement cycle is consistent with the"Shapiro step" along the current-voltage curve ane then the frequency of the tested signal will be as follows:

$$\omega = U(t_N) * 2e/\hbar |_{\Delta U = (U(t_N) - U(t_{N-1}))} \quad (6).$$

There also exists an approach when two parallely connected (by power circuits) Josephson junctions are used as a measuring element; at one of those of is provided the reference frequency of a standard generator and on its I-U characteristics there is formed a net of standard frequencies .The other one has been irradiated by the tested microwave signal and "Shapiro steps" form marks on its current-voltage curve when feeding linearly increasing bias voltage $U(t)$ (Fig. 3).

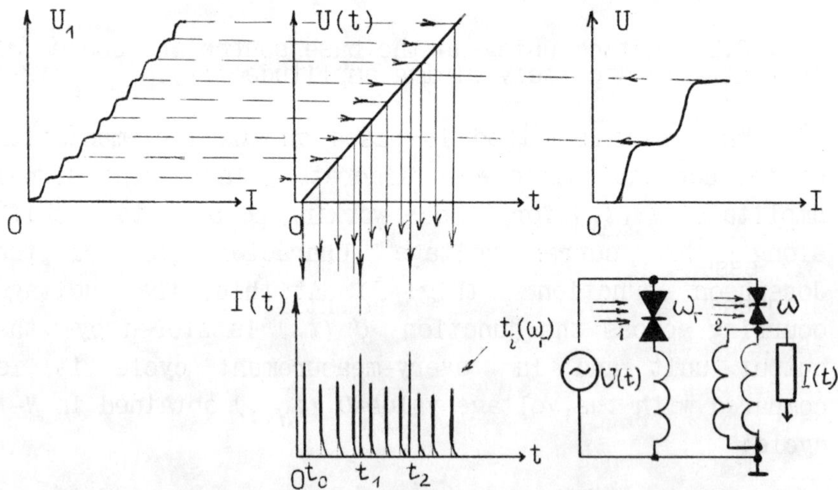

Fig. 3.The method based on the use of two parallely connected Josephson junctions one of which is irradiated by the microwave signal of the studied frequency, while the other one is irradiated by the signal of the standard generator frequency.

The output unit effects counting of reference

frequency pulses fallen in the interval between the pulses-marks (so called gate) and so the desired frequency can be expressed as:

$$\omega -N*\omega_r \quad (7).$$

All the above approaches were tried experimentally on acting sample models including the partial use of High-T_c superconductive microbridges of 1-2-3 system type. Upon our estimations at nitrogen cooling temperatures in the near future it can be feasable to use bridge structures out of the High-T_c superconductive materials for frequency measurement with the following possible parameters:

- measured frequency band-up to 800 GHz;
- start power sensitivity over the whole band not less than 10^{-8} W;
- frequency measurement precision up to 10^{-5} relative units;
- speed (the time of one measurement cycle) not more than 10^{-3}sec.

There exists a number of problems which so far essentialy limit the possibilities of Josephson frequency meters. Practical scientific interest lies in the following:

- analysis and small signal processing against noise background;
- development of ultrawideband path of the transmission of microwave signal to Josephson junction;
- electro-dynamic match of Josephson junction with the path;
- increase of the measurement precision, which is limited by the unstability of the reference-voltage sources.

We have obtained the following results on the sample model, realizing the first and the second described above methods of frequency measurement with the use of a Josephson effect:

- measured frequency band - 30 - 300GHz;
- sensitivity not less than 10^{-6} W over the whole band;
- measurement error 10^{-4} relative units;
- digital monitoring of the measured frequency.

References:

1. Бароне А., Патерно Дж. Эффект Джозефсона : физика и применение. Пер. с англ.-М. : Мир, I984

2. Лихарев К.К., Ульрих Б.Т. Системы с джозефсоновскими контактами.-М. : Изд-во МГУ, I978.

3. Mc.Donald D.G., Kose V.E., et al. Appl. Phys. Letts., No15, p.121 (1979).

4. Werthamer N.R., Shapiro S., Phys. Rev., No164, p.523 (1967).

5. Hinken J., Niemeyer J., Computer-controlled mm- and sub-mm wave Josephson spectrometer with a planar integrated front end. //Conference Proccedings 18th European microwave conference, 1988, Stockholm.

8. BUSINESS, MARKETS AND LEGAL

DEVELOPMENT AND MARKETING OF SUPERCONDUCTOR TECHNOLOGIES

R.C. Ropp, Vice President for Technology
International Superconductor Corp.
138 Mountain Ave.
Warren, NJ 07059-5260

ABSTRACT

International Superconductor Corp.(ISC) has, as its sole product, patents concerning ceramic superconductors available for license and/or sale. A survey of the current market potential for high-T_c superconductor products is presented in terms of the R & D currently being conducted on devices by private Industry, Government Labs., and funding by the Federal Government. Specific areas are discussed in some detail, including Physics Machines such as the Superconducting Supercollider. Targeted markets and general licensing procedures are also explored in relation to the business market discerned by ISC.

Three (3) points will be addressed in this presentation. They are: 1) The Superconductor Business in the United States, past and present; 2) Target markets in this business; and 3) Licensing strategies. But, before we can do this, we need to examine the history of our company in relation to the high T_c business.

International Superconductor Corp. (ISC) was formed in 1987 in response to a perceived need that can be concisely stated by a quote from the Wall Street Journal -

"Companies that control crucial patents....could freeze out competitors and some day dominate a multi-billion dollar industry ......".

Our marketing philosophy is: "We are searching for firms who will take over a crucial superconductor patent and diligently **use** it to manufacture and sell products in the international marketplace. *We do not intend to market products ourselves, only patents.*

In order to elucidate this philosophy further, we need to explore the business background of ISC. In early 1988, a consortium of 40 + Scientists was formed by the company.

The inducement to join continues to be:

"We are serving as an outlet for the individual inventor who might not otherwise patent his invention and, more importantly, market it".

By assuming all legal, marketing and developments costs, ISC provides a unique service to inventors wherein future revenues become possible for technologies that might otherwise **not** be commercially developed and possibly would be abandoned by the Inventor.

Mode of Operation: ISC pays all costs for filing the patent including payment of monies and stock to the Inventor for:

a) write-up of patent submission,
b) when accepted
c) when filed,
d) when patent issues
e) a percentage of royalties generated upon license or sale of patent. This may include a one-time payment up to $20,000.
f)Inventor is available as CONSULTANT to licensing company to ensure proper commercialization of his invention.

From our Consortium, we solicited, and obtained patent disclosures which were written into patent applications by our patent attorney. Most of these were filed in late 1988 and early 1989.

In December, 1989, we completed a public offering of stock, and began day-to-day operations the following month.

Products and Services: ISC has a Brochure available which describes "Superconductor Inventions for License or Sale". Five (5) categories are listed:

i) Superconducting (SC) materials & processes
ii) Non-destructive evaluation (SQUIDS)
iii) Electronic applications
iv) Military and space applications
v) Industrial applications

The technological advances introduced by these proprietary properties include:

i) New design for a SQUID
ii) Self-cooled SC wire design
iii) Electronic modulation of a magnetic field
iv) Ultra-high energy storage capacitors
v) Switchable superconducting mirrors and pixel arrays
vi) New design for superconducting gears and clutches
vii) "∂ singlet oxygen generation" and apparatus

viii) Rail gun space launchers
ix) Flux Motion Bolometer

For example, in the **Flux Motion Bolometer** invention, the detector does not have to be kept midway between its superconducting and normal state during its functional phase. This simplifies operation which is based on detecting an induced fluxoid motion following a temperature rise induced by an impinging infra-red photon. Additionally, in the **Electronic Modulation of a Magnetic Field** invention, a method of shaping and modulating a magnetic field is described. The modulated magnetic field is made possible by suitable utilization of the Meissner effect. Whether the modulation rate can reach into the radio-frequency range has yet to be determined.

Additionally, several grants and solicitations have been made by private industry, based upon our patents, to wit:

i) SBIR Phase 1- "Thermal Shielding in Space Applications", Foster-Miller, Waltham, Mass.

ii) SBIR Proposal- "Variable Cutoff Filter for Far Infra-red", Janos Technology, Townsend, Vt.

iii) SBIR Proposal- "Miniature Wide-Band Electric and Magnetic Field Sensors", Wake-Forest College, Winston-Salem, N.C.

iv) SBIR Proposal-"Passive Self-Charging Superconductor Magnetic Bearing", University of Wisconsin Center for Superconductivity, Madison, Wisc.

We welcome inquiries of submissions for funding to demonstrate feasibility based upon our patents to any interested party.

Let us now turn to a survey of the market potential of the Superconductor Industry, our **first point** in this profile.

The **present market** is a result of the past 25 years of development of low T_c materials and devices. As most of you already know, these are based upon niobium (Nb) wires and alloys such as Nb(Ti). Some of the significant advances made include:

I. "Maglev" high-speed trains
II. Superconducting Supercollider
III. Magnetic Resonance Imaging
IV. Tokamac fusion research
V. Practical "SQUIDS" for radiation and field detection
VI. RF Cavities and MMIC's

VII. Energy storage and transmission
VIII. Magnetohydrodynamics
IX. Bubble chambers

However, for all of these applications, operation at 4 °K (liquid helium temperatures) is mandatory. Probably the largest present market is that of magnetic resonance imaging (MRI). Although funding of the Superconducting Supercollider has begun, this physics machine is still in the conceptual stage. The actual MRI market is difficult to estimate, since the market is perceived differently by various parties. For example, the Nb-wire manufacturer sees a much smaller market than a manufacturer of the actual MRI machine which is purchased by Medical consortia for use in medical diagnosis (typical cost per machine = \$2-3 x 10^6). About 1000 of these machines will be built this year.

For the most part, the second largest market lies in supplying low T_c coils and solenoids for use in R&D of large devices, including Maglev trains, Tokamac fusion, linear particle accelerators, and bubble chambers.

The current market for each of the above 9 categories has been variously estimated to lie in the \$ 10^6 - \$ 10^9 range, but some of these are still in the planning stage.

<u>Current R & D work</u> being carried out in various laboratories throughout the US on high-T_c materials includes: 1) Processing; 2) Research on Devices; & 3) work connected with Physics Machines.

This work may be summarized as:

I. **Processing** - thin film deposition techniques
- methods of forming "wires"
- reproducible compositions

In this Conference, you undoubtedly have heard many presentations concerning advances achieved in processing of materials. Some of the more relevant included work @ University of Houston wherein neutron irradiation produced a J_c of 2.5 x 10^7 a./cm^2 (@ zero field -this is equivalent to Nb-based alloys), and the paper on optimization of processing variables by Seigal of Bell Labs. Thin film materials are now being made which are usable in current devices in the Microelectronics Industry.

II. **Device Research**

Many industrial labs. are currently pursuing development of:

A. Electromagnetic Launchers and Rail Guns
B. Electro-optics
C. Levitated bearings
D. Magnet design

E. Hybrid Thin-film Microstrip Circuits
F. RF Cavity Design
G. Optical sensors
H. Proton therapy for carcinogenic patients
I. Superconducting antennas
J. Superconducting motor design

Much of the work is directed towards feasibility studies, with specific products in mind. For example, magnet design studies are directed toward sale of superconducting magnets for motors, linear accelerators, magnetic resonance imaging, and construction of large-scale Physics Machines. Some of the Physics Machines currently being built, planned, or scheduled, include:

III. **Physics Machines**

A. Superconducting Supercollider (SSC)

B. Continuous Electron Beam Accelerator Facility(CEBAF)

C. Superconducting Magnetic Energy Storage Facility (SMEF)

The projected funding for the SSC is currently about $ 2 x 10^9. The project will use about 10,000 magnet dipoles and quadrupoles (~ 6.6 T), with an equal number of trim coils. Both magnet designs were originally set @ 40 mm., with a length of about 17 meters. Most recently, the dipole bore was changed to 80 mm. with the quadrupole diameter remaining the same. Each magnet will probably cost between $10,000 & $100,000. There will be 10 cryogenic units, each of which will be 8.06 kilometers in length. The total track that a particle will traverse is therefore 80.6 kilometers. **At present, only Nb-alloy wire- based magnets are under consideration.**

The CEBAF installation at Newport News, Va. is currently under design and the low T_c rf-cavities will be used to generate three (3) simultaneous beams of electrons, varying in energy from 0.5 to 4.0 Gev. @ 200 μamps. The cavities will operate @ 1500 Mhz. and will be arranged in a recirculating "racetrack" configuration. There will be two (2) segments, one on each side, each having a length of 235 meters. Each segment will be composed of 25 cryomodules of 8 cavities/module, or 200 modules/segment. **At this time, only rf-cavities based on Nb-alloy wires are under consideration.**

In regard to the SMEF installation, I was not able to get detailed information, except that any planned installation will use low T_c based wires and coils.

In addition to work on coils, solenoids and magnets, considerable work in Industrial Labs. is being directed toward development of a usable SQUID (superconducting quantum interference device). The major problem encountered has been use of a suitable substrate for the device.

However, the actual products being offered in the marketplace continues to be limited. Some of these include:

IV. **Actual Products Being Marketed**

A. Powders, rods and wires - both low Tc and high Tc.
B. Cables and solenoids - low Tc
C. Hybrid superconducting films and circuits - high Tc.
D. Evaporation and sputtering targets- high Tc
E. Electro-optic devices - high Tc
F. Liquid cryogen level sensors - both low & high Tc superconductors

We may summarize all of the above as: " In the *short run* , high Tc material forms and wires are being sold while research continues to refine the production of reproducible compositions and to optimize manufacturing processes. In the *long run* , it can be expected that markets where low Tc materials are being used will gradually be overtaken by superior high-Tc compositions, forms and devices".

The current market has been estimated to be as much as $ 350 million for this year, with most of the sales involving low T_c Nb-alloys and forms.

Let us now concern ourselves with the **future status** of the high T_c market. Obviously, this will depend upon the progress made in R&D in various laboratories. Room temperature superconductivity is not likely to become a reality in the near future. At present, what is currently desired is a thin film with a J_c near to 150 °K. so as to minimize noise. Wire current-carrying capacity and formability are, and will be, deciding factors for high-T_c ceramic forms. Another factor will be the availability and cost of liquid helium.

As far as the **present view** of the probability of replacement of low T_c materials in their various contemporary applications, the following is a reasonable summary:

I. Magnetic Resonance Imaging: no advantage seen. Room temperature operation would save about 12%/scan. If high-Tc materials can be made more cheaply than low-Tc SC's, they may become used in this Industry.

II. RF Cavities are a good bet for high-Tc materials

III. Magnetohydrodynamics & Nuclear Fusion: coils and magnets are the main product used here. Refrigeration is about 2% of cost. Thus, high-Tc materials will be used only if lower in cost, but the subject of helium availability and cost remains an unknown.

IV. SSC and Particle Accelerators: the same comment applies. the amount of actual helium used in such installations may seriously limit its availability for other uses.

V. Power storage and transmission: good probability for high-Tc coils and magnets. This includes both DC & AC applications.

The future market for **all** superconductors has been estimated to lie in the $ 5-6 billion range, but with a time frame of at least 10 years.

We can summarize the **current attitudes** of those in the Industrial Marketplace, as follows:

i) Large Firms: have reduced R&D efforts but continue work on processing and processes; some work on devices is ongoing.

ii) Small Firms: most are marketing superconductor forms for ongoing research in Academia and Industry. Some are developing products for sale in the marketplace.

iii) Government: supports SBIR contracts which are getting preference over non-superconductor projects.

iv) National Labs.: mandated by Congress to support technology transfer on high-Tc materials and to aid in R&D efforts by small firms.

With this in mind, let us summarize where ISC fits into the marketplace:

i) We have patents available which lend themselves to development of **immediate** products for the marketplace. Our patents deal with innovative advancements in many of the areas where both small and large firms are carrying out research on devices for manufacture.

ii) The inducement continues to be that our technological innovation will result in an exclusive, marketable product.

The **second point** to be examined is the **TARGET MARKETS** that ISC has defined for itself, notably GOVERNMENT AND INDUSTRY:

1. **Government**

i) Large Physics Machines

I. Superconducting Supercollider
II. CEBAF
III. Tokamac

All of the above are now committed to low Tc coils and solenoids. As such, they require high quality wire and cable for construction. Coil sales are expected to amount to billions of dollars. A major advance in high-Tc wire quality and cost is necessary before it could expect to replace low-Tc wire. Special magnet design is a prerequisite because of the exaggerated Meissner effect found in some high-T_c materials.

ii. Space applications

I) Antennas
II) Thermal insulation and isolation
III) Levitated bearings
IV) Microwave generators
V) Superconducting microstrip integrated circuits, including hybrid circuits.

In general, Government is funding developmental activities directed toward space applications via: SBIR grants, research contracts and directed research. There appears to be a sizable market for the future.

b. **Industry**

i) Product development is primary activity. This includes materials processing and processes for manufacturing devices.
ii) Industry continues to develop high Tc materials in:

I. Device Research

A. Electromagnetic Launchers and Rail Guns
B. Electro-optics
C. Levitated bearings
D. Magnet design
E. Hybrid Thin-film Microstrip Circuits
F. RF Cavity Design
G. Optical sensors
H. Proton therapy for carcinogenic patients
I. Superconducting antennas
J. Superconducting motor design
K. Energy storage

II. Coil Development for:

A. Physics machines
B. Magnetic resonance imaging

III. Space Applications

A. Energy storage
B. Levitated bearings
C. Microwave power transmission
D. Antennas

Most of this work is R&D being carried out in private Labs. on product design for the Electronics and Space Industry.

The **third point** I wish to consider is the various ramifications of possible Licensing procedures with our customers. One definition of licensing is:

"Licensing is the means of transferring Intellectual Property from one organization to another". Its scope includes nine (9) areas:

* Funding the R&D of Intellectual Property

* Evaluating and packaging licensable Intellectual Property

* Identifying potential markets and licenses

* Protecting the Intellectual Property to be licensed

* Determining what Intellectual Property rights should be licensed

* Negotiating reasonable terms beneficial to the licensor and licensee and drafting an appropriate license

* Monitoring the flow of Intellectual Property and the payment of royalties

* Enforcing the license

The activity of licensing is primarily a business activity and the agreed, final written record forms a legal document called a **Licensing Agreement**. The license is a contractual agreement between a seller who, for financial gain, allows the buyer the right to use the seller's Intellectual Property rights to sell a product in the marketplace.

Because Mr. Alan Gordon, Esq, of Arnold, White & Durkee P. A. (Houston, Tx.) will present a detailed description of the licensing process, I will keep my description to a minimum.

We can list at least two types of *licensing activities*, to wit:

1) **Non-exclusive**: Licensee can use patent to develop and manufacture devices for sale in the marketplace on a competitive basis.

Advantage: Seller - can sell to several buyers at the same time. Competition may enhance final success for product in the marketplace; Buyer - pays less for license.

Disadvantage: Seller- Has more responsibility for enforcement of diligence; Buyer- final success may depend upon speed of getting product to marketplace in competition with others.

2) **Exclusive**: Licensee has sole use of patent to develop product for marketplace:

Advantage: Seller - can choose buyer more carefully who will enhance final success for product in the marketplace;

Buyer- has less competition, particularly if patented product is innovative and without equal in marketplace.

Disadvantage: Seller- Initial payment is likely to be less, but long term prospects may be better.

Buyer- likely to pay more initially.

The third category consists of the outright sale of Intellectual Property, and may be summarized as:

3. **Sale of Intellectual Property**

i) Outright Sale: Buyer owns patent and all rights thereto

Advantage: Buyer can develop products as he wishes.

Seller realizes profits immediately

Disadvantage: Buyer has sole responsibility for product development.

Seller has obligation to aid Buyer in his product development efforts

ii) Restricted Sale: Buyer obtains patent exclusively for a limited term. If products are not developed and sold in the marketplace after a specified time, the patent reverts to the Seller.

Advantage:	Seller can oversee development of products for the marketplace; Buyer still has an exclusive right, but for a more limited time.
Disadvantage:	Seller has more responsibility for product development; Buyer agrees to a time limit for product development.

Obviously, the final agreement can be quite flexible.

The final point I want to make in this section concerns **Joint Ventures.** Here, a definitive advantage accrues to both parties in that each side brings its most advantageous activity to the joint effort.

In summary, I have described the following areas:

a. Company History
b. Products and Services
c. Potential of the Superconductivity Industry
d. Target Markets
e. Licensing vs: Sale of Superconducting Properties

It should be clear that the market for high T_c materials and devices is likely to be evolutionary, and that as R&D continues to improve material properties, including chemical, physical and mechanical ones, the market for devices utilizing ceramic superconductors will continue its slow growth. The most likely "instant" use will come in the Microelectronics Industry, followed by SQUIDS and then the Space Industry.

A room temperature superconductor will make some impact, but not as much as has already appeared in the news media.

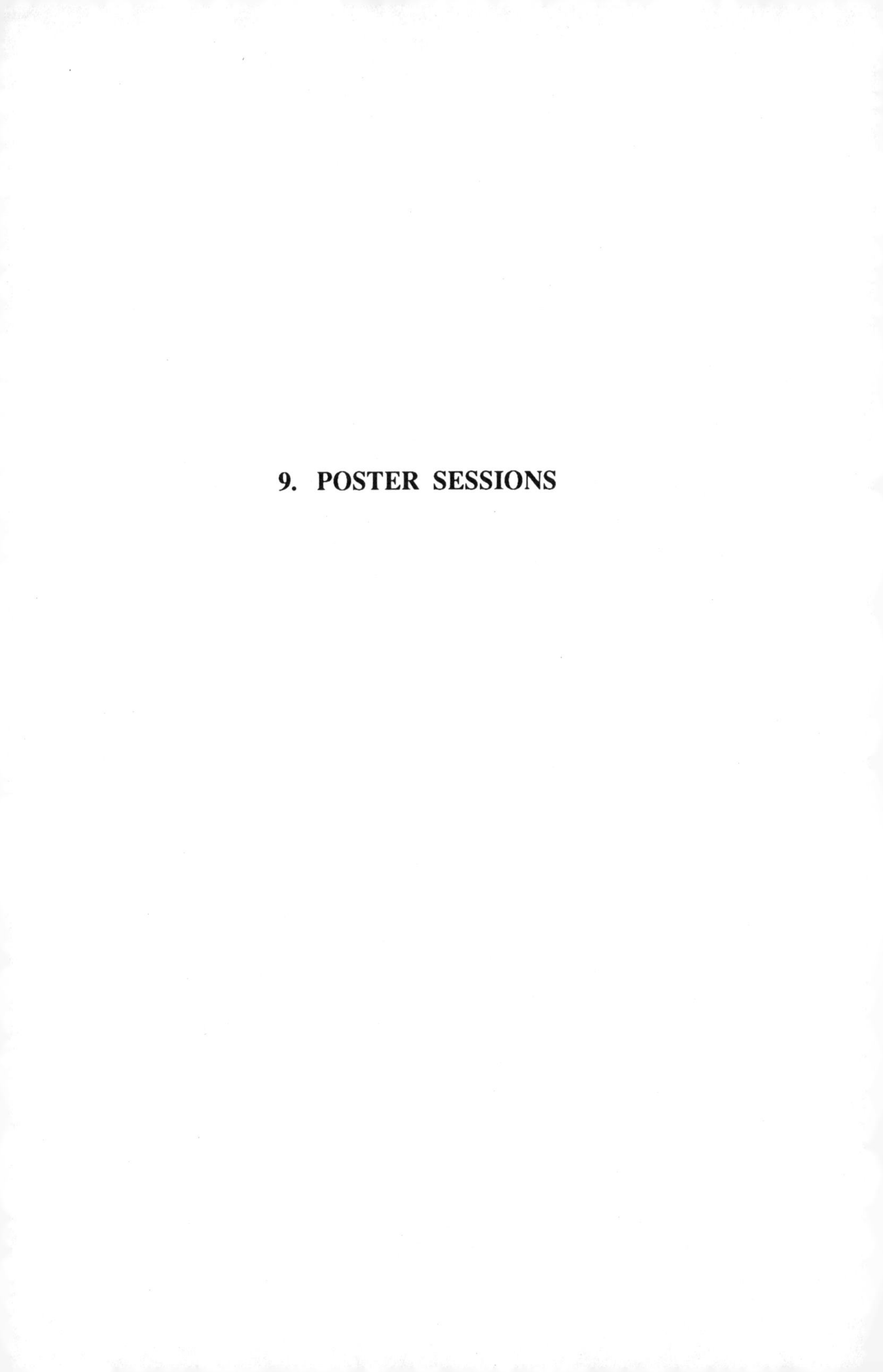

9. POSTER SESSIONS

Processing of Textured BiSCCO Materials for Improved Critical Currents

I E Denton, A Briggs, J A Lee, J Moore*, L Cowey#
H Jones#, C R M Grovenor* and A Hooper

Harwell Laboratory, AEA Technology, Oxon. OX11 0RA, UK.
* Department of Metallurgy and Science of Materials
# Clarendon Laboratory
University of Oxford, Oxford OX1 3PH, UK.

ABSTRACT
Work is described on the fabrication of textured 2212 and 2223 BiSCCO materials, using mechanical alignment and partial melt processes. In bulk samples, measured critical current densities are around 10^3 Acm^{-2} at 77K in zero field, and at 4K up to 15T. Melt-processed 2212 films exhibit improved values of T_c above 85K and tape specimens have J_c values up to 6,500 Acm^{-2} at 4K.

1. INTRODUCTION

As part of the Harwell-based consortium* on high temperature superconductivity, work has been carried out on the fabrication of BiSrCaCuO 2212 and 2223 superconductor samples with morphology and microstructure optimised for high current applications. Both bulk and tape samples have been produced with strong texture to improve grain connectivity and current carrying capacity, and a study of the sintering process for obtaining the maximum values of T_c and J_c has been undertaken.

---

* The Harwell-Industry Superconducting Ceramics Club comprises Air Products, BICC, BOC, Ford, Johnson Matthey and Oxford Instruments and is also supported by the UK Department of Trade and Industry.

This paper reports some of the findings of this programme, including the effect of sintering time on the J_c values measured at 77K, 63K and at 4K in fields up to 15T. Problems experienced in making these measurements at high current densities and high field are also described. The measurements are correlated with microstructural observations on the grain structure, impurity phase content and lattice defect density.

Finally, some preliminary observations are made on the relatively simple process of melt casting of the 2122 phase, which can have some beneficial effects on its superconducting properties.

2. SAMPLE FABRICATION

2212 powder was obtained from BDH Ltd. Pb-doped 2223 powder was prepared by repeated sintering and regrinding treatments starting with a stoichiometric powder mix.

Textured bulk material was obtained by pressing bar-shaped samples from pre-reacted 2223 powder at 640 MPa and then giving a final sinter at 840°C for times between 16 and 150 hours. These samples show transitions to zero resistance at temperatures in excess of 100K.

Figure 1 shows an SEM micrograph of the fracture surface of a typical specimen illustrating the partial alignment of the plate-shaped grains; an observation confirmed by the XRD pattern. A typical R/T curve is also shown.

Textured 2212 and 2223 tapes were prepared by rolling on silver backing sheets, and giving a final anneal above 800°C. The superconductor layers produced in this process are typically 20 - 50 μm thick.

The 2212 tapes exhibit T_c values in the range 70-85K, as shown in Figure 2(a). Figure 2(b) is an SEM micrograph of the top surface of a 2223 tape, showing the well aligned plate morphology, and an XRD pattern illustrating the textured nature of these samples.

Melt-processed 2212 material has been prepared by melting at 1000-1100°C, casting the melt onto a variety of metallic and ceramic substrates, and giving a final anneal above 800°C. This simple process generally results in T_c values between 85-90K, Figure 3.

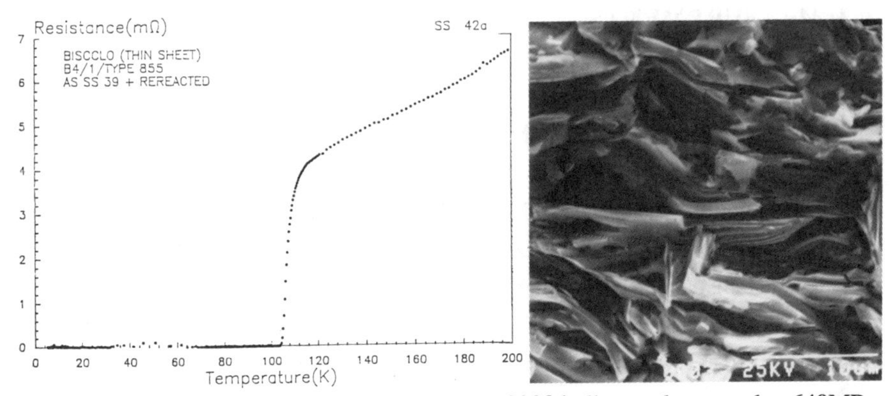

Fig.1 An SEM micrograph and R/T curve from a 2223 bulk sample pressed at 640MPa.

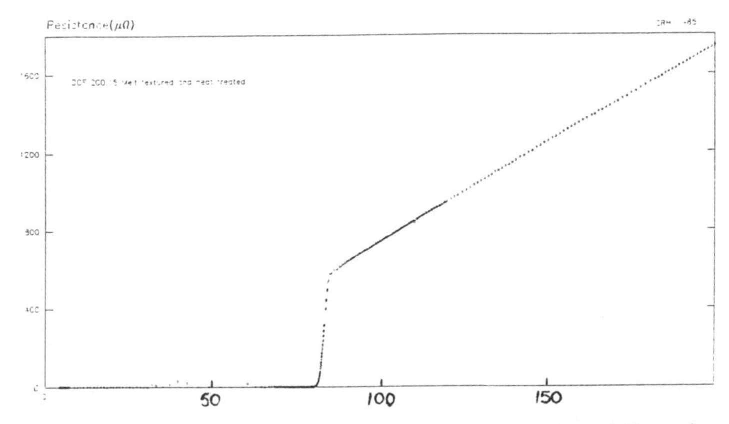

Fig. 2a Typical R/T curve for a 2212 tape rolled onto silver and partially melt-processed.

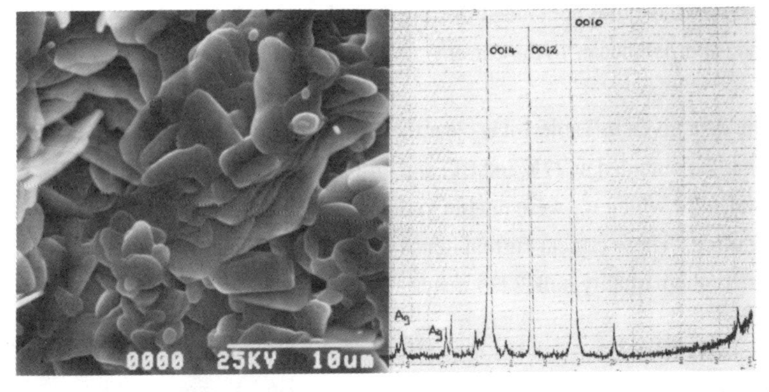

Fig.2b SEM micrograph and XRD spectrum showing the strong texture achieved in a rolled and annealed 2223 tape on silver

3. J_c MEASUREMENTS

3.1 Bulk Pressed 2223 Samples.

These samples have been tested in the high field facility of the Clarendon Laboratory. Typical sample dimensions are bars 20 by 1 by 1 mm, which are reasonably sturdy to handle and minimise the currents which must be passed through the indium contacts. However, numerous samples have been found to fail during high current testing by a localised melting phenomenon either at a contact or in the bulk of the sample. This failure of the bulk samples remains to be overcome, and has restricted the number of results obtained at low fields and temperatures where the critical current density is highest.

Figure 4 compares a typical set of data of J_c versus B at temperatures between 77-63K and at 4K. The current carrying capacity of this 2223 sample is rapidly quenched on the application of magnetic field at 77K, and this decrease is still marked at 63K. By contrast, at 4K, the rate of decrease of J_c with field is much slower, presumably as a result of increased pinning of the flux lines.

We have also observed quite different behaviour of the samples depending on the orientation of the plate-shaped grains with respect to the applied magnetic field. Comparing the data at 77K in Figure 4 with that from the same specimen in Figure 5 shows that $J_c(B)$ decreases much more rapidly in Figure 5. The Figure 5 data were obtained with the sample oriented at 90° with respect to its position for the data in Figure 4. We have thus concluded that the degradation in $J_c(B)$ is because the field direction is perpendicular to the current carrying plane in the second set of data.

From the data given in Table 1 it is clear that critical current densities greater than 800 A/cm^2 can be obtained at 77K and 0T, and around 1000 A/cm^2 at 4K and 15T. Many of the samples for which we were unable to obtain J_c values were carrying higher currents before failure. These critical current density values are in line with much of the published literature on high quality bulk 2223 material.

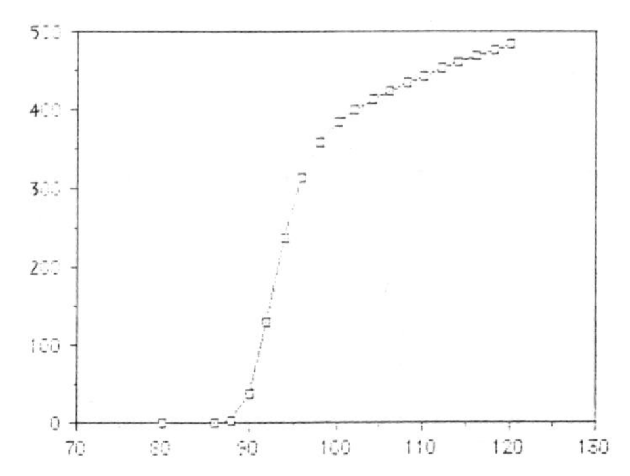

Fig. 3 Typical R/T curve for 2212 melt-cast material.

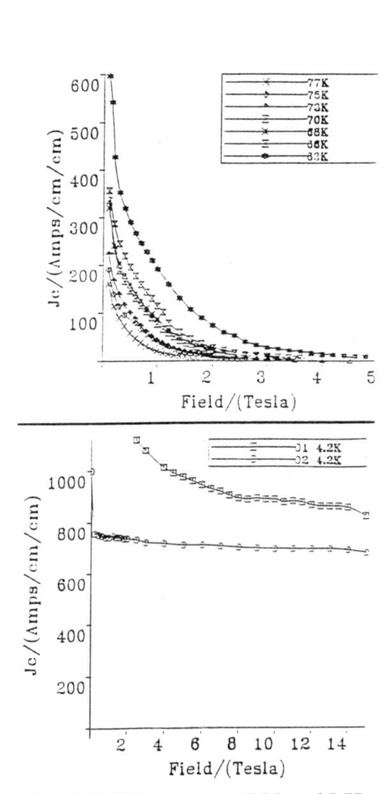

Fig. 4 Jc(B) curves at LN and LHe temperatures for bulk pressed 2223 samples.

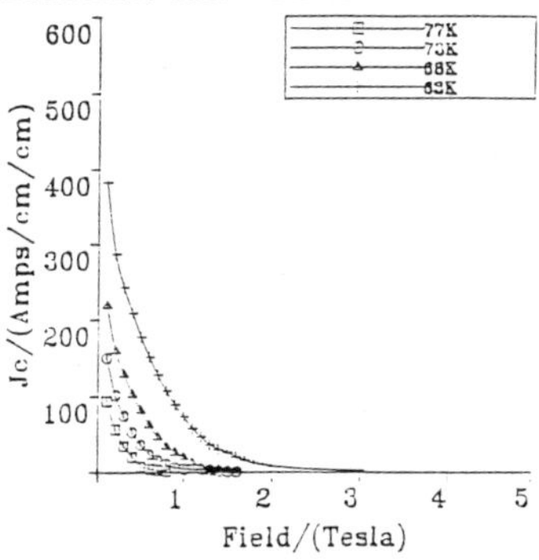

Fig.5 Jc(B) curves at LN temperature for the same sample as Fig.4 but rotated 90° with respect to the applied field.

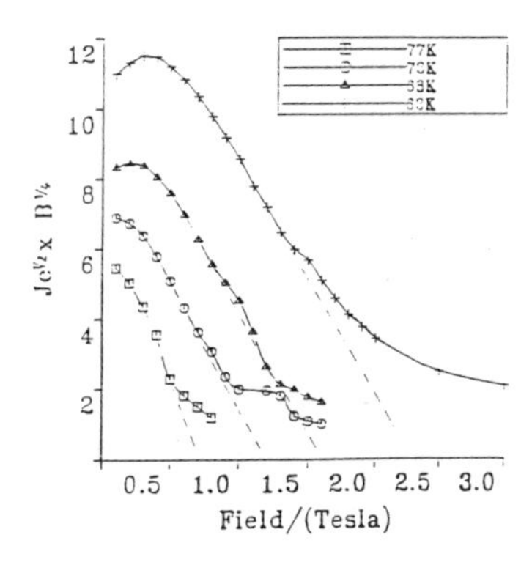

Fig.6 Family of Kramer plots from bulk pressed 2223 from which values of Bc2 can be extracted.

Table 1

A summary of J_c measurements on a set of pressed 2223 bars annealed at 840°C for times between 16 and 150 hours.

HOURS AT 840°C	MEASUREMENT TEMP.	77K		63K		4K	
	MEASUREMENT FIELD	0T	1T	0T	1T	5T	15T
16		321	4.8	F	60	1018	712
24		F	6.8	F	190	1087	912
35		900	NS	1400	NS	F	>502
48		F	10	F	80	-	-
150	SEVERELY DEGRADED PROPERTIES						

F SAMPLE FAILED DURING TESTING
NS SAMPLE CARRIES NO SIGNIFICANT CURRENT AT THIS FIELD

These measurements, on samples annealed at times much less than the frequently reported 50-100 hours, and the poor performance of the samples heated for 150 hours, illustrate that it is not necessary to use very long final anneals to obtain high quality bulk 2223 superconducting material.

This bulk 2223 material exhibits saturated flux pinning at temperatures between 63 and 77K; i.e. there is a peak in the volume pinning force at approximately 0.2 B_{c2}. This is analogous to the behaviour of conventional superconductors, eg. Nb_3Sn. For these types of materials, plots of $J^{0.5}B^{0.25}$vs B will generally show good linearity in the high field regime. Figure 6 is a family of these 'Kramer' plots for this material at liquid nitrogen temperatures. Exploitation of the linear regime yields $B_{c2}*$, an effective B_{c2} for high transport current materials, and an important parameter for flux pinning studies.

4. MICROSTRUCTURAL STUDIES

4.1 Bulk Pressed 2223 Material

This material is essentially phase-pure (as determined by XRD and EPMA analysis), and so the microstructural features of most interest have been defects introduced during the deformation process. Figure 7 shows a conventional TEM image of the basal plane dislocations found in almost every grain of the pressed and annealed material, and an HREM image showing the bending of the lattice planes and the insertion of wedge-shaped regions of different orientation. This material is thus highly strained and distorted, which may add pinning sites to a compound in which intrinsic pinning at 77K is very weak.

4.2 Rolled 2212 Tape

We have studied the phase purity of this material with XRD and EPMA. Even when XRD patterns show no sign of phases other than 2212, impurity phases are easily identified in the EPMA. These are generally CuO, (CaSr)CuO and Sr-rich particles. Figure 8 shows EPMA maps for Ca, Cu, Bi and Sr from one of these tapes. The non-uniformity of the elemental distribution is clear, as is the presence of a variety of impurity phases. In a tape only 30 μm thick, these phases can have a significant effect on the superconducting properties and only by very careful processing can they be avoided.

5. MELT PROCESSED 2212

Melt processing of 2212 results in the formation of a mixture of 2201, (CaSr)CuO and a Bi-rich phase, Figure 9(a). Heating this mixture at 820°C for 16 hours results in the growth of large 2212 grains, Figure 9(b), and this material is very dense and almost phase-pure. The T_c of samples treated in this way is always above 85K, but the bulk J_c values at 77K are poor.

More interestingly, the cation ratios are quite different from those found in sintered specimens. This is most easily demonstrated by plotting typical EPMA figures for the Ca/Sr and Cu/Sr ratios from bulk sintered and melt cast materials, Figure 10. The Sr level is far higher in the cast sample.

Fig.7 Conventional TEM image [left] of basal plane dislocations in bulk pressed 2223 and a high resolution image of the same material showing the remaining intercalation defects and low angle boundaries.

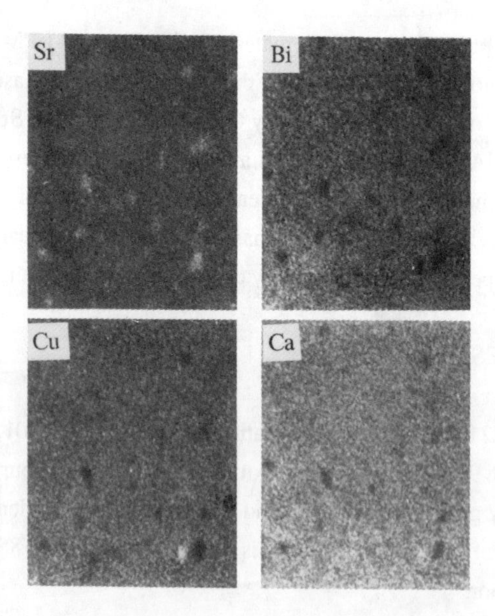

Fig.8 EPMA elemental maps for Sr, Bi, Cu, and Ca, illustrating the existance of impurity phases which can limit current flow in thin tape samples. [each picture area is approx 200 by 250 microns]

Fig.9 Optical micrographs of (a) melt-quenched 2212 showing a distribution of needle-like phases, and (b) the formation of large grained 2212 after annealing.

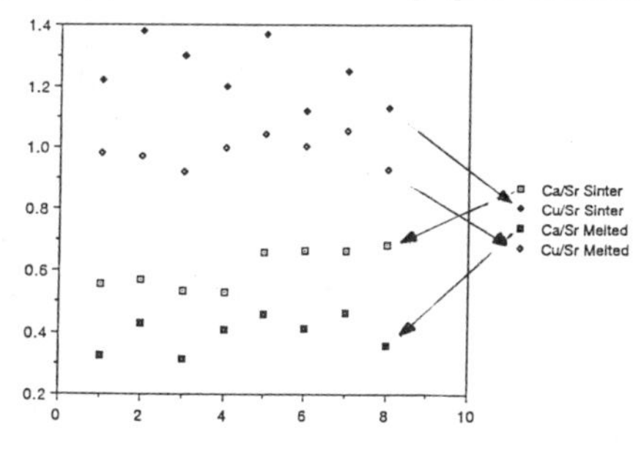

Fig. 10 Comparison of Ca/Sr and Cu/Sr ratios from ordinary sintered and melted 2212 material. The high Sr levels in the melted material are clearly seen.

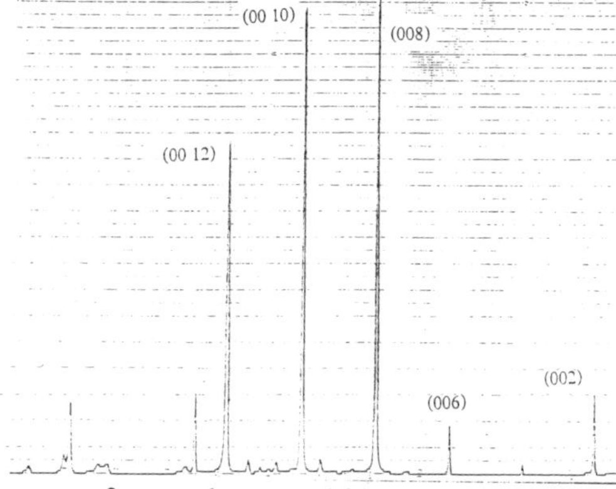

Fig. 11 An XRD pattern from melt-textured 2212 on polycrystalline MgO. The (00l) peaks are very prominent, indicating a strong texture.

Casting 2212 onto MgO substrates can result in the formation of a highly textured material as illustrated in the XRD pattern in Figure 11. The T_c of this material is above 80K after annealing at 850°C, but the critical current density values at 77K are very low; around 10A/cm^2. However, at 4.2K and 10T, J_c values in excess of 10^3 A/cm^2 have been measured.

Samples of rolled 2212 tape on silver substrates have been tested at 4K, after melt processing. They have shown J_c values of up 6500 A/cm^2 in zero field.

6. CONCLUSIONS

(1). Mechanical alignment and partial melting processes have been used to prepare bulk and tape BiSCCO material with a high degree of texture.

(2). High values of critical current density have been measured in bulk samples: above 800 A/cm^2 at 77K and 0T, and around 1000 A/cm^2 at 4K and 15T.

(3). Melt texturing of 2212 samples on a variety of substrates has been used to improve values of T_c above 85K. J_c values remain low at 77K, but values as high as 6500 A/cm^2 at 4K have been measured in tape specimens.

TEXTURE ANALYSIS OF BULK $YBa_2Cu_3O_x$ BY NEUTRON DIFFRACTION

A. C. Biondo,[1,2] J. S. Kallend,[1] A. J. Schultz,[2] and K. C. Goretta[2]

[1]Illinois Institute of Technology, Chicago, Illinois 60616
[2]Argonne National Laboratory, Argonne, Illinois 60439

ABSTRACT

Neutron diffraction has been used to generate Orientation Distribution Functions for two sinter–forged $YBa_2Cu_3O_x$ specimens. Sinter forging imparted a strong texture, with *c* axes of crystals preferentially aligned parallel to the forging direction. The distribution of *a* and *b* axes was not uniform, which may have implications to critical current density.

INTRODUCTION

Superconducting properties of all high–temperature superconductors are highly anisotropic.[1] Unless a random orientation of grains exists in the material, bulk properties will generally show anisotropic behavior. Few processing routes produce a random grain orientation. It is very likely, therefore, that some preferred orientation, or texture, will exist.

In high–temperature superconductors, superconducting properties are excellent in the *a–b* plane of the crystals and considerably poorer along the *c* axis. Thus, in order to maximize the critical current density (J_c) of a bulk superconductor, the alignment of the *c* axis perpendicular to the direction of current is desired. Because the presence of high angle grain boundaries may lower J_c,[2] a goal of bulk processing is to minimize the misorientation between grains. A method of identifying the distribution of crystal orientations is needed in order to make property predictions as a function of the deviation from perfect crystal alignment. A method of obtaining this information is to obtain pole figures from diffraction data and to then calculate an Orientation Distribution Function (ODF). An ODF is a complete statistical description of grain orientations within the volume examined.[3,4] In this study, neutron diffraction has been used to examine the textures of sinter–forged $YBa_2Cu_3O_x$ specimens. Neutrons enabled the full volume of the specimens to be examined. A brief description of the ODF method will be provided and the results for the sinter forgings will be presented and discussed.

ORIENTATION DISTRIBUTION FUNCTIONS

In standard X–ray analysis by a diffractometer method, a sample is held in a given orientation and a scan of crystal d-spacings is made. The reflecting planes examined are restricted to those parallel to the sample surface. To gain information on the volume fraction of crystals in arbitrary orientations, more information is required.

Pole figures are a measure of the diffracted intensity from a given crystallographic plane as a function of orientation. They typically give the distribution of plane normals in the sample as a function of orientation on a 5° x 5° grid in a stereographic projection. Because pole figures locate plane normals only, no information is given about the rotation of a plane about its normal. If a complete description of crystal orientations is desired, three dimensions must be used. The function that describes grain orientations in three dimensions is called an Orientation Distribution Function.[3,4]

An ODF gives the volume fraction of crystals in a given orientation with respect to some external set of coordinates. In rolling operations, for example, the external coordinates may be the rolling, transverse, and normal directions. In order to relate the three–dimensional crystal axes to the three dimensional sample axes, one uses Euler angles. Figure 1 shows a diagram that explains the angles used in the Roe convention of ODF mapping as they appear in three dimensions.[3-5]

In the Roe convention, the Euler angles are Ψ, Θ, and Φ, where Ψ and Θ locate the crystal z axis and Φ locates the crystal x and y axes. When plotting an ODF, constant Φ sections are used. Figure 2 shows a diagram of the Euler angles in two dimensions.

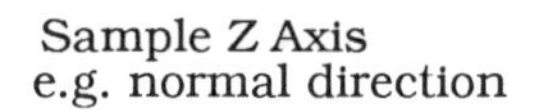

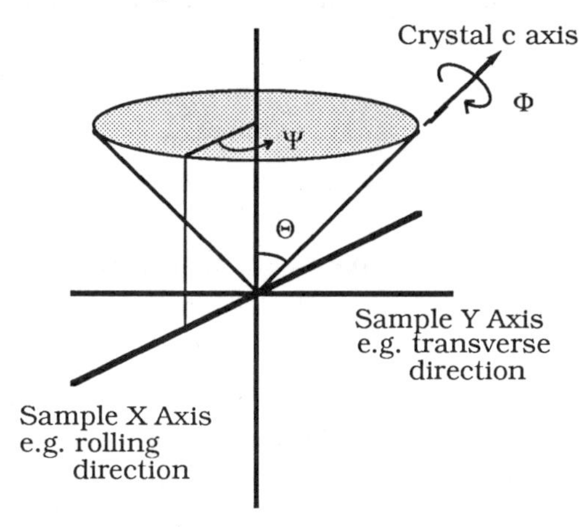

Figure 1. Description of ODF angles in three dimensions.

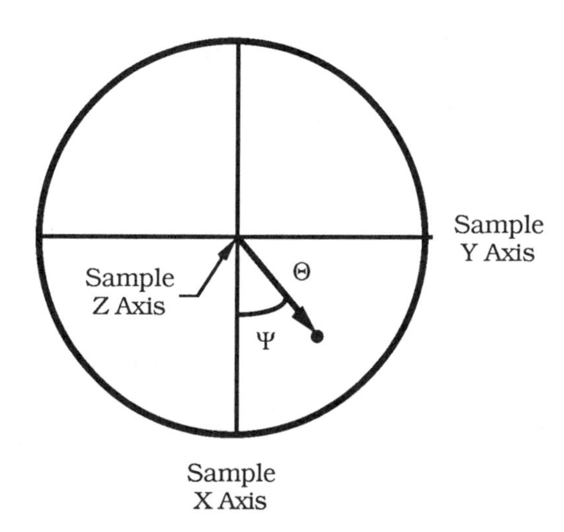

Figure 2. Description of ODF angles in two dimensions.

Orientation Distribution Functions can be measured experimentally by methods such as TEM; however a very large number of grains must be tested to make the result statistically meaningful, and thus a very long time would be required. In general, these functions are not measured directly, but are calculated by a mathematical method that requires as input the information in pole figures. For high–symmetry crystal systems, the calculations can be made from 2 or 3 pole figures. For low–symmetry crystal systems, more pole figures are required.

Once the pole figures are measured, they are mathematically manipulated by the Williams–Imhof–Matthies–Vinel (WIMV) algorithm.[6] Each point on a pole figure represents a projection along a path in the ODF. This projection path depends upon the crystal geometry and the diffracting plane. The WIMV algorithm makes an initial estimate at the ODF by assigning a value to each ODF cell. This value is given by the geometric mean of the values in all the pole figure cells to which it contributes. The pole figures that result from the estimated ODF are compared with the actual data and the discrepancies are used to refine the estimate. This procedure is repeated until the match between the recalculated pole figures and the actual data is considered satisfactory. In practice, the algorithm converges rapidly, and 6 to 12 iterations are generally sufficient for a good solution.

EXPERIMENTAL PROCEDURES

Two sinter forgings[7,8] were made from $YBa_2Cu_3O_x$ powder synthesized by solid–state reaction of Y_2O_3, $BaCO_3$, and CuO.[9] The powder was pressed at a pressure of about 100 MPa into pellets 22 mm in diameter and 13 mm in height. The pellets were compressed in air at 930°C between Al_2O_3 rams. ZrO_2 felt was placed on the ram faces;[8] Pt foil separated the ZrO_2 from the specimens. Forging was conducted at a constant displacement rate that ranged from 1.7×10^{-4} to 3.3×10^{-4} mm/s. Final specimen heights were about 7 mm and final densities were greater than 97% of theoretical.

The sinter–forged samples were studied at the Intense Pulsed Neutron Source at Argonne National Laboratory. The time–of–flight single–crystal diffractometer was used to gather the data.[10] Thirteen histograms were collected covering one quarter of orientation space. From these data pole figures were derived and expanded into four quadrants; orthorhombic sample symmetry was assumed in the calculations. To calculate the ODFs, pole figures were obtained for the following planes: (001), (200), (103), and (012).

RESULTS AND DISCUSSION

Optical microscopy of the cross sections of the two sinter forgings are shown in Fig. 3. Although the starting particle size for each specimen was approximately 3–5 μm, the second specimen exhibited much larger grains because of grain growth during forging. This grain growth is associated with formation of a small amount of liquid phase due to a lack of phase purity in the starting powder.[11] It is noted that it is difficult to assess optically the extent of texture for either specimen.

Figure 3. Optical micrographs of sinter forgings (forging direction is parallel to micron markers).

The (001) and (200) pole figures for the second sinter forging are shown in Figure 4. It is evident that strong *c*–axis alignment exists and that the distribution of *a*–*b* planes is not uniform. ODFs for the sinter forgings are shown in Figures 5 and 6. They display, as expected,[8,9] a strong *c*–axis normal texture. Both of the ODFs from the sinter forgings exhibit distinct regions of the off–normal orientation (i.e., for $\Psi \approx 30°$ and $\Psi \approx 60°$). The large–grained specimen has a slightly stronger, *c*–axis–normal texture. This stronger texture is probably a result of the grain growth. It is noted that, as were the pole figures, the ODFs are not uniform about the origin. This implies that the distribution of *a* and *b* axes in the forgings are not random. Therefore, this technique may prove useful in determining whether alignment of *a* and *b* axes is needed to produce high J_c values.[2]

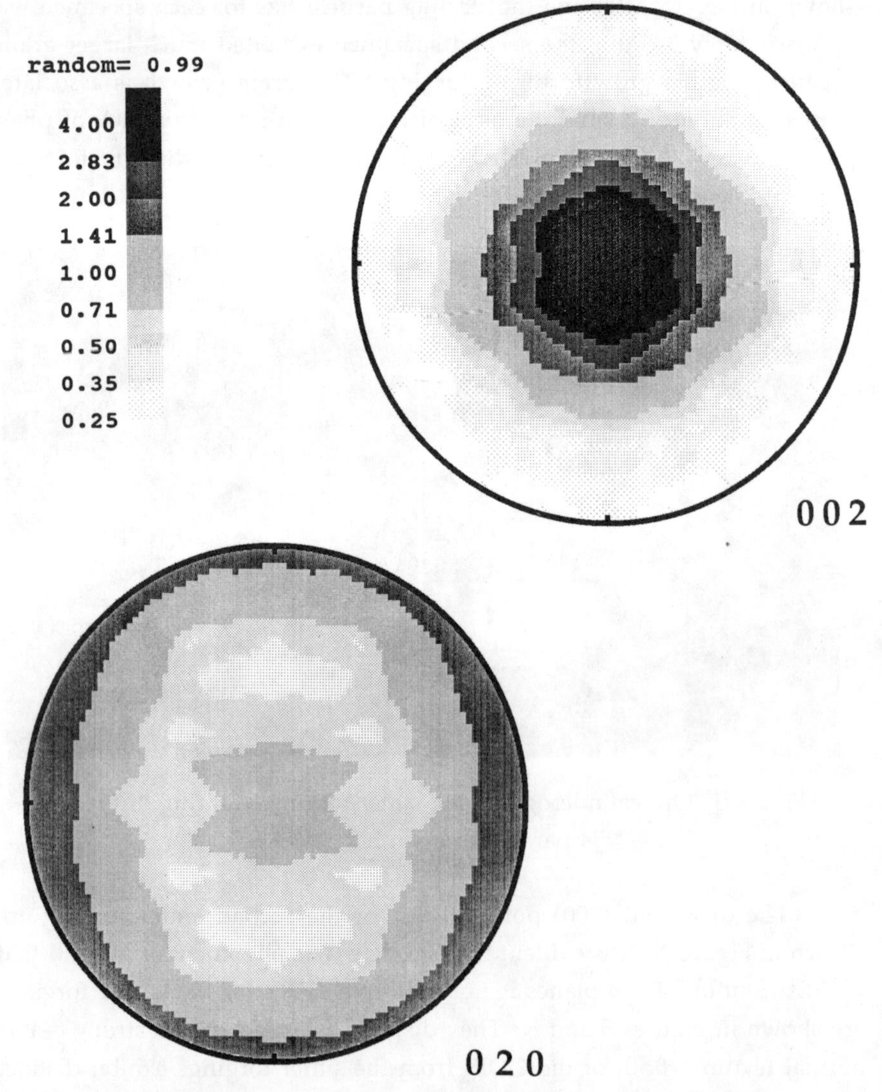

Figure 4. Representative pole figures for first sinter forging.

Favorable texture in bulk $YBa_2Cu_3O_x$ has been shown to produce increases in J_c values.[8,12] Texture of $YBa_2Cu_3O_x$ has been studied extensively.[13,14] Use of neutrons and ODFs provides the opportunity to quantify the textures, including differentiation of *a* and *b* axes. Work is now underway to examine systematically effects processing on texture and of texture on superconducting properties.

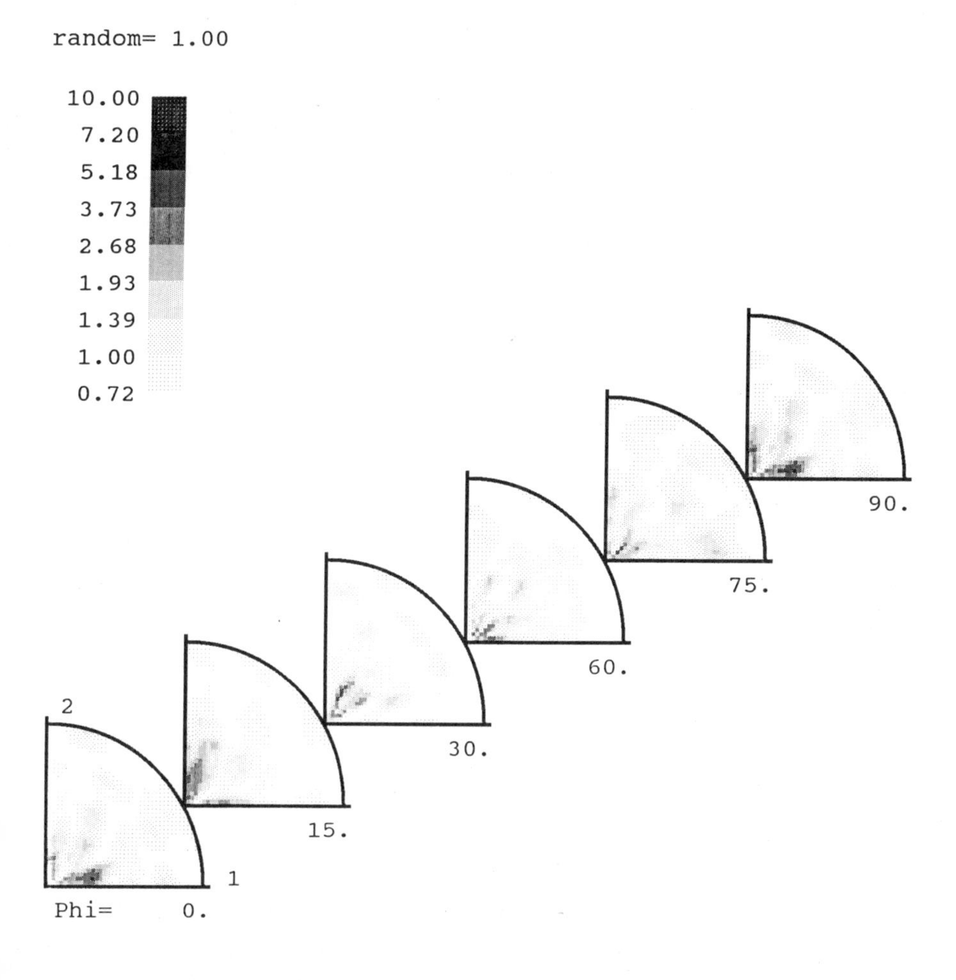

Figure 5. ODF of the first sinter forging.

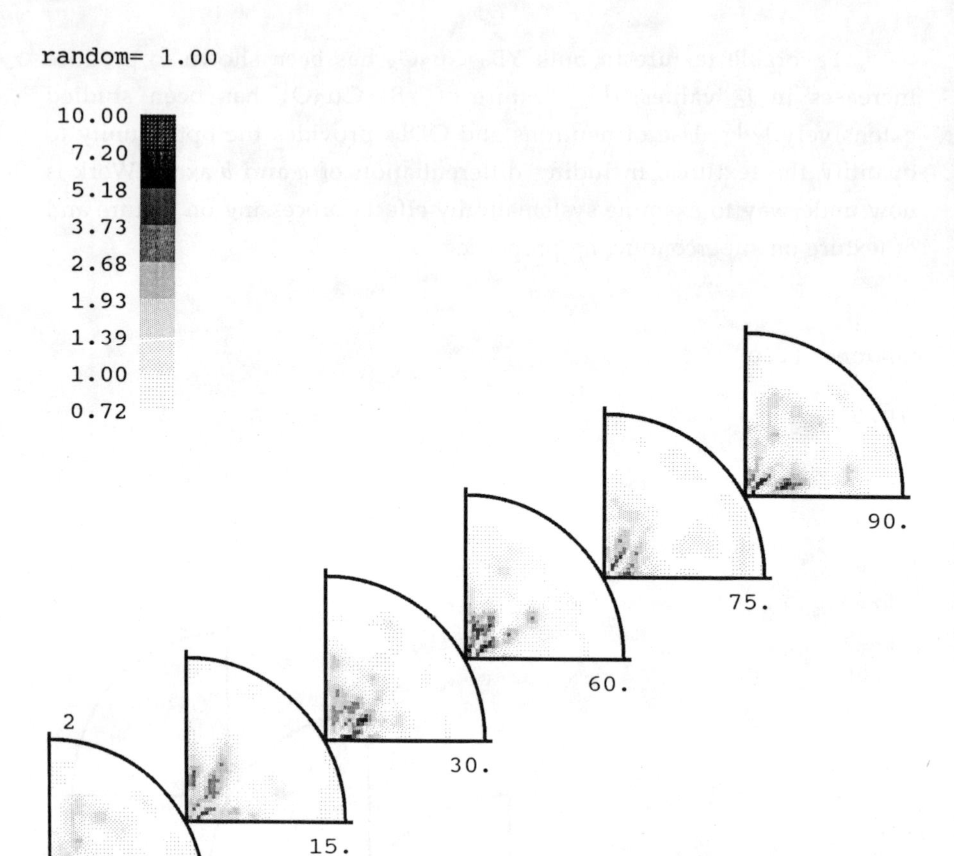

Figure 6. ODF of the second sinter forging.

CONCLUSION

Neutron diffraction was used to generate Orientation Distribution Functions of sinter–forged $YBa_2Cu_3O_x$. The sinter forgings had strong textures, with *c* axes being preferentially aligned with the forging direction. Grain growth during forging appeared to enhance texture slightly. The ODFs reveal that the *a* and *b* axes of the sinter forgings are distributed nonrandomly.

ACKNOWLEDGMENTS

This work was supported by the U. S. Department of Energy (DOE), Conservation and Renewable Energy, as part of a DOE program to develop electric power technology, and by Basic Energy Sciences–Materials Science, under Contract W–31–109–Eng–38. The work of A.C.B. was done in partial fulfillment of the requirements for the Ph.D. degree at the Illinois Institute of Technology, Chicago, IL, and was supported by the National Science Foundation–Office of Science and Technology Centers, under Contract STC–8809854.

REFERENCES

1. Iye, Y., Int. J. Mod. Phys. B 3, 367 (1989).
2. Dimos, D., et al., Phys. Rev. Lett. 61, 219 (1988).
3. Bunge, H. J., Int. Mater. Rev. 32, 265 (1987).
4. Bunge, H. J., Textures and Microstructures 8 & 9, 55 (1988).
5. Roe, R. J., J. Appl. Phys. 36, 2024 (1965).
6. Matthies, S., in Proceedings of the 8th ICOTOM, edited by J. S. Kallend and G. Gottstein (Warrendale, PA, The Metallurgical Society, 1988), p. 37.
7. Robinson, Q., et al., Adv. Ceram. Mater. 2, 380 (1987).
8. Grader, G. S., et al., Appl. Phys. Lett. 52, 1831 (1988).
9. Goretta, K. C., et al., Mater. Lett. 7, 161 (1988).
10. Schultz, A. J., Trans. Am. Cryst. Assoc. 23, 61 (1987).
11. Shi, D. et al., Mater. Lett. 6, 217 (1988).
12. Chen, K., et al., Appl. Phys. Lett. 55, 289 (1989).
13. Choi, C. S., et al., J. Appl. Cryst. 22, 465 (1989).
14. Knorr, D. B. and Livingston, J. D., Supercond. Sci. Technol. 1, 302 (1989).

SILVER SHEATHING OF HIGH–T_c SUPERCONDUCTOR WIRES

C.–T. Wu,[1] M. J. McGuire,[2,3] G. A. Risch,[2] R. B. Poeppel,[2] K. C. Goretta,[2] H. M. Herro,[4] and S. Danyluk[3]

[1] Illinois Institute of Technology, Chicago, Illinois 60616
[2] Argonne National Laboratory, Argonne, Illinois 60439
[3] University of Illinois–Chicago, Chicago, Illinois 60680
[4] Nalco Chemical Company, Naperville, Illinois 60566

ABSTRACT

The properties of Ag sheaths on high–temperature superconductors are examined. Ag is chemically compatible with $YBa_2Cu_3O_x$ and Bi–based superconductors and can be safely coprocessed with them. Residual stresses created by differences in thermal expansion coefficients are favorable and can be controlled by proper annealing. Although Ag forms low–resistance contacts with high–temperature superconductors, it is not certain that effective cryogenic stabilization by Ag can occur at 77 K and above.

INTRODUCTION

Long lengths of high–temperature superconductor wires have been made by plastic extrusion,[1,2] hot extrusion,[3,4] pyrolysis of solutions,[5,6] and powder–in–tube processing. Of these methods, only processing within a metallic tube has been used successfully with each of the high–temperature superconductors: $YBa_2Cu_3O_x$,[7,8] Bi(Pb)–Sr–Ca–Cu–O,[9–12] and Tl–Ba–Ca–Cu–O.[13] In addition, multifilament wires have been produced only in metallic tubes.[11,14]

High critical current density values have been obtained by processing high–temperature superconductors in Ag tubes.[7–14] Although it may be possible to use other metals as sheaths,[15,16] virtually all current approaches make use of Ag. Effective metallic sheaths must have several properties. In this paper, for Ag–sheathed wires made from $YBa_2Cu_3O_x$ and Bi–based compounds, the following considerations are examined: corrosion and chemical compatibility; coprocessing constraints; and electrical contacts and cryogenic stabilization. Use of Ag as a sheath may be limited by its cost, but this possibility will not be addressed here.

CHEMICAL CONSIDERATIONS

$YBa_2Cu_3O_x$ and Bi–based superconductors are highly susceptible to attack by moisture.[17,18] Our studies have shown that $Bi_2Sr_2CaCu_2O_x$ is more stable in water than $YBa_2Cu_3O_x$ is, but both readily corrode (Fig. 1). It has been reported[19] that because of the high oxidation state of Cu in $YBa_2Cu_3O_x$, electrochemical corrosion occurs in addition to chemical attack. Polarization studies have shown, however, that the chemical attack is far more severe than is the electrochemical attack.[20]

Figure 1. $Ba(OH)_2$ corrosion products on $YBa_2Cu_3O_x$ exposed to Chicago tap water at 25°C.

A metallic sheath offers protection of the underlying superconductor. Ag is a good choice for the sheath because it is relatively compatible with high–temperature superconductors.[7–14,21,22] X–ray[23] and synchrotron–radiation[24] photoemission studies of Ag/super–conductor interfaces reveal that Ag affects the electronic structures of both $YBa_2Cu_3O_x$ and $Bi_2Sr_2CaCu_2O_x$. Slight reduction of Cu^{2+} to Cu^{1+} occurs near the surface, and changes in oxygen core levels are observed.

The effects on the superconductors from Ag are, however, less than those observed for all other metals examined.[23–25]

Corrosion of an Ag sheath due to atmospheric exposure should be limited.[26] Ag is virtually inert in near neutral pH waters. It can be attacked, however, by waters containing hydrogen sulfide, ammonia, or ozone. The principal concern for atmospheric exposure of Ag is galvanic corrosion. Ag is cathodic to all metals except Au and the Pt metals; and in seawater, Ti is cathodic to Ag. Because of the inherent nobility of Ag, exposed brazes or contacts may be susceptible to corrosion.

PROCESSING CONSIDERATIONS

Processing superconductor powder in an Ag tube involves both mechanical and diffusional considerations. The tube must be worked, either at room or elevated temperature, and post–working heat treatment will create residual stresses. Oxygen may be lost during heat treatment, and if insufficient oxygen is present at elevated temperature, the superconductors may decompose.[27]

Mechanical Effects

Residual stress analysis Residual stress will be created during cooling from the sintering temperature because of differences in thermal expansion coefficients. Stress–relief processes for Ag–sheathed superconductors were examined by estimation of X–ray diffraction line broadening[28] on Ag that was deformed by rolling. A 99.9% pure Ag rod 6.35 mm in diameter was rolled with a reduction of 0.5 mm per pass to a strip about 0.5 mm thick. Small specimens cut from strip were heated at 10°C/min to annealing temperatures of 200, 400, 500, 600, 700, 800, or 900°C, held for 4 h, and then cooled at 3°C/min to 25°C. X–ray diffraction on the specimens annealed at various temperatures was performed to estimate the amount of residual stress that had been relieved by recovery and/or recrystallization processes. Diffraction peaks at (111), (200), (220), and (311) were chosen to be measured by Cu Kα radiation. Widths of peaks were measured at mid–height for the various annealing temperatures. Optical micrographs were taken to determine grain size for each annealing temperature. Hardness (HV) was measured by a Vickers indentation tester, with 1 kg load and 10 s loading time.

Figure 2a illustrates line sharpening of rolled Ag after annealing. (Conversely, when an annealed metal is deformed, the X–ray lines progressively broaden.) The variation of peak width as a function of annealing temperature is shown in Fig. 2b. It was found that the width decreased with increasing annealing temperature.

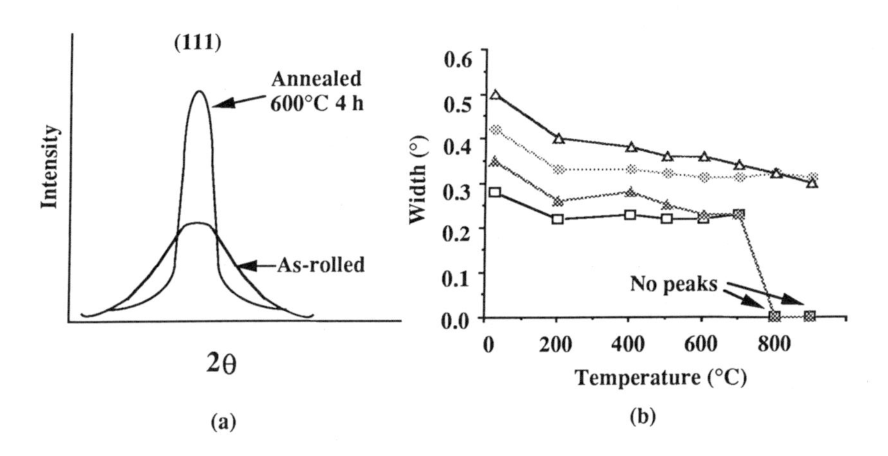

Figure 2. X–ray diffraction of annealed Ag sheet. (a). Line sharpening after annealing at 600°C for 4 h; (b) Peak width versus annealing temperature: square is (111), gray triangle is (200), gray circle is (220), and open triangle is (311).

Figure 3 shows the hardness as a function of annealing temperature. A sharp decrease of hardness at about 200°C indicates the onset of recrystallization. Microstructures of Ag for the as–rolled condition and annealed conditions are shown in Fig. 4. The as–rolled grain size is extremely fine. Significant grain growth took place at 400°C and higher temperatures.

A simple model was used to calculate the residual stress that was eliminated by annealing. This residual stress can be generated in many ways. Anisotropic thermal contraction of differently oriented grains in a polycrystalline metal or unequal contraction of adjacent phases in an alloy or composite can generate residual stress. Inhomogeneities in the internal stress pattern cause a range of lattice spacings, d, consequently,

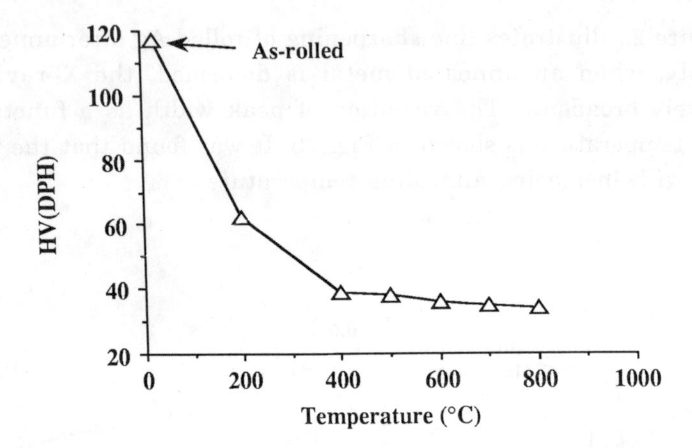

Figure 3. Relation between hardness and annealing temperature.

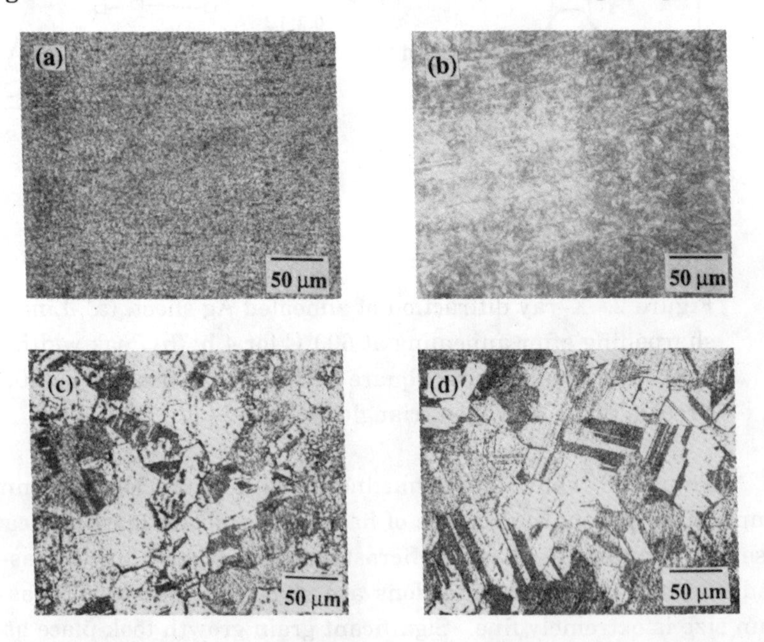

Figure 4. Microstructures of Ag: (a) as–rolled, (b) 200°C anneal, (c) 400°C anneal, and (d) 600°C anneal (bars are 50 μm).

a range of Bragg angles in diffraction and, therefore, a broadening of Debye–line width and position. A relation between the peak broadening produced and the nonuniformity of the internal stress (strain) is obtained by differentiating the Bragg law:

$$b = \Delta 2\theta = -2\,(\Delta d/d)\tan\theta \ , \qquad (1)$$

where b is the extra broadening due to a fractional variation in plane spacing, $\Delta d/d$, and θ is the angle of diffraction. This equation allows the variation in strain, $\Delta d/d$, to be estimated from the observed broadening. The strain so found can then be multiplied by the elastic modulus , E, to give the maximum stress present (Fig. 5). Two assumptions were made in this model: (1) that the nonuniformity of internal stress (strain) produced by cold work was the major cause of broadening; (2) that the tensile and compressive strains shown in Fig. 5c were equal.

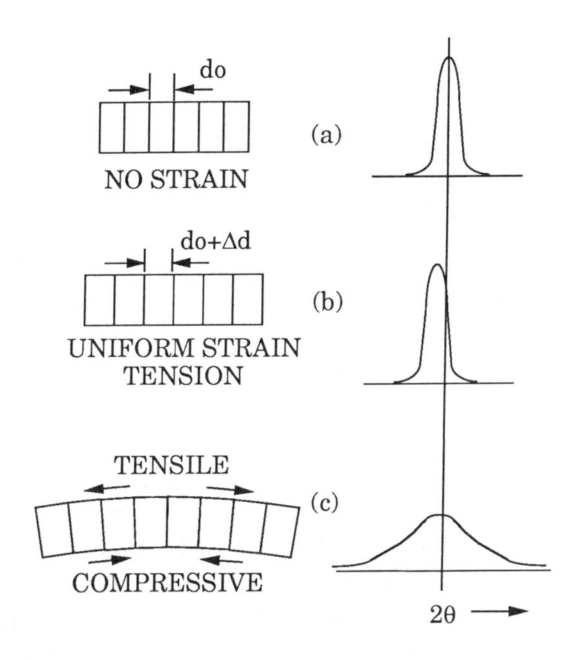

Figure 5. Line broadening due to plastic deformation: (b) uniform strain and (c) inhomogeneous strain.

From peak width data and Equation 1, one can estimate the amount of residual stress that was relieved by annealing. Table 1 shows the calculated results. Because preferred orientation was generated by rolling, it is not sufficient to calculate the overall residual stress of the bulk specimens simply by summing the value for each diffraction peak. Pole figures and Orientation Distribution Functions may aid in calculation of

Table 1. Residual stress relieved by annealing

Temperature (°C)	$\sigma_{(111)}$ MPa	$\sigma_{(200)}$ MPa	$\sigma_{(220)}$ MPa	$\sigma_{(311)}$ MPa
25	–	–	–	–
200	32	72	57	89
400	22	44	72	128
500	32	89	89	174
600	32	128	107	174
700	14	128	107	228
800	*	*	89	288
900	*	*	107	288

* No peaks were observed

residual stress for bulk specimen. It is, however, clear that the residual stresses in Ag generated during cold working and by differences in thermal expansion coefficients can be relieved by recovery and recrystallization processes, possibly at low temperatures.

Strength of Ag To evaluate the mechanical properties of Ag at high temperature, constant strain rate compression tests were employed. Ten pieces of 99.9% Ag with diameter of 6.35 mm and length of 10 mm were made. Tests were performed in an Instron Model 1125 universal testing machine. One specimen was compressed at room temperature in air; the other specimens were compressed in a 10^{-7} Torr vacuum from 800 to 950°C at an initial strain rate of 10^{-5} sec^{-1}. Load–versus–time data were recorded and stress–strain curves were generated.

A typical curve is shown in Fig. 6. Since Ag was compressed at high temperature (> 0.8 T_m), dynamic recrystallization took place. The softening produced by dynamic recrystallization is followed by renewed hardening and a cyclic flow curve is traced out. Table 2 lists maximum and average effective flow stresses (average of maximum and minimum stress) of Ag for different test temperatures. Softening of the Ag and occurrence of dynamic recrystallization imply that the workability of Ag at high temperature is very high.

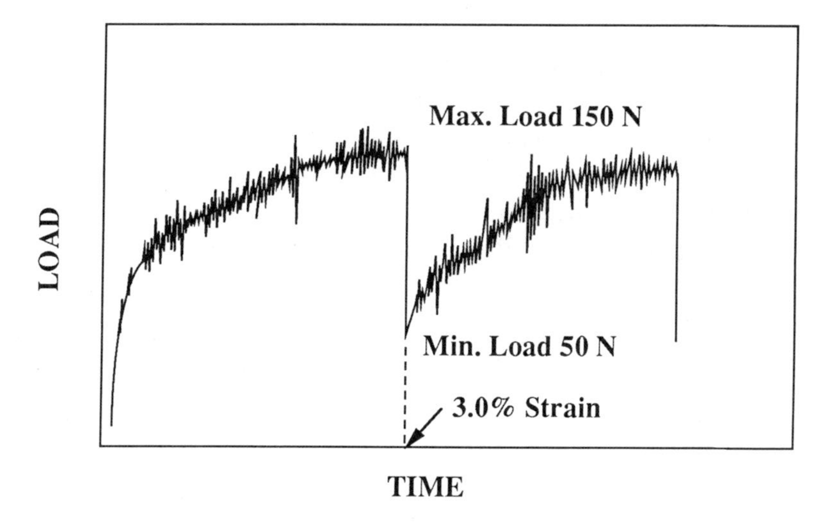

Figure 6. Flow curve of Ag at 920°C for a strain rate of 10^{-5} s^{-1}.

Table 2. Mechanical properties of Ag at temperature

T (°C)	25	800	850	900	920	940	950
Max. σ (MPa)	137	5.37	5.37	4.74	4.42	3.47	2.84
Ave. σ (MPa)	--	3.79	3.79	3.10	2.84	2.37	1.90

Diffusional Effects

Superconducting properties of $YBa_2Cu_3O_x$, $Bi_2Sr_2CaCu_2O_x$, and $(Bi,Pb)_2Sr_2Ca_2Cu_3O_x$ are influenced by their oxygen contents.[29,30] The compounds lose oxygen upon heating and require optimal concentrations for best properties. The effects of oxygen on $YBa_2Cu_3O_x$ are most widely studied, and reoxygenation annealing occurs at a low temperature. Hence, a $YBa_2Cu_3O_x$/Ag oxygen diffusion couple will be examined in some detail.

Chemical diffusion of oxygen through Ag has been studied over a wide range of temperature.[31,32] In considering a diffusion couple, both solubilities of oxygen and oxygen diffusion rates must be considered. $YBa_2Cu_3O_x$/Ag tubes are sintered near 900°C. Diffusion data at that temperature indicate that oxygen diffuses very rapidly through Ag[31,32] and $YBa_2Cu_3O_x$.[33,34] Equilibrium oxygen contents are established within several minutes.[35] Since oxygen will be lost from the superconductor at elevated temperature, it therefore becomes import to examine reoxygenation of $YBa_2Cu_3O_x$ during cooling. A schematic of the diffusion couple is shown in Fig. 7.

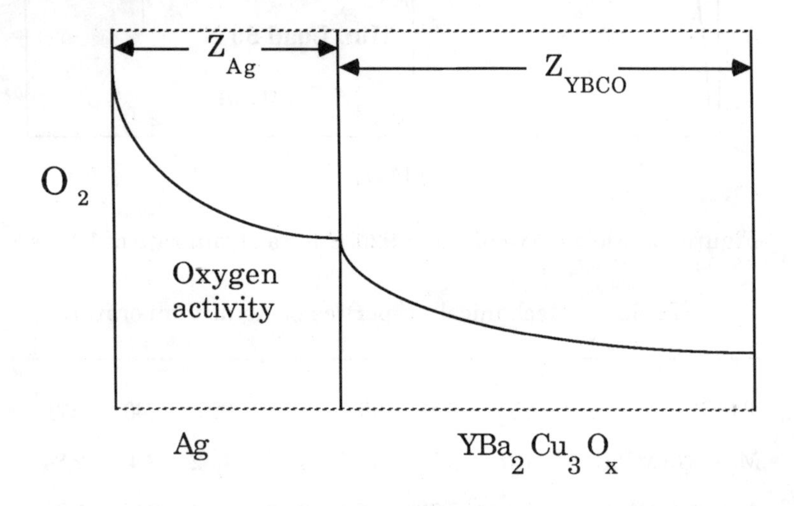

Figure 7. Schematic figure of oxygen activity in Ag and $YBa_2Cu_3O_x$ during a reoxygenation anneal.

This problem presents a few complications. First, chemical diffusion in $YBa_2Cu_3O_x$ is a function of both temperature and oxygen content, x. For a given value of x, diffusion rates decrease as temperature decreases. For a given temperature, diffusion rates increase as x increases. As temperature decreases, the equilibrium value of x increases. One is thus faced with competing mechanisms: although slower diffusion is expected as temperature decreases, the expected decrease is offset by changes in stoichiometry. As a consequence, reoxygenation can be accomplished most expeditiously in stages, over a range of temperatures rather than at about 450°C only.[35]

To approach diffusion of oxygen through Ag into $YBa_2Cu_3O_x$, it is assumed that the atmosphere is oxygen, the $YBa_2Cu_3O_x$/Ag interface is perfect, and the flux of oxygen across the interface is continuous. By Fick's first law, flux, F, can be related to the chemical diffusion coefficient of oxygen, D, and the concentration gradient of oxygen, $\Delta C/\Delta z$:

$$F = -D_{Ag}\,(\Delta C/\Delta z)_{Ag} = -D_{YBCO}\,(\Delta C/\Delta z)_{YBCO}\,, \qquad (2)$$

where the subscripts denote the two materials.[36]

The question is whether the presence of an Ag sheath adds *substantial* time to the anneal. A simplifying assumption to address this question is that any oxygen that passes through the Ag is immediately taken up by the $YBa_2Cu_3O_x$. This assumption implies that the effective concentration of oxygen in the Ag *at the interface* is zero. Saturation concentrations of oxygen in each material at any temperature can be calculated from published data.[31-34] Chemical diffusion rates can be related approximately to diffusion distance, z, by:[36]

$$D \approx (4Dt)^{1/2}\,, \qquad (3)$$

where t is time. In practice, the Ag and the $YBa_2Cu_3O_x$ can be any thickness. Thus, a family of solutions for various geometries exists for the time needed to reoxygenate $YBa_2Cu_3O_x$. Calculations show that for typical thicknesses z_{Ag} and z_{YBCO}, the time needed to diffuse oxygen through Ag is appreciable. In practice, it has been found that the value of x in $YBa_2Cu_3O_x$ can be increased to over 6.9 by annealing at 450°C for times a factor of a few greater than those for unsheathed $YBa_2Cu_3O_x$.[37]

Mention must be made of a severe constraint to coprocessing Bi–based superconductors with Ag at elevated temperature. It has been reported that the 110–K, (Bi,Pb)–Sr–Ca–Cu–O phase can be destabilized by Ag in the presence of a highly oxidizing atmosphere. No degradation of T_c is observed, however, if the oxygen pressure is about 6.7×10^3 Pa.[38] This restriction does not appear to apply to the 90–K, Pb–free phase.[39]

ELECTRICAL CONSIDERATIONS

Ag forms low–resistance contacts with high–temperature superconductors. Resistivities of Ag contacts on $YBa_2Cu_3O_x$ have been reported[40] to be as low as $10^{-10}\ \Omega\ cm^2$ at 77 K. Ag would appear therefore to be an ideal sheath material and to offer hope of effective cryogenic stabilization. Little work has been reported on this possibility. A recent paper,[41] however, casts serious doubt on whether Ag can stabilize $YBa_2Cu_3O_x$ against a temporary loss of superconductivity. Computer modelling of a two–coil solenoid system indicated that $YBa_2Cu_3O_x$ is effectively self stabilizing only below temperatures of about 20 K. Ag is not expected to allow for stabilization at temperatures up to 77 K.[42] This finding has yet to be confirmed.

SUMMARY

Ag is compatible with $YBa_2Cu_3O_x$ and Bi–based superconductors and is itself corrosion resistant. Ag is highly ductile and can readily be processed with high–temperature superconductors. Residual stresses between Ag and the superconductors are favorable: Ag is in tension and the superconductor is in compression. The magnitude of these stresses can be minimized by low–temperature annealing. Ag forms low–resistance contacts with high–temperature superconductors, but preliminary analyses have cast doubt on whether Ag can effectively stabilize $YBa_2Cu_3O_x$ at 77 K.

ACKNOWLEDGMENTS

This work was supported by the U.S. Department of Energy (DOE), Conservation and Renewable Energy, as part of a DOE program to develop electric power technology, and by Basic Energy Sciences–Materials Science, under Contract W–31–109–Eng–38; by the National Science Foundation–Office of Science and Technology Centers, under Contract STC–880954; and by the Illinois Department of Energy and Natural Resources Grant No. SWSC2. Helpful discussions were held with J. L. Routbort and J. R. Hull (Argonne) and S. Shim (Nalco).

REFERENCES

1. Lanagan, M. T., et al., J. Less–Common Met. 149, 305 (1989).
2. Enomoto, R., et al., Jpn. J. Appl. Phys. 28, L1207 (1989).
3. Chen, I.–W., et al., J. Am. Ceram. Soc. 70, C–388 (1987).
4. Samanta, S. K., et al., J. Appl. Phys. 66, 4532 (1989).
5. Wang, J. G. and Yang, R. T., J. Appl. Phys. 67, 2160 (1990).

6. Chien, J. C. W., et al., Adv. Mater. 2, 305 (1990).
7. Okada, M., et al., Jpn. J. Appl. Phys. 27, L185 (1988).
8. Osamura, K., et al., Supercond. Sci. Technol. 2, 111 (1989).
9. Hikata, T., et al., Jpn. J. Appl. Phys. 28, L82 (1989).
10. Mimura, M., et al., Appl. Phys. Lett. 54, 1582 (1989).
11. Sekine, H., et al., J. Appl. Phys. 66, 2762 (1989).
12. Heine, K., et al., Appl. Phys. Lett. 55, 2441 (1989).
13. Okada, M., et al., Jpn. J. Appl. Phys. 27, L2345 (1989).
14. Shi, D., et al., Mater. Lett. 9, 1 (1989).
15. Sadakata, N., et al., Mater. Res. Soc. Symp. Proc. 99, 293 (1988).
16. Kammlott, G. W., et al., Appl. Phys. Lett. 56, 2459 (1990).
17. Trolier, S. E., et al., Am. Ceram. Soc. Bull. 67, 759 (1988).
18. Yoshikawa, K., et al., Jpn. J. Appl. Phys. 27, L2324 (1988).
19. Lyon, S. B., et al., Supercond. Sci. Technol. 2, 107 (1989).
20. Shim, S., unpublished information.
21. Singh, J. P., et al., J. Appl. Phys. 66, 3154 (1989).
22. Hoshino, K., et al., Jpn. J. Appl. Phys. 27, L1297 (1988).
23. Meyer III, H. M., et al., J. Appl. Phys. 65, 3130 (1989).
24. Wagener, T. J., et al., Phys. Rev. B 38, 232 (1988).
25. Meyer III, H. M., et al., Phys. Rev. B 38, 6500 (1988).
26. R. H. Leach, in The Corrosion Handbook, ed. H. H. Uhlig (John Wiley, New York, 1948) p. 314.
27. Bormann, R. and Nölting, J., Appl. Phys. Lett. 54, 2148 (1989).
28. Cullity, B. D., Elements of X-ray Diffraction (Addison-Wesley, New York, 1967) p. 431.
29. Jorgensen, J. D., et al., Physica C 153-155, 578 (1988).
30. Tallon, J. L., et al., Nature 333, 153 (1988).
31. Steacie, E. W. and Johnson, F. M. G., Proc. Roy. Soc. A 112, 542 (1926).
32. Eichenauer, W. and Müller, G., Z. Metallk. 53, 321 (1962).
33. Matsui, T., et al., Presented at NATO Meeting on Nonstoichiometric Compounds, Tegersee, FRG, 1988.
34. Kishio, K., et al., J. Sol. State Chem. 82, 192 (1989).
35. Goretta, K. C., et al., in Proceedings First International Ceramic Science and Technology Congress (American Ceramic Society, Westerville, OH, 1990) in press.
36. Shewmon, P. G., Diffusion in Solids (McGraw-Hill, New York, 1963).
37. Goretta, K. C. and Risch, G. A., unpublished information.
38. Dou, S. X., et al., Appl. Phys. Lett. 56, 493 (1990).
39. Jin, S., et al., Appl. Phys. Lett. 52, 1628 (1988).
40. Ekin, J. W., et al., Appl. Phys. Lett. 52, 1819 (1988).
41. Iwasa, Y. and Butt, Y. M., Cryogenics 30, 37 (1990).
42. Hull, J. R., unpublished information.

HIGH-J_c SUPERCONDUCTING Y-Ba-Cu-O PREPARED BY RAPID QUENCHING (RQ) AND DIRECTIONAL ANNEALING (DA)

T. Yamamoto,[a] T.R.S. Prasanna, S.K. Chan, J.G. Lu, and R.C. O'Handley

Department of Materials Science and Engineering
Massachusetts Institute of Technology
Cambridge, Massachusetts 02139, USA

[a]Permanent Address: Mitsui Mining & Smelting Co., LTD, Corporate R&D Center, 1333-2 Haraichi, Ageo-shi Saitama, 362 Japan

ABSTRACT

Textured high-T_c superconducting bulk ceramics Y-Ba-Cu-O were produced by rapid quenching (RQ) and directional annealing (DA) processes. $YBa_2Cu_3O_7$ prepared by conventional ceramics method was rapidly quenched with a twin roller apparatus (5.8-11.5m/s) and directionally annealed by slowly (2.5mm/h) moving the specimen through temperature gradient of 45°C/cm from 1020 to 900°C. The resulting ceramic samples showed T_c (ρ=0) of 91.9K and a sharp superconducting transition, ΔT_c = 0.2K. Enhancement of the transport critical current density up to 4.3×10^4 A/cm^2 (77K, in 1T) has been demonstrated on the bulk ceramics prepared by the above procedures. The microstructure is strongly textured $YBa_2Cu_3O_7$ with a 20% volume fraction of Y_2BaCuO_5 and a small amount of CuO.

1. INTRODUCTION

Enhancement of critical current densities (J_c) of high T_c bulk superconducting $YBa_2Cu_3O_7$ (1:2:3) is actively pursued.[1-14)] These early published results indicate several factors favorable to large critical current in the magnetic field; (1) large superconducting grains oriented with a-b planes lying in the superconducting current direction, (2) fine Y_2BaCuO_5 (2:1:1) and CuO precipitates, (3) low density of microcracks, and relaxation of stress concentrations, (4) clean grain boundaries, and (5) high density.[4,8,10-13] It has also been reported[15] that high J_c values could be achieved in 1:2:3 thin films by neutron irradiation to induce a dispersion of very fine crystal defects (10-100 Å) which serve as flux pinning centers. The beneficial effects of RQ and post-annealing processes are apparent in the fine microstructure, high density, and chemical homogeneity they produce.[16,17] RQ and post-annealing processes are well suited to controlling the formation and growth of flux-pinning 2:1:1 precipitates in Y-Ba-Cu-O superconductors.[18] This paper focusses on enhancing transport critical current density by growing highly oriented superconducting grains in a homogeneous, RQ precursor by directional annealing (DA) in a sharp temperature gradient.

2. EXPERIMENT

The starting powders, Y_2O_3, $BaCO_3$ and CuO had a nominal purity level of 99.99%. Sintered pellets of 1:2:3 phase were prepared according to the conventional ceramics processing methods described in our previous paper.[18] These pellets were melted with an H_2-0_2 torch (0_2-rich flame) and the droplets were rapidly quenched in a high speed, 5.8 to 11.5 m/s twin-rolling mill. The flakes produced by twin-roller-quenching were ground into powders and isostatically pressed into a bar under 40 MPa pressure. (Typical dimensions of bar specimen are 6 cm in length and 0.5 cm in diameter). A horizontal tubular furnace was used for directional annealing the bar specimen. The hot zone temperatures are between 1100 and 1020° C. The sample is held here for 30 min. then transported slowly (2.5 mm/h) through a temperature gradient which could be varied between 40 and 50° C/cm in a distance of about 3cm. It is here that directional annealing has taken place. The specimen was then cooled slowly at 60° C/hr. through the tetragonal-to-orthorhombic transition in flowing oxygen. We moved the specimen at a speed of 2.5 mm/h to expose its entire length sequentially to the sharp temperature gradients.

Microstructure characterization was carried out by X-ray diffractometry XRD (Rigaku RU300), scanning electron microscopy SEM and energy dispersive X-ray analysis EDXA (Cambridge Instruments). Transport critical current was measured by the four-point method with both DC (low current densities) and pulse technique (10 μs in width, 200 ms in duty cycle) using the 2-6 μV/mm (1.2-3.6 x 10^{-10} ohm·cm) voltage criterion. Leads were attached by means of baked on silver paint and solder.

3. RESULTS AND DISCUSSION

3.1 Rapid Quenching and Post Annealing

Sintered Y1:2:3 pellets prepared by conventional ceramics processing were rapidly quenched at various surface speeds between 5.8 and 11.5 m/s (2000 to 4000 rpm). The rapidly quenched materials were in the form of flakes, some as large as 1.5-1 cm in length, 0.5 cm in width, and 50 to 100 μm in thickness. The faster the cooling rates are, the thinner the flakes become. The flakes were glossy black.

Figure 1 shows XRD patterns of rapidly quenched Y1:2:3. The results show the presence of Y_2O_3 as a matrix phase with the exception of the peak marked with the arrow (2θ = 36°). Some amorphous phase (Ba and Cu oxides) also appears at this stage. There is no difference between the XRD patterns of rapidly quenched materials with the surface speeds of 5.8 to 11.5 m/s. It has been found that the twin-roller surface speed of 5.8 m/s was enough for obtaining the Y_2O_3 and amorphous phase of Ba and Cu oxides.

The rapidly quenched material was pressed into pellets then annealed at a temperature within the range of 1060° C to 1100° C for 0.5 hour. In this stage, Y_2BaCuO_5(2:1:1) was formed from Y_2O_3 and liquid Ba and Cu oxides. A larger amount of Y_2O_3 exists at

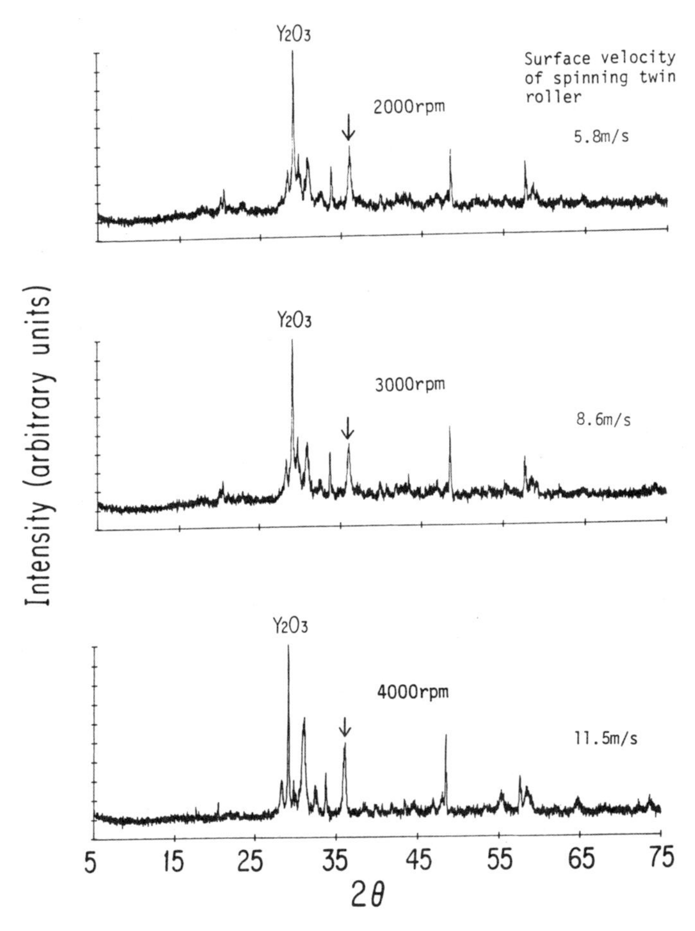

Fig. 1 X-ray powder diffraction patterns of rapidly quenched $YBa_2Cu_3O_7$ at various surface velocities of spinning twin-roller.

lower temperature as Fig. 2(a) and (b). It has been found that temperatures higher than 1060° C promote the formation of 2:1:1 phase from Y_2O_3 and liquid Ba and Cu oxides.

3.2 Directional Annealing

Following the rapid quenching the samples are drawn through

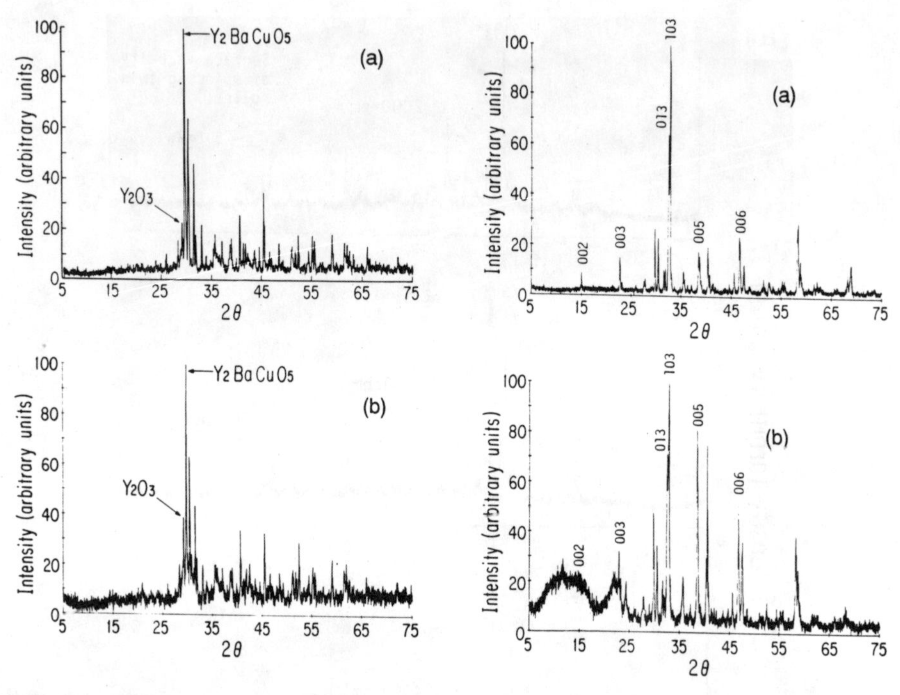

Fig. 2 X-ray powder diffraction patterns of post-annealed materials after rapid quenching. (a) 1100° C (b) 1060° C

Fig. 3 X-ray diffraction patterns of directionally annealed materials (a) powder sample (b) bulk sample cut along the longitudinal direction.

the temperature gradient which constitutes the directional annealing (DA) process. Figure 3 shows XRD patterns of DA powders (a) and of the bulk DA surface along the longitudinal direction of specimen (b). The material is predominantly 1:2:3 phase with a 20% volume fraction of 2:1:1 phase as shown in Fig. 3(a). The (001) peaks are more intense in the DA bar (Fig. 3(b)) than in the powdered DA sample (Fig. 3(a)). This indicates that the c axes are strongly oriented normal to the rod length.[11,19]

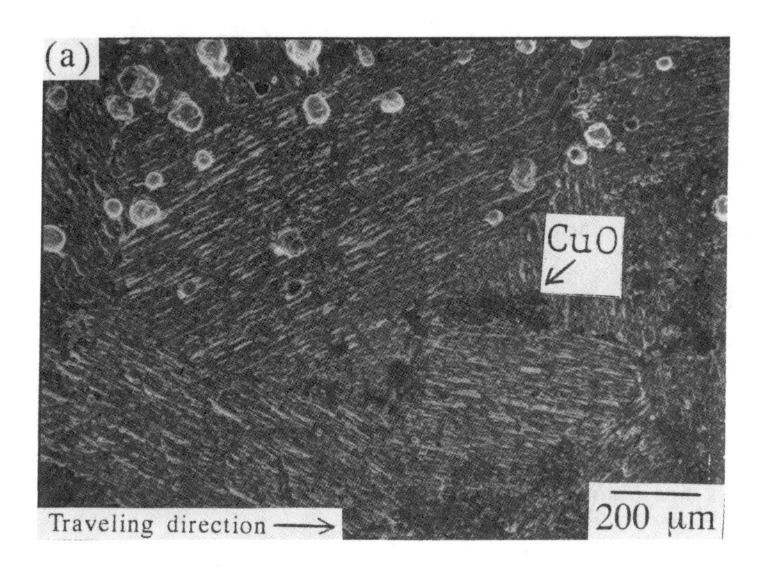

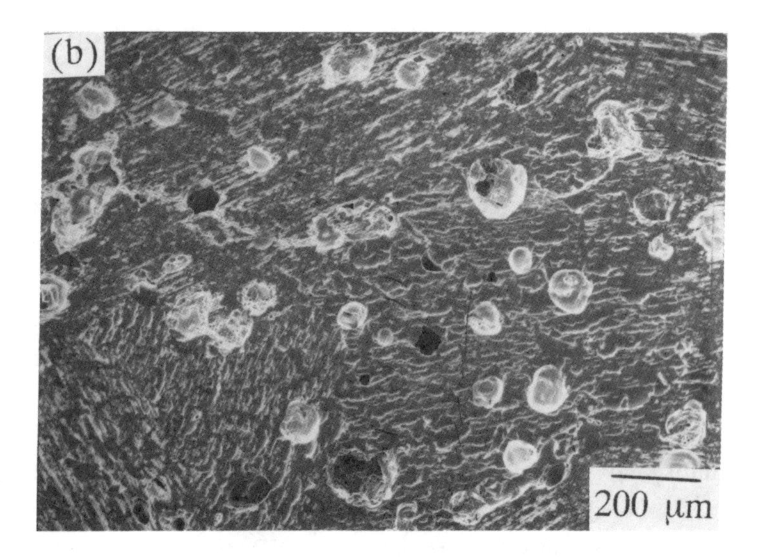

Fig. 4 SEM micrographs of polished surfaces of directionally annealed specimen.
(a) surface parallel to traveling direction,
(b) surface perpendicular to traveling direction.

Figure 4 shows SEM micrographs of polished surfaces of a directionally annealed specimen. The surfaces are longitudinal (a) and perpendicular (b) to the bar axis, respectively. The, surface along the longitudinal direction shows longitudinally textured microstructure (Fig. 4(a)). By contrast, the perpendicular face shows more conventional ceramic microstructure (Fig. 4(b)). These SEM results are consistent with the XRD results of Fig. 3. EDX analysis indicates that copper oxides are precipitated mainly at grain boundaries as marked in Fig. 4(a), and the grains are strongly connected (i.e. grain boundaries are pore-free). Precipitates of 2:1:1 phase in the matrix (evident in Fig. 3(a)) could not be detected with SEM and EDX probably because their diameters are below the resolution of EDX analysis.

3.3 Superconducting Properties

The typical textured material showed T_c (ρ=0) = 91.9 K. Transport current densities were measured from V-I characteristic curves using the 2-6 μV/mm voltage criterion at 77 K in a magnetic field of 1.06 T perpendicular to the current direction. The current, current densities, and cross-sectional areas are summarized in Table 1. The transport J_c ranged from 1.6×10^4 to 4.3×10^4 A/cm^2. Sample A was still superconducting at 3.3×10^4 A/cm^2. These values are higher than any others reported for ceramic Y1:2:3 at 77 K and 1T. Thus a significant enhancement of transport critical current density has been demonstrated in the samples prepared by RQ and DA processes (cf. Fig. 5).

Additional efforts are needed to determine the most effective directional annealing parameters for enhancement of the critical current density of bulk superconducting $YBa_2Cu_3O_7$. However, it is clear that large grains, significant grain orientation, formation of a large amount of 2:1:1 phase, and perhaps some CuO precipitates, as achieved by the DA process shown here combine to increase J_c compared to that obtained by conventional sintering or by molten oxide processing technique.

Table 1 Net Current supplied, cross-sectional areas and current densities. (77K, 1T)

Sample	Net current supplied (A)	Cross-sectional area (cm^2)	Current density (A/cm^2)
A	24.8	0.029X0.058	32900
B	28.0	0.030X0.058	16100*
C	72.0	0.026X0.029	42800*

*Critical current densities

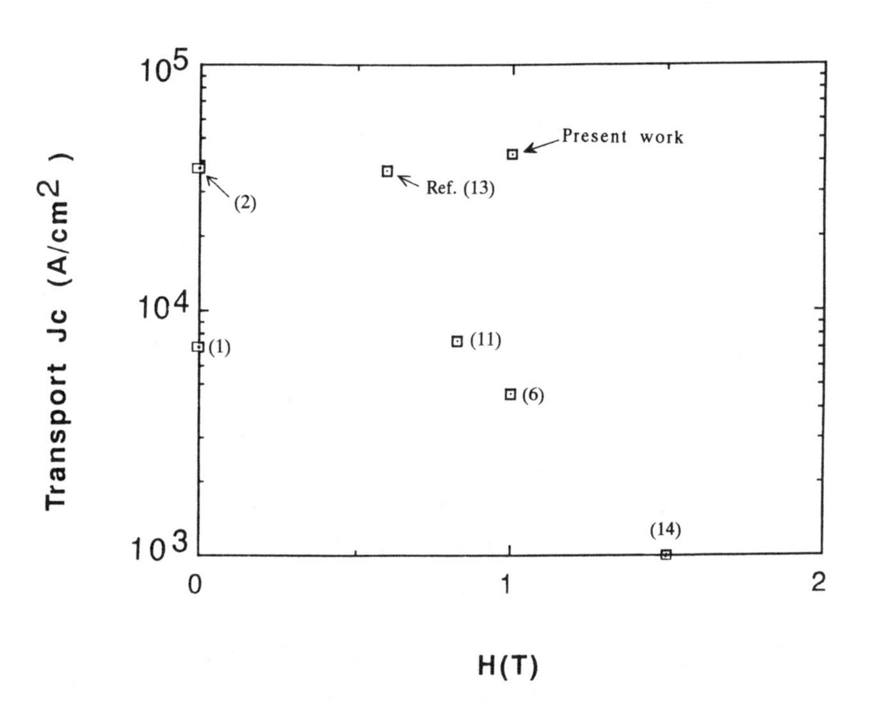

Fig. 5 Transport J_c at 77K in bulk superconducting $YBa_2Cu_3O_7$

4. SUMMARY

Enhancement of the transport critical current density up to $4.3 x 10^4$ A/cm^2 in 1.06 T at 77 K has been demonstrated on bulk superconducting Y-Ba-Cu-O prepared by rapid quenching and directional annealing processing. The material is predominantly a strong textured $YBa_2Cu_3O_7$ with a 20% volume fraction of Y_2BaCuO_5.

ACKNOWLEDGEMENT

The authors wish to thank Dr. K. Kitazawa of the University of Tokyo for helpful discussions and encouragement of this work. This reserach is supported by Mitsui Mining and Smelting Co., Ltd., and by the U.S. Department of Energy grant DE-FG02-84ER45174.

REFERENCES

1. Alford, N.M. and Burton, T.W., J. Appl. Phys. 66, 5930 (1989).
2. Chen, K., Hsu, S.W., Chien, T.L., Lan, S.D., Lee, W.H., and Wu, P.T., Appl. Phys. Lett. 56, 2675 (1990).
3. Jin, S., Tiefel, T.H., Sherwood, R.C., Davis, M.E. Van Dover, R.B., Kammlott, G.W., and Fiastmacht, R.A., Phys. Rev. B 13, 7850 (1988).
4. Jin, S., Tiefel, T.H., Sherwood, R.C., Davis, M.E. Van Dover, R.B., Kammlott, G.W. Fiastmacht, R.A., and Keith, D., Appl. Phys. Lett 52, 2074 (1988).
5. Jin, S., Sherwood, R.C., Gyorgy, E.M., Tiefel, T.H., Van Dover, R.B., Nakahara, S., Schneemeyer, L.F., Fiastmacht, R.A., and Davis, M.E., Appl. Phys. Lett. 54, 584 (1989).
6. Kase, J., Shimoyama, J., Yanagisawa, E., Kondoh, S., Matsubara, T., Morimoto, T., and Suzuki, N., Jpn. J. Appl. Phys. 29, L277 (1989).

7. Matsushita, T., Murakami, M., Morita, M. Miyamoto, K., Saga, M., Matsuda, S., and Tanino, M., Jpn. Jl. Appl. Phys. 28, L1545 (1989).
8. McGinn, P.J., Chen, W., and Black, M.A., Physica C 161, 198 (1989).
9. McGinn, P.J., Black, M.A. and Valenzuela, A., Physica C 156, 57 (1989).
10. McGinn, P.J., Chen, W., Zhu, N., Balachandran, V., and Lanagan, M.T., Physica C 165, 480 (1989).
11. Meng, R.L., Kinalidis, C., Sun, Y.Y., Gao, L., Tao, Y.K., Hor, P.H., and Chu, C.W., Nature 345, 326 (1990).
12. Murakami, M., Morita, M., Doi, K., and Miyamoto, K., Jpn, Jl. Appl. Phys. 28, 1189 (1989).
13. Salama, K., Selvamanickam, V., Gao, L. and Sun, K., Appl. Phys. Lett. 54, 2352 (1989).
14. Shi, D., Tang, M., Chang, Y.C., Jiang, P.Z., Vandervoot, K., Malecki, B., and Lam, D.J., Appl. Phys. Lett. 54, 2358 (1989).
15. Kirk, M.A. et al., High T_c Update, 4 [3] 2, (1990).
16. McHenry, M.E., McKittrick, J., Sasayama, S., Kwapong, V., O'Handley, R.C., and Kalonji, G., J. Appl. Phys. 63, 4229 (1988).
17. McKittrick, J., Sasayama, S., McHenry, M.E., Kalonji, G., and O'Handley, R.C. J. Appl. Phys. 65, 3662 (1989).
18. Yamamoto, T., Chan, S.K., Stubicar, M., Prasanna, T.R.S., and O'Handley, R.C., to be submitted to Mat. Lett. (1990).
19. McCallum, R.W., Verhoeven, J.D., Novack, M.A., Gibson, E.D., Laabs, F.C., Finnemore, D.K., and Moodenbaugh, A.R., Adv. Ceram. Matter. 2, 388 (1987).

FABRICATION AND PROPERTIES OF Ag-CLAD Bi-Pb-Sr-Ca-Cu-O WIRES

H.K. Liu, S.X. Dou, W.M. Wu, K.H. Song, J. Wang and C.C. Sorrell

School of Materials Science and Engineering
University of New South Wales
Kensington, NSW 2033, Australia

L. Gao

Texas Center for Superconductivity
University of Houston
Houston, Texas 77204-5506, USA

ABSTRACT

Ag-clad Bi-Pb-Sr-Ca-Cu-O wires and coils have been fabricated using the powder-in-tube technique. A J_c of 1.2×10^4 A/cm^2 at 77 K and zero field and 3×10^3 A/cm^2 at 77 K and 4000 Oe has been achieved for Ag-clad tape. The J_c for a coil of 1 m length was 2×10^3 A/cm^2 at 77 K and zero field. The high J_c of Ag-clad Bi-Pb-Sr-Ca-Cu-O tape is attributed to a combination of elimination of the poisoning effect of Ag on superconductivity through low partial oxygen atomosphere heat treatment, grain alignment through cold rolling, and enhancement of flux pinning through control of microstructure, and composition.

INTRODUCTION

Owing to the brittleness and reactivity of oxide ceramic superconductors, it is advantageous to make metal/ceramic composites and metal-clad wires, tapes, and multifilaments, where the metal's malleability compensates for the ceramic's brittleness. Furthermore, the metal provides a good means of thermal dissipation. This stabilizing effect is of fundamental importance for type II superconductors, in which undesirably large local rises in temperature can develop through flux jumping in the mixed state. Unfortunately, these ceramics, in particular Bi-Pb-Sr-Ca-Cu-O, react with nearly every metal with which they come into contact.

Ag-clad Bi-Pb-Sr-Ca-Cu-O wires have been fabricated using the powder-in-tube technique [1,2]. Critical current densities (J_c) of 10^4 A/cm^2 at 77 K and zero field have been achieved in these wires. In particular, the J_c does not drop significantly up to 25 T at 4.2 K [3]. In this paper, we report the properties of Ag-clad Bi-Pb-Sr-Ca-Cu-O wire fabricated by using the powder-in-tube technique. Some contributing factors to the high J_c are discussed.

EXPERIMENTAL PROCEDURE

Powders having the cation ratios Bi:Pb:Sr:Ca:Cu = 1.6:0.4:1.6:2.0:3.0 and 1.8:0.4:2.0:2.2:3.0 were prepared by freeze drying [4]. The powders were calcined at 830°C for 10 h, pressed into pellets, sintered at 850°C in air for 20 to 190 h and crushed. Nearly pure (2223) phase was identified by X-ray diffraction (XRD) patterns obtained using a Philips Type PW 1140 powder diffractometer with CuKα radiation. The powder was packed into a silver tube of 10 mm outer and 8 mm inner diameter. The composite was then drawn to a final outer diameter of 1 mm. The wires were rolled into tapes of total thickness ~ 0.2 mm (~ 0.1 mm for the superconductor) and width ~ 2-3 mm. The tapes were heat treated at 820°C for 80 to 190 h in a mixture of oxygen and nitrogen at $P_{O_2} = 5.0 \times 10^3$ Pa. To make coils, Ag/superconductor tapes of ~ 1 m length were wound onto ceramic tubes of 12 and 35 mm outer diameter. The coils were heat-treated under the same conditions.

Microstructural and compositional analyses were performed with a CAMSCAY scanning electron microscope (SEM) and a JEOL FX2000 transmission electron microscope (TEM), both of which were equiped with Link Systems AN10000 energy dispersive spectrometers (EDS).

Electrical characterisation was performed using the standard four-probe d.c. technique. The transport critical current density was determined from the current-voltage curve under magnetic fields varying from 0 to 1.1 T using a 1 μv/cm criterion.

RESULTS

The J_c for short samples was found to be reproducible at a level of 10^4 A/cm^2 at 77 K and zero field. The highest value of it was 1.2×10^4 A/cm^2 for a tape sample treated in 5×10^3 Pa O_2 for 150 h.

A J_c of 2.0×10^3 A/cm^2 at 77 K and zero field was achieved for coils of 35 mm diameter when the coil was heat treated in 5×10^3 Pa O_2. The coil of 12 mm diameter had over 40 A•turn at 77 K. The tape from which the coil was made had a cross sectional area of 0.16 mm^2 and a length of 1000 mm. The critical current density was measured over the length of the coil. It should be pointed out that, although the J_c of the coil was lower than that of the straight tape, the fact that the degree of the bending is substantial for what is presumably a uniform cross section demonstrates the feasibility of the production of ceramic superconducting oxide coils.

Figure 1 shows typical results for the dependence of the J_c on magnetic field for Ag-clad tapes and sintered pellet samples. The sintered samples were from (2223) powder and were pressed at 3.5×10^3 kg/cm^2 into pellets and sintered at 845°C for 60 h followed by cooling at a rate of 60°C/h. The final density was 4.2 g/cm^3. At low magnetic field (H < 300 Oe), the J_c in both the sintered sample and Ag-clad tape dropped rapidly. The rapid decrease in the J_c in this region has been interpreted by a Josephson weak link model [5]. It is noticed, however, that the J_c for the sintered sample loses more than 99% of its zero field value within this region, whereas the J_c of the Ag-clad tape drops only 50%, indicating a significant improvement in the weak link structure by cold rolling. At fields higher than 300 Oe, the J_c of the rolled tape reached a plateau region and exhibited pronounced anisotropy in relation to the direction of the applied magnetic field.

The J_c under magnetic field perpendicular to the *a* axis had a value of 2.0×10^2 A/cm^2 at 77 K and 4000 Oe. The slow decrease in the J_c in this region is interpreted by a strongly coupled percolation path regime [6]. The J_c in this regime is controlled by the flux pinning within grains. The magnitude of this regime is strongly dependent on the orientation of the grains relative to the magnetic field. The plateau region for the tape under magnetic field perpendicular to the *c* axis extends to 1.1 T (the measurement limitation of the available magnet), whereas a rapid drop in the J_c under a magnetic field parallel to the *c* axis begins at 0.3 T. It is evident that the grain alignment significantly improves the J_c-H characteristics when H is perpendicular to the *c* axis.

DISCUSSION

The high J_c and the improved J_c-H charateristics of Ag-clad tape may be attributed to the following factors.

1. Suppression of the Ag Poisoning Effect on Superconductivity

Ag as a cladding material has been found to be compatible with and non-poisoning to the superconducting systems Y-Ba-Cu-O [7], Bi-Sr-Ca-Cu-O [8], and Tl-Sr-Ca-Cu-O [9]. It has been reported that the presence of Ag in the Bi-Pb-Sr-Ca-Cu-O system has been found to degrade the superconductivity when the composites were treated in oxygen or air. However, when the composites were treated in low oxygen partial pressure (5×10^3 Pa), no such degradation was observed [10]. The mechanism for the degradation of superconductivity by Ag addition is the formation of a low-temperature melting eutectic liquid of Ag-CuO-PbO, which affects the composition of the superconducting phase and degrades the superconductivity. This eutectic may be suppressed by reducing the oxygen partial pressure. Figure 2 shows X-ray diffraction patterns for Ag/$Bi_{1.6}Pb_{0.4}Sr_{1.6}Ca_{2.0}Cu_{3.0}O_y$ samples heat heated in air for 50 h. It may be seen that the undoped sample consisted of nearly single high-T_c phase (2223), while the (2223) phase is degraded with increasing Ag content in the samples. At a level of 30 wt% Ag addition, the 85 K phase (2212) becomes the major phase and the 7 K phase (2201) has appeared. However, when the same series of samples was treated in low oxygen partial pressure (5.0×10^3 Pa), the X-ray diffraction patterns remained the same as for undoped samples, as shown in Figure 3.

It is, therefore, essential that the Ag-clad Bi-Pb-Sr-Ca-Cu-O wires are treated in low oxygen partial pressure in order to eliminate the interaction of Ag with the superconductor.

2. Alignment of the Superconductor Grains

It is well know that a large anisotropy in the J_c is one of factors that limits the transport current in randomly oriented polycrystalline materials. This difficulty can be overcome through cold drawing and rolling owing to the plate-like morphology of Bi-Pb-Sr-Ca-Cu-O grains. The grain alignment in the tape was confirmed by X-ray diffraction patterns (Figure 4), which were obtained from the polished surface of an Ag-clad tape by grinding away the cladding silver. It is seen that only the (001) peaks were observed when the incident X-rays were normal to the face of the tape, indicating that the *a-b* plane is aligned in the plane of the tape. This alignment allows the optimal conduction path along *a-b* plane to be utilised.

3. Enhancement of the Flux Pinning

As shown in Figure 1, the plateau region in the J_c-H curve indicates strong pinning within the grains. It has been found that a small amount of highly dispersed Ca_2CuO_3 may act as pinning centres in Bi-based superconductors [11]. The starting compositions which have slightly higher content of Ca and Cu relative to stoichiometric (2223), allow the formation of excess Ca_2CuO_3 during sintering. Through freeze drying and cold rolling, the excess Ca_2CuO_3 was high dispersed and deformed into an elongated shape along the rolling direction, parallel with the *a-b* plane.

It is also evident from TEM studies on the rolled tapes that the deformation through rolling creates massive defects, such as dislocations, stacking faults, and other types of planar defects. Figure 5 is a micrograph showing a high density of dislocations, which show more or less oriented patterns and have 5-10 dislocation in lines per 100 nm. This density of dislocations is comparable with the magnetic flux lines. For example, in a field of 2 T, the flux lines in the high-T_c superconductor are estimated at a spacing of 30 nm [12]. Therefore, it is expected that the dislocations and stacking faults may act as effective pinning centres. On the other hand, the defect regions may cause Josephson weak links within the grains owing to the extremely small coherence length [13]. The fact that the density of the dislocations and stacking faults is high suggests that these defects are aligned through rolling. Hence, the defects, which act as a s weak links, are located along the *c* axis, and they play a role in effective pinning in the *a-b* direction [14].

4. Other Factors

Other factors, such as highly reactive, uniform, and nearly single phase powders derived from freeze drying, and high density, and good connectivity between grains through rolling and pressing, also contribute to the improvement in the J_c and the J_c-H dependence.

REFERENCES

1. Hikata T, Nishikawa T, Mukai H, Sato K, and Hitotsuyanagi H, Jap. J. Appl. Phys. 28, L288(1989).
2. Dou S X, Liu H K, Apperley M H, Song K H, and Sorrell C C, Supercon. Sci. Tech. 3, 138(1990).
3. Tenbrink J, Heine K, Krauth H, Szuluzyk A, and Th ner M, VDI Berichte NR. 733, 399(1989).
4. Song K H, Liu H K, Dou S X, and Sorrell C C, J. Amer. Ceram. Soc. 73, 1771(1990).
5. Peterson R L and Ekin J W, Physca C 157, 325(1989).
6. Ekin J W, Larsin T M, Hermann A M, Sheng Z Z, Togano K, and

Kumakura H, Physica C 160, 489(1989).

7. McCallum R W, Verhoeven J D, Noack M A, Gibson E D, Laabs F C, Finnemore D K, and Moodenbaugh A R, Adv. Ceram. Mater. 2, 388 (1987).
8. Jin S, Sherwood R C, Tiefel T H, Kammlott G W, Fastnacht R A, Davis M E, and Zahurak S M, Appl. Phys. Lett. 52, 1628(1988).
9. Dou S X, Liu H K, Bourdillon A J, Tan N X, Savvides N, Andrikidis C, Roberts R B, and Sorrell C C, Supercon. Sci. Tech. 1, 83(1988).
10. Dou S X, Song K H, Liu H K Sorrell C C, Apperlly M H, and Savvides N, Appl. Phys. Lett. 56, 493(1990).
11. Dou S X, Guo S J, Liu H K, and Easterling K E, Supercond. Sci. Technol. 2, 308 (1989).
12. Kitazawa K, Ceram. Bulletin 68, 880(1989).
13. Deutscher G, Vide Coaches Minces (France), 43(241), Suppl. pp91 (1990).
14. Dou S X, Liu H K, Wang J, Apperley M H, Loberg B, and Easterling K E, Physica C, in press.

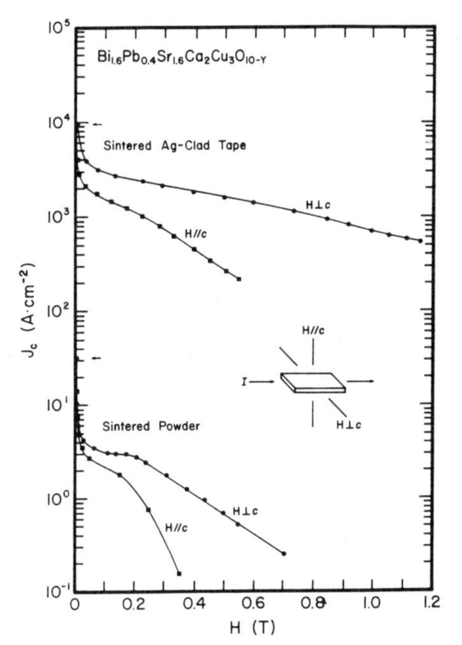

Fig.1 J_c dependence on magnetic field for the sintered pellet and Ag-clad BPSCCO wire

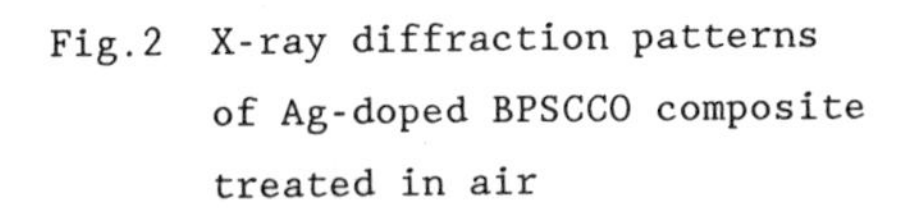

Fig.2 X-ray diffraction patterns of Ag-doped BPSCCO composite treated in air

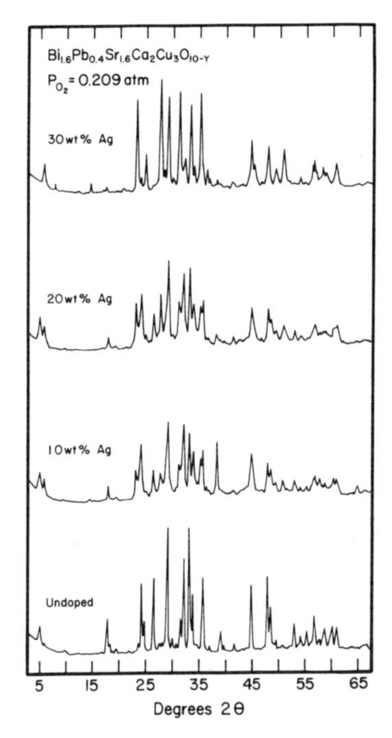

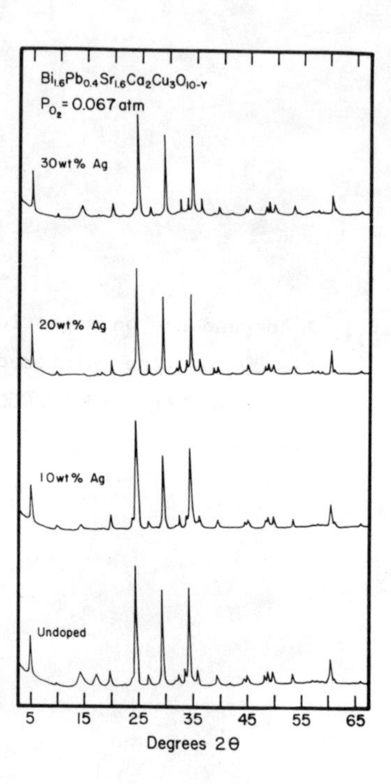

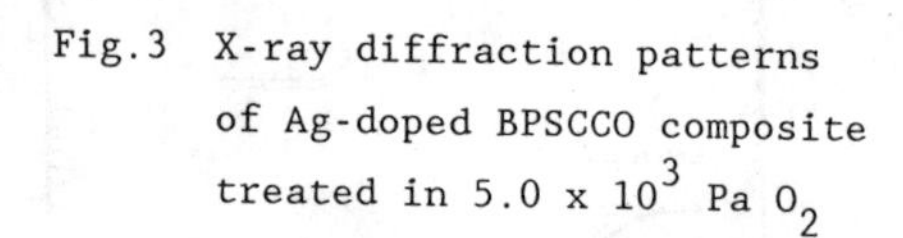
Fig.3 X-ray diffraction patterns of Ag-doped BPSCCO composite treated in 5.0×10^3 Pa O_2

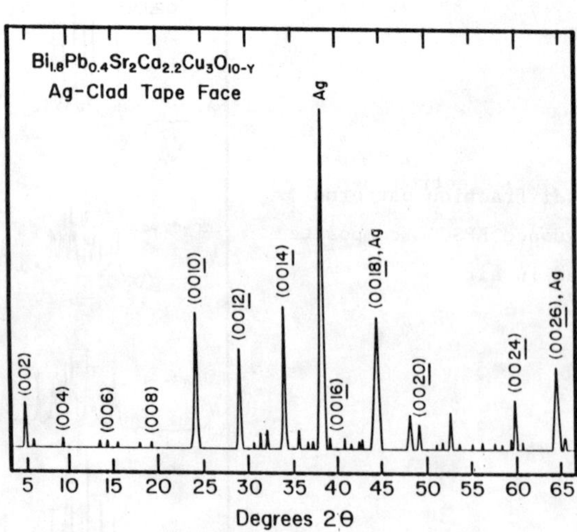

Fig.4 X-ray diffraction patterns of Ag-clad BPSCCO tape showing the grain alignment achieved through rolling

Fig.5 Transmission electron micrograph showing a high density of dislocations in the rolled BPSCCO tape

THICKNESS DEPENDENCE OF THE LEVITATION FORCE IN SUPERCONDUCTING $YBa_2Cu_3O_x$

T.H. JOHANSEN, H. BRATSBERG AND Z.J. YANG

Department of Physics, University of Oslo, P.O. Box 1048 Blindern, 0316 Oslo 3, Norway

Abstract: Using a 0.1 mg resolution electronic balance we have studied the levitation force on a small spherical magnet placed above superconducting $YBa_2Cu_3O_x$ planar disks. The force was measured as function of distance for many thicknesses of the superconducting sample. The sample thickness was reduced from 6.6 to 0.9 mm by sanding. We find that during the first approach, the behaviour of the force at small distances is essentially independent of the thickness, and can be described by an exponential force law, $F \propto \exp(-\alpha z)$. At large distances the force is highly thickness dependent. In the thick sample limit, however, the far away region can also be described by an exponential relation but with a different α. The distinct crossover is observed when the maximum field from the magnet on the YBCO surface exceeds 80 G. We believe this is due to the onset of intragranular flux penetration near H_{c1}.

Introduction

One of the most fascinating properties of superconducting materials is their ability to levitate small permanent magnets. Classically, this phenomenon is explained as a consequence of perfect diamagnetism possessed by the superconducor. Surface currents are induced so that the external magnetic field is expelled from the interior of the sample (Meissner effect). When this repelling interaction balances the weight of the magnet it can levitate.

The discovery of the high-T_c oxide superconductors has made it possible to study the levitation force quite conveniently at liquid nitrogen temperature. During the last few years several articles have reported experiments that address this subject.[1–5] It is found that the force as function of separation is highly hysteretic, which demonstrates that one easily gets flux penetration and considerable pinning in these materials. Quantitative results using large planar samples also show that the image dipole interaction, which is expected from the Meissner effect alone, does not fully describe the observed behaviour. Instead of an inverse fourth power dependence, one finds that the force-separation relation is approximately exponential. At present, there exists no theory to explain this behaviour.

Quite recently, Chang et al.[6] reported that for $YBa_2Cu_3O_x$ (YBCO), the force-separation curve is essentially independent of the thickness of the sample. In that work sample thicknesses down to 5.4 mm were investigated. In an earlier investigation Hellman et al.[7] described measurements of the levitation height of a magnet over YBCO slabs of thickness varying between 2.8 and 0.15 mm. They found that the sample thickness is a vital parameter in determining the levitation position. In fact, with the thinnest sample the magnet did not levitate at all. The thickness dependence was interpreted in terms of the the energy cost of putting flux vortices into the superconductor.

The purpose of the present investigation is to perform measurements over a large thickness interval, so that the existence of the two qualitativly different types of behaviour can be verified and studied on the same sample.

Experimental

The measurements were performed using an experimental arrangement shown in Fig.1. The superconducting disk was placed on an aluminum base in a polystyrene container filled with liquid nitrogen. The magnet was glued onto the lower end of a thin cylinder which hangs vertically down from the force sensor, a Mettler AE 260 Delta Range electronic balance with a resolution of 0.1 mg. In order to avoid condensation of water on the magnet and its support a moisture shield is surrounding the setup as illustrated in the figure.

The distance between the magnet and the superconductor is controlled by the motion of a step motor, which can displace the polystyrene container vertically by 1/800 mm per step. The force versus distance measurements are fully automated using a PC-AT computer. In the experiments described here the vertical force was sampled at intervals of 0.125 mm displacement.

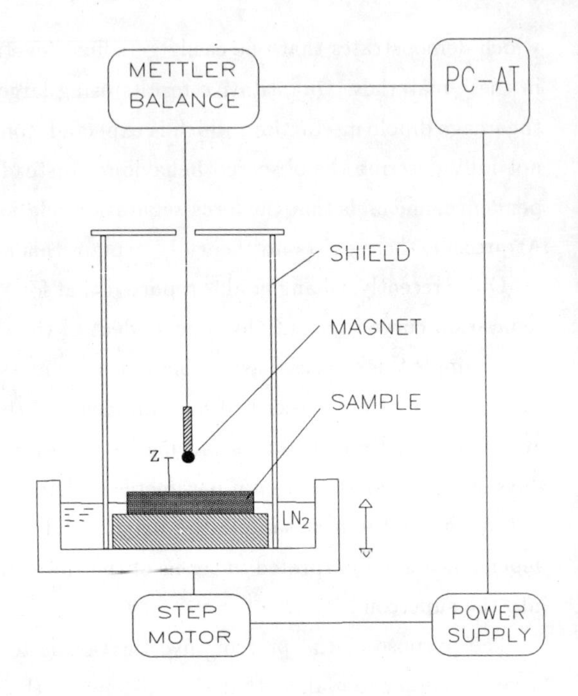

Figure 1: *Experimental arrangement used in the force versus displacement measurements.*

In order to have precise knowledge about the magnetic field applied to the superconductor the Nd-Fe-B magnet was shaped as a sphere using a ball mill. A separate experiment was performed to determine the dipole moment of the 2.1 mm diameter magnet. We used the same measurement equipment described above where the superconductor interaction was replaced by the force produced by a circular current loop. The coil consisted of 30 turns and carried a current of 1.0 A per turn. Figure 2 shows the force on the magnet as it was displaced along the coil axis over a distance of 20 mm starting from the center point. The sphere was mounted with the magnetization pointing in the vertical direction. The full line is the fitted curve corresponding to the force on a point dipole positioned along the field gradient of a circular current loop. The only fitting parameter here is the dipole moment, which was found to be $M = 5.3\ 10^{-3}$ Am^2. The high quality of the fit permits us to use the theoretical flux density from a point dipole,

$$\mathbf{B} = \frac{\mu_0}{4\pi}\left(\frac{3\mathbf{M}\cdot\mathbf{r}}{r^5}\,\mathbf{r} - \frac{\mathbf{M}}{r^3}\right), \tag{1}$$

to determine the field at the superconducting surface.

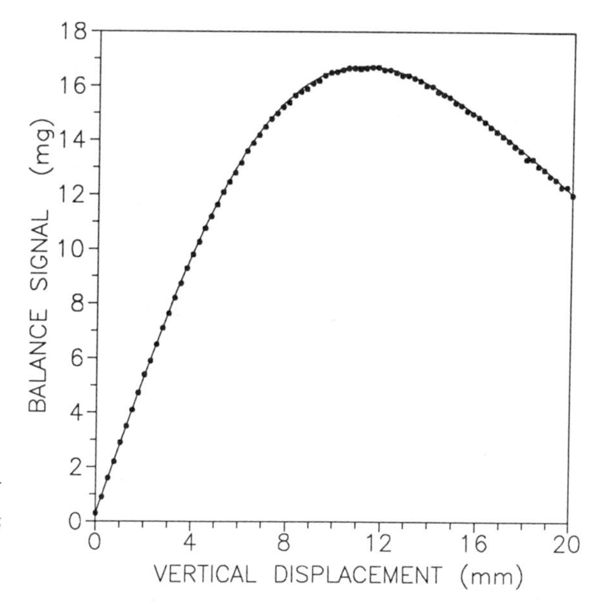

Figure 2: *Force on the magnet from a circular current loop.*

The superconducting disks studied in this work were all taken from the same YBCO mother sample, which was 25 mm in diameter and 6.6 mm thick. The thickness was reduced by attaching the sample by wax onto a grinding jig for sanding. Alcohol was constantly supplied to the ground surface during the process. By careful handling of the sintered material, the thickness could be made as small as 0.9 mm without any appearance of visual cracks.

The measurements of the superconductor-magnet interaction were also performed with the magnetic moment pointing vertically. The YBCO disks were always cooled to the superconducting state while the magnet was located at a large distance (zero-field cooling). In this way essentially no flux was frozen into the material initially.

Results and Discussion

Figure 3 shows the behaviour of the levitation force, F_z, as the magnet approaches from an initial distance of 13.3 mm above the superconductor. The distance is measured from the center of the magnet ball to the top surface of the YBCO disk. When the magnet is withdrawn after a closest approach of 1.2 mm, the force curve lies below the curve from the initial descent. This is the commonly observed hysteretic behaviour, which shows that irreversibly trapped flux results in an attractive contribution to the total force. In fact, after about 4 mm withdrawal the net force is totally attractive.

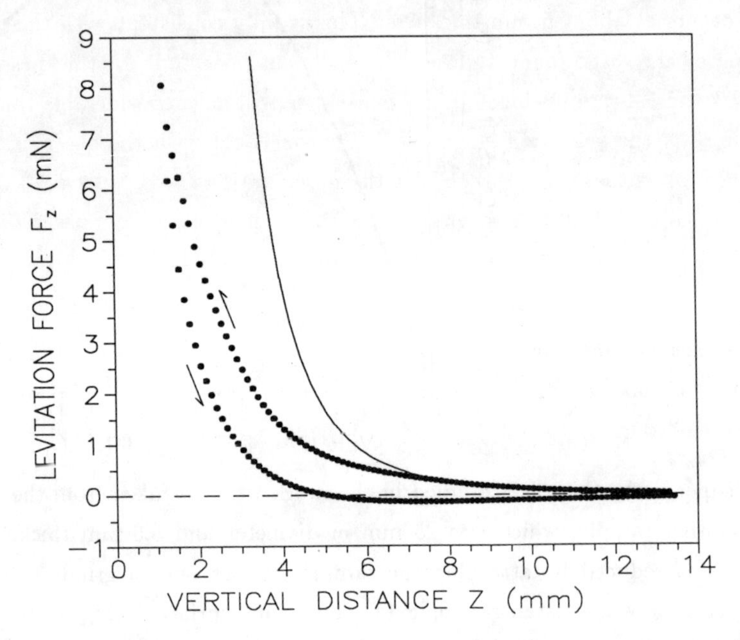

Figure 3: *Repelling force F_z on a 5.3 10^{-3} Am^2 magnetic dipole as function of distance. The full line corresponds to the dipole - image dipole interaction.*

For comparison, the figure also includes the reversible levitation force resulting from a complete Meissner effect, i.e., $F_z \propto M^2/z^4$, which follows from a dipole - image dipole interaction. The observed force is seen to be considerably smaller that what the image model predicts, in particular when the magnet is close to the sample. This has also been reported by other authors,[3] who have studied thick disk samples. The results in Fig.3 were obtained using a maximum sample thickness of 6.6 mm. Thus, our YBCO material displays the typical force characteristics.

The thickness dependence of the levitation force is displayed in Fig.4. All the curves were obtained during the initial approach after zero-field cooling of the sample. As one sees, the curves for the two largest thicknesses are essentially identical. As the thickness is reduced further (below 5.6 mm) the magnitude of the force is gradually reduced. These results show that there exists a thick-limit value for the force profile, where nothing is gained in levitation strength by increasing the thickness of the YBCO disk. We find that a crossover to a thickness dependent regime occurs at about 5 mm thickness. This is fully consistent with the results of Chang et al.[6], who found thickness independent behaviour for thicknesses between 9.3 and 5.4 mm. Below 5 mm thickness we will have considerable thickness dependence of the levitation height, which is consistent with the observations of Hellman et al.[7] However, we note that the observed thickness dependence of the levitation height will be influenced by the actual weight of the levitated body.

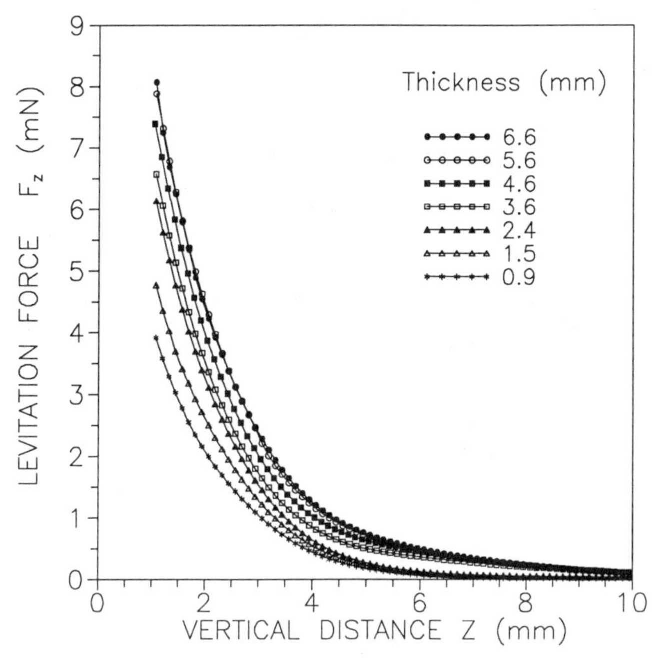

Figure 4: *Levitation force during initial descent for samples with thickness ranging from 6.6 to 0.9 mm.*

A different view of thc behaviour can be obtained by displaying the data in a semi-logarithmic plot, see Fig.5. Exponential dependence upon the distance,

$$F_z \propto \exp(-\alpha z),$$

should here appear as a linear curve. It is clear from this plot that non of the force curves follows a simple exponential law over the entire interval investigated. The thick sample limiting curve is instead well described by two straight line segments. The far away region, $z > 5$ mm, is characterized by $\alpha = 0.36$ mm^{-1}, while the steeper slope in the close viscinity region, $z < 5$ mm, gives $\alpha = 0.63$ mm^{-1}. The broken lines in Fig.5, which have these two slopes, serve as a guide to the eye.

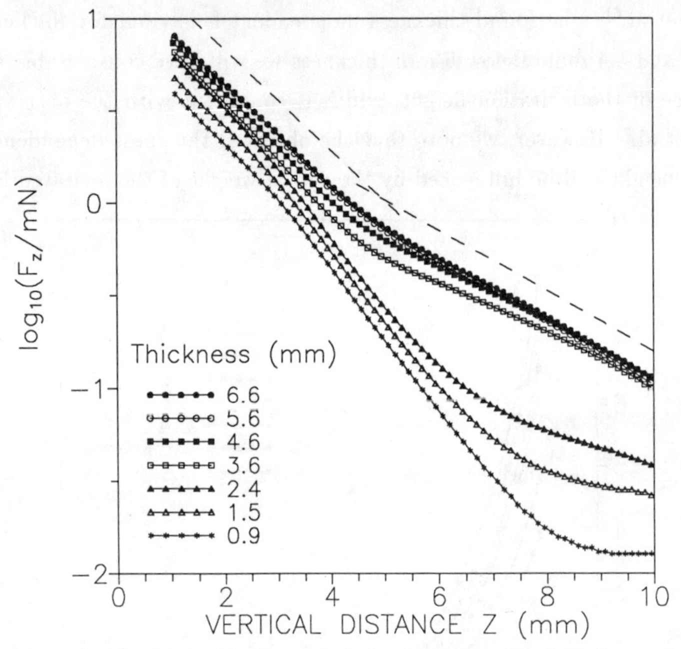

Figure 5: *Semi-logarithmic plot of the levitation force F_z for various thicknesses.*

It is clear that the levitation force is basically the response of the sample to the field applied from the magnet. A central point is therefore; what is the applied field when the force suddenly starts to increase more rapidly as the magnet approaches? The maximum field on the superconductor surface was in the present experiments directly below the magnet. From Eq.(1) it follows that

$$B(z) = \frac{\mu_0}{4\pi}\frac{2M}{z^3}$$

along the line defined by **M**. At a distance of 5 mm, the flux density amounts to 8.5 mT. This corresponds to a field strength close to the value reported for the critical field of YBCO. A plausible explanation for the change in slope may be found by taking into account the granularity of the specimen. At very low fields

(< 1 Oe) the flux will penetrate into regions between the grains due to the low critical field there, and a high demagnetization factor as given by the dimensions of the superconducting disk. The individual grains have a much higher critical field and a much smaller demagnetization factor. Thus, the field will start to penetrate into the grains first at a field strength slightly below H_{c1} of the superconductor. The slope in each regime will depend on both the energy cost of introducing a flux line, and the energy associated with flux pinning.

The strong thickness dependence of the force in the low field region, can be understood in the following way. The force from the Meissner effect in individual grains will significantly contribute to the total force. The sum of these Meissner forces will depend on the number of grains, i.e. the sample thickness. When the thickness of the sample is sufficiently large to screen the field completely, the thickness becomes unimportant, and the levitation force saturates.

Acknowledgment

The authors wish to thank G. Helgesen and A.T. Skjeltorp for helpful discussions. The financial support from NAVF – The Norwegian Council of Natural Sciences – is gratefully acknowledged.

References

1. F.C. Moon, M.M. Yanoviak and R. Ware, Appl. Phys. Lett. **52** 1534 (1988).
2. F.C. Moon, K-C. Weng, and P-Z. Chang, J. Appl. Phys. **66** 5643 (1989).
3. D.B. Marshall, R.E. DeWames, P.E.D. Morgan, and J.J. Ratto, Appl. Phys. A48, 87 (1989).
4. D.E. Weeks, Rev. Sci. Instrum. **61** 197 (1990).
5. B.R. Weinberger, L. Lynds and J.R. Hull, Supercond. Sci. Technol. **3** 381 (1990).
6. P-Z. Chang, F.C. Moon, J.R. Hull, and T.M. Mulcahy, to appear in Jap. J. Appl. Phys.
7. F. Hellman, E.M. Gyorgy, D.W. Johnson Jr., H.M. O'Bryan, and R.C. Sherwood, J. Appl. Phys. **63** 447 (1988).